IMAGING THE EARTH'S INTERIOR

Imaging the Earth's Interior

Jon F. Claerbout

Department of Geophysics
Stanford University

BLACKWELL SCIENTIFIC PUBLICATIONS
Oxford London Edinburgh
Boston Palo Alto Victoria

© 1985 Blackwell Scientific Publications

Editorial offices:

Osney Mead, Oxford, OX2 OEL
8 John Street, London, WC1N 2ES
23 Ainslie Place, Edinburgh, EH3 6AJ
52 Beacon Street, Boston, Massachusetts 02108
667 Lytton Avenue, Palo Alto, California 94301
107 Barry Street, Carlton, Victoria 3053, Australia

Distributors:

USA and Canada
  Blackwell Scientific Publications
  P.O. Box 50009
  Palo Alto, CA 94303

Australia
  Blackwell Scientific Book Distributors
  31 Advantage Road
  Highett, Victoria 3190

Europe and United Kingdom
  Blackwell Scientific Publications Ltd.
  Osney Mead, Oxford OX2 OEL

First published 1985

Library of Congress Cataloging in Publication Data
  Claerbout, Jon F.
    Imaging the earth's interior.

    Bibliography: p.
    Includes index.
    1. Geophysics 2. Exploration Geophysics
  3. Seismic prospecting 4.Imaging systems in seismology
  5. Seismic reflection method 6. Acoustic Imaging
  7. Seismic waves. I. Title.
  QE539.C53 1985   551'.028   84-28397
  ISBN 0-86542-304-0

This book is dedicated to Cecil and Ida Green who enjoyed the excitement of being among the pioneers, had the foresight to garner a fortune, and now have the fun of giving it away to universities throughout the world.

Contents

Preface

Some History

Reflection seismologists make images of the earth's interior. Through the 1960s this was done in an ad hoc fashion. Between 1968 and 1972, I conceived and field tested a new method of image making based directly on the wave equation of physics. Previously the wave equation had been used to predict observations starting from simplified, hypothesized models. It was not used in routine data analysis. My imaging method using finite differences soon came into widespread use in the petroleum exploration industry. Many other people quickly became involved and made important improvements. The earlier ad hoc methods were reinterpreted and they too improved in the light of wave theory.

An industrial affiliates group, known as the Stanford Exploration Project (SEP), was founded at Stanford University to pursue the new developments. Of the forty-eight sponsoring organizations, many have substantial research departments of their own. Thus began the decade of the 1970s, in which much progress was made. In the 1980s progress continues to be made at a rapid pace.

The Place for This Book

This textbook was born of the need to teach the best of the many new ideas to those entering the industry. Because so many people enter geophysics from outside the field, I have kept specialized geophysical terminology to a minimum and defined everything. So this book should be useful, not only to those interested in petroleum exploration, but also to professionals in all disciplines in which waves are analyzed.

My previous book, *Fundamentals of Geophysical Data Processing* (FGDP), was published in 1976 and has recently been reissued. It covered more basic aspects of reflection seismic data processing such as Z-transforms, Fourier transforms, discrete linear system theory, matrices, statistics, and theory of the stratified earth. FGDP also introduced wave-equation imaging, but extensive supplements became necessary. The supplements evolved into this book. The two books are about ninety percent different, the ten percent overlap being necessary to keep this book fully self-contained.

This book provides a coherent overview of the whole field of data processing as it is used in petroleum exploration, and it is the basic textbook in exploration geophysics at Stanford University. I make no claim, however, that the book is encyclopedic in scope. Some important processes such as *deconvolution* and *statics* are lightly sketched, as are the experimental applications of *tomography,* while other techniques such as *ray tracing* (see Cerveny [1977]) and many kinds of *modeling* are omitted altogether. Regrettably, the migration literature alone has grown so large that significant contributions such as the theoretical side of Kirchhoff migration (see Berkhout[1980]) are omitted.

Organization

Seismic imaging is a subject that draws much from mathematics and physics. These subjects build one idea upon another in a logical progression. I chose to organize this book likewise. This organization favors the new student who wants to understand the material thoroughly. A detailed index and more than a hundred cross references are included to link the practical topics that a logical organization has left somewhat scattered about.

More encyclopedic, topic-oriented textbooks on reflection seismology are those of Waters [1981] and Sengbush [1983]. Books focused on petroleum prospecting that treat reflection seismology descriptively are Sheriff [1980] and Anstey [1980]. Complementary books on earthquake seismology are Aki and Richards [1980] and Kennett [1983].

I have also tried to reach readers who want to learn concepts while skimming the mathematics. Individual sections (which are lectures) carry practical and descriptive matters as far as possible before the mathematical analysis. The chapters themselves are also organized in this way, so for example, when you get to the middle of Chapter 1, you can skip forward to Chapter 2.

As it happens, waves are marvelously geometrical objects, and much can be learned with little mathematical analysis. But you should begin the book having previous familiarity with calculus, complex exponentials, and Fourier transformation.

Philosophy

There is always a gap between theory and practice. Many books give you no clue as to the size and location of the gap — even books in exploration geophysics. This gap is nothing to be embarrassed about. It represents the current life of the subject — the current life of any science. It is a moving target, and its size is a matter of opinion. So it is risky for me to tell you what works, what doesn't, what is important, and what isn't. Opinions go beyond facts. Your knowledge won't be complete if you don't know some opinions as well as the facts. You will be getting opinions as well as facts when I explain the discrepancies between theory and industrial practice, and when I explain what should work, but doesn't seem to.

Thanks

This book is less my personal creation than was FGDP. I am indebted to many people. My colleagues Francis Muir and Fabio Rocca collaborated extensively in research. The following Stanford students contributed figures, exercises or ideas: David Brown, Junyee Chen, Robert Clayton, Steve Doherty, Raul Estevez, Paul Fowler, Bob Godfrey, Alfonso Gonzalez, Dave Hale, Bill Harlan, Bert Jacobs, Einar Kjartansson, Walter Lynn, Larry Morley, Dave Okaya, Richard Ottolini, Don Riley, Shuki Ronen, Dan Rothman, Chuck Sword, Jeff Thorson, John Toldi, Oz Yilmaz, and Li Zhiming. Industrial teaching helped me clarify the material in this book. Many useful ideas on presentation came from those who assisted me in industrial teaching, namely, Rob Clayton, Walt Lynn, Einar Kjartansson, Rick Ottolini, Fabio Rocca, Dave Hale, and Dan Rothman. Thanks are also due those who helped arrange the classes: Phil Hoyt, Lee Lu, Andre Lamer, Aftab Alam, Gary Latham, and Mike Graul.

JoAnn Heydron was first my secretary, and at a later date, the editor. Terri Ramey provided drafting and graphics assistance. Pat Bartz maintained order in ways too numerous to mention. I did most of the typing myself, as computerized typesetting has lightened that load and allows a more heavily revised and debugged final product. Peter Mora, Stew Levin, Dan Rothman, and Bill Harbert assisted in final proof reading.

Major thanks are due to Stanford University and the sponsoring organizations. Without them, much less would have been achieved. A list of sponsor names is given in an appendix. Many sponsors also donated data for our research. This is acknowledged in the figure caption. Many figures were based on data from a world-wide assortment prepared for academic use by Yilmaz and Cumro [1983]. Where figures were prepared by students, I have also acknowledged them in the caption. I prepared most of the figures myself often using outstanding graphics software utilities prepared by Rick Ottolini (*Tiplot, Movie*), Shuki Ronen (*Thplot*), Rob Clayton (*pen*), Joe Dellinger (*Ipen*), and Dave Hale (*Graph*).

Special thanks must also go to Western Electric Company, who donated the typesetting software that I used to prepare my lecture notes at Stanford. This software enabled me to go through many generations of improvements, eradicating scores of errors, rarely introducing new ones, all the while seeing (and having the students see) the material in almost final form. I am confident that this book has many fewer errors than did my previous book even after its second printing. When these lectures are in final form, I will simply mail a camera-ready copy and you my readers should have the book within eight weeks. So long, I'll see you at the SEG meeting.

Jon F. Claerbout

Introduction

Seek truth from facts. - Deng Xiaoping

Prospecting for oil begins with seismic soundings. The echoes are processed by computer into images that reveal much geological history. Worldwide, echo sounding and image making constitute about a four-billion-dollar-per-year activity.

Meaning of the Measurements

The presence of oil and gas has little *direct* effect on seismic reflections. The volume of rock is much larger than the volume of hydrocarbon. The reflections are well correlated, however, with interfaces between various rock types. In porous rocks, the hydrocarbons are free to flow. Fluids tend to rise. The shapes of rock interfaces tell us where hydrocarbons may accumulate. The discovery of oil and gas in the middle of the North Sea is a remarkable success story for the reflection seismic method. As the first exploratory wells, located by reflection seismology, were being drilled, it was impossible to predict whether they would hit oil. But if oil was to be found anywhere under the North Sea, there was great confidence that these initial drill sites were much more favorable than random locations. And as it turned out, of course, oil was soon found.

After a well has been drilled and logged, the reflection images become even more valuable, because then it is known what rock type corresponds to each echo. Seismology is usually able to provide a remarkably accurate mapping of rock types at some distance from the well. It is particularly valuable to know in which direction the rocks tilt upward, and where the strata are broken by faults. Seismology provides this information at a much lower cost than more drilling. When petroleum prospecting moves offshore the cost of seismology goes down by an order of magnitude, while the cost of drilling goes up by an order of magnitude.

Interested readers can purchase three volumes of reflection seismic images of the earth from the American Association of Petroleum Geologists (Bally [1983]).

Reproducibility of the Data

Reflection seismic data is voluminous. It is not like pencil marks on a sheet of paper. It is row upon row of high density magnetic tapes. Much seismic data is readily comprehensible. But much remains that is not, especially on the first try. Although much data is incomprehensible and seems noisy and random, it is remarkable that the data is experimentally repeatable. And we find that by working with this data we learn more and more, and are encouraged to continue. Because there is so much still hidden in the data which is routinely collected, this book concentrates on the mainstream data geometry: single survey lines with conventional near-surface sources and receivers. Experimental techniques are indicated but not examined.

Computers as Imaging Devices

Philosophers ask the question, "What is knowledge?" As technologists, our answer is that there is a real world and there is also an image of it in our minds. Knowledge means that the two are similar. To help form images we use imaging devices, such as microscopes, telescopes, cameras, television, etc. In this book computers are imaging devices for seismic echo soundings.

As an imaging device, a computer is in many ways ideal. A telescope is limited by the quality of its components. The image created by a computer is limited more by our understanding of mathematics, physics, and statistics, than by limitations inherent in the computer. For imaging radar or ultrasonics, computer capacity would be a real problem. It happens that the information content (bandwidth) of seismic echo soundings is about matched by today's computers.

Why is it Fun?

Many young people seem to enjoy tackling tough theoretical problems, but when the time comes for application they are often disappointed to find that the theory is in some ways irrelevant or inadequate to the problem at hand. At first this causes a diminished interest in practical problems. But eventually many come to see the real problems as more interesting than the original mathematical models. Why is this?

Maybe life is like a computer game. I have noticed that the games students like best are not those with a predetermined, intricate logical structure. They like the games which allow them to gradually uncover the rules as they play. It is really fun when a period of frustration with a game is ended by some application of a personal idea. But to be fun, a game must *have* rules, and you must *be able* to uncover them with a reasonable amount of effort. Luckily, reflection seismology, together with modern computers, provides us with a similar environment. Sometimes a game can be too frustrating, and you need a hint to get you over some obstacle into a new and deeper level. Reading this book is not like playing the game. It is more like being given a collection of hints, a bag of tricks, to help you to the deeper level.

These tricks are mostly new, many being less than ten years old. They have been selected because they really work, not always, but often enough. I have repressed the urge to include many promising tricks which have not been sufficiently tested.

Practical problems are not only deeper than theoretical problems, but ultimately they yield more interesting theory. For example, in freshman physics laboratory I learned to deduce Newton's laws of motion from simple experiments. I should have found experimentally that force equals mass times acceleration. Of course I didn't find exactly that. The experiment didn't seem to work out too well because of friction. *Friction,* now there's a really interesting subject for you. Physicists, chemists, metallurgists, earth scientists, they all know Newton's laws but wish they understood friction!

The theoretical book you are now holding wouldn't have been written except that two earlier theoretical approaches, (1) the theory of mathematical physics in stratified media, and (2) time series analysis, couldn't touch some of the most interesting aspects of our data. Some people thought we just had dirty data! Reflection seismic data are repeatable. Most of our problems really arise from the theory, not the data.

Computers and Movies

This book includes a number of computer programs. These programs are used for illustration and as exercises, but they should also be useful for reference. While they cannot be guaranteed, they worked for me when I generated many of the figures in this book, and they should work for you. You will notice that they are in a language similar to Fortran. The language is described at the beginning of Section 1.7. Since everyone has different facilities for graphical output, to use these programs yourself, you will have to understand them well enough to direct their outputs into your plotting equipment.

A movie is really a stack of pictures. In a computer it is just a three-dimensional matrix of floating point numbers which must somehow be converted to brightness *pixels* (*picture elements*). At the time of writing, few people are equipped to directly convert such a three-dimensional matrix into a movie. In our laboratory (Ottolini et al. [1984]) this is done on a high-quality video computer terminal (AED 512). Movie capability is a valuable asset. It enhances our understanding of our data and of the processing. Students are inspired by seeing their programming work result immediately in a movie, which is easily videotaped. Compared to other graphical devices this one is easy to maintain. It is used by both the research students and the students in the master's degree program, who use it for homework exercises.

The cost of such equipment, including the direct memory access (DMA) computer interface, is less than $10,000. For a really good experience with movies, you should also have physical control of a computer with a memory greater than a few megabytes. If you don't already have this, the price (1985) increases by about a factor of ten.

Will There Be Jobs ?

The main use of reflection seismic imaging is petroleum prospecting. Unlike nuclear energy, hydrocarbons are a non-renewable resource and there is evidence that petroleum production must decline during the lifetimes of the young generation. Does this mean that young people should avoid these studies? I think not. Taking the long-range view, with the population of the earth continuing to increase, it is not easy to imagine people losing interest in the earth's crust. A view of more intermediate range is that as the resource declines in abundance, there will be greater efforts to seek it. A short-range view is that workers are needed today, and there are no coal- or nuclear-powered airplanes. In any case, the skills developed in this book, computer implementations of concepts from physics, will always be of general utility.

Guide to This Book

Chapters 1 and 3 describe the basic concepts of imaging in reflection seismology. Chapters 2 and 4 cover computer techniques for the analysis of observed wavefields. Chapter 5 describes advanced imaging concepts. At Stanford University, Chapters 1-3 are taught to master's students in a course that runs for one quarter. These students also take a class from FGDP either before or after the class from this book.

You may want to understand concepts without learning about techniques. You could try reading just Chapters 1 and 3, but Chapter 2 will increase your comprehension because of its concrete nature and the examples it includes. Chapter 4 is for craftsmen who want to know what is involved in a high-quality implementation, or for unusually skilled interpreters who wish to understand artifacts and the accuracy limitations of various techniques. Chapter 5 describes seismic imaging concepts that are novel and seem correct in principle, but for various reasons (many unknown to me) have not yet come into widespread practical use. Interpreters who can stand the mathematics may appreciate Chapter 5 because it claims to explain how and why things often work out the way they do in practice. But the main attraction of Chapter 5 will be for those who wish to develop new echo imaging techniques.

1

Introduction to Imaging

1.1 Exploding Reflectors

The basic equipment for reflection seismic prospecting is a source for impulsive sound waves, a geophone (something like a microphone), and a multichannel waveform display system. A survey line is defined along the earth's surface. It could be the path for a ship, in which case the receiver is called a hydrophone. About every 25 meters the source is activated, and the echoes are recorded nearby. The sound source and receiver have almost no directional tuning capability because the frequencies that penetrate the earth have wavelengths longer than the ship. Consequently, echoes can arrive from several directions at the same time. It is the joint task of geophysicists and geologists to interpret the results. Geophysicists assume the quantitative, physical, and statistical tasks. Their main goals, and the goal to which this book is mainly directed, is to make good pictures of the earth's interior from the echoes.

A Powerful Analogy

Figure 1 shows two wave-propagation situations. The first is realistic field sounding. The second is a thought experiment in which the reflectors in the earth suddenly explode. Waves from the hypothetical explosion propagate up to the earth's surface where they are observed by a hypothetical string of geophones.

Notice in the figure that the raypaths in the field-recording case seem to be the same as those in the exploding-reflector case. It is a great conceptual advantage to imagine that the two wavefields, the observed and the hypothetical, are indeed the same. If they are the same, then the many thousands of experiments that have really been done can be ignored, and attention can be focused on the one hypothetical experiment. One obvious difference between the two cases is that in the field geometry waves must first go down and then return upward along the same path, whereas in the hypothetical experiment they just go up. Travel time in field experiments could be divided by two. In practice, the data of the field experiments (two-way time) is analyzed

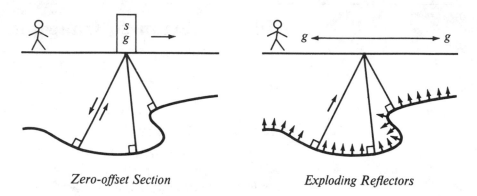

FIG. 1.1-1. Echoes collected with a source-receiver pair moved to all points on the earth's surface (left) and the "exploding-reflectors" conceptual model (right).

assuming the sound velocity to be half its true value.

Huygens Secondary Point Source

Waves on the ocean have wavelengths comparable to those of waves in seismic prospecting (15-500 meters), but ocean waves move slowly enough to be seen. Imagine a long harbor barrier parallel to the beach with a small entrance in the barrier for the passage of ships. This is shown in figure 2. A plane wave incident on the barrier from the open ocean will send a wave through the gap in the barrier. It is an observed fact that the wavefront in the harbor becomes a circle with the gap as its center. The difference between this beam of water waves and a light beam through a window is in the ratio of wavelength to hole size.

Linearity is a property of all low-amplitude waves (not those foamy, breaking waves near the shore). This means that two gaps in the harbor barrier make two semicircular wavefronts. Where the circles cross, the wave heights combine by simple linear addition. It is interesting to think of a barrier with many holes. In the limiting case of very many holes, the barrier disappears, being nothing but one gap alongside another. Semicircular wavefronts combine to make only the incident plane wave. Hyperbolas do the same. Figure 3 shows hyperbolas increasing in density from left to right. All those waves at nonvertical angles must somehow combine with one another to extinguish all evidence of anything but the plane wave.

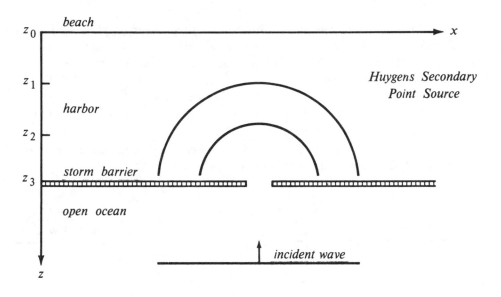

FIG. 1.1-2. Waves going through a gap in a barrier have semicircular wavefronts (if the wavelength is long compared to the gap size).

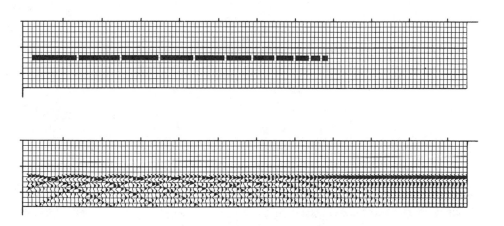

FIG. 1.1-3. A barrier with many holes (top). Waves, (x, t)-space, seen beyond the barrier (bottom).

A Cartesian coordinate system has been superimposed on the ocean surface with x going along the beach and z measuring the distance from shore. For the analogy with reflection seismology, people are confined to the beach (the earth's surface) where they make measurements of wave height as a function of x and t. From this data they can make inferences about the existence of gaps in the barrier out in the (x, z)-plane. Figure 4a shows the arrival time at the beach of a wave from the ocean through a gap. The earliest arrival occurs nearest the gap. What mathematical expression determines the shape of the arrival curve seen in the (x, t)-plane?

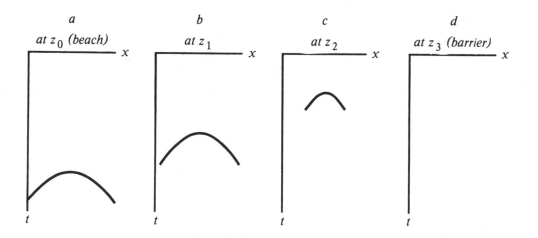

FIG. 1.1-4. The left frame shows the hyperbolic wave arrival time seen at the beach. Frames to the right show arrivals at increasing distances out in the water. (The x-axis is compressed from figure 2). (Gonzalez)

The waves are expanding circles. An equation for a circle expanding with velocity v about a point (x_3, z_3) is

$$(x - x_3)^2 + (z - z_3)^2 = v^2 t^2 \tag{1}$$

Considering t to be a constant, i.e. taking a snapshot, equation (1) is that of a circle. Considering z to be a constant, it is an equation in the (x, t)-plane for a hyperbola. Considered in the (t, x, z)-volume, equation (1) is that of a cone. Slices at various values of t show circles of various sizes. Slices of various values of z show various hyperbolas. Figure 4 shows four hyperbolas. The first is the observation made at the beach $z_0 = 0$. The second is a hypothetical set of observations at some distance z_1 out in the water. The third set of observations is at z_2, an even greater distance from the beach. The fourth set of observations is at z_3, nearly all the way out to the barrier, where the hyperbola has degenerated to a point. All these hyperbolas are from a family of hyperbolas, each with the same asymptote. The asymptote refers to a wave that turns nearly 90° at the gap and is found moving nearly parallel to the shore at the speed dx/dt of a water wave. (For this water wave analogy it is presumed —incorrectly— that the speed of water waves is a constant independent of water depth).

If the original incident wave was a positive pulse, then the Huygens secondary source must consist of both positive and negative polarities to enable the destructive interference of all but the plane wave. So the Huygens waveform has a phase shift. In the next section, mathematical expressions will be found for the Huygens secondary source. Another phenomenon, well known to boaters, is that the largest amplitude of the Huygens semicircle is in the direction pointing straight towards shore. The amplitude drops to zero for waves moving parallel to the shore. In optics this amplitude dropoff with angle is called the *obliquity factor*.

Migration Defined

A dictionary gives many definitions for the word *run*. They are related, but they are distinct. The word *migration* in geophysical prospecting likewise has about four related but distinct meanings. The simplest is like the meaning of the word *move*. When an object at some location in the (x, z)-plane is found at a different location at a later time t, then we say it *moves*. Analogously, when a wave arrival (often called an *event*) at some location in the (x, t)-space of geophysical observations is found at a different position for a different survey line at a greater depth z, then we say it *migrates*.

To see this more clearly imagine the four frames of figure 4 being taken from a movie. During the movie, the depth z changes beginning at the beach (the earth's surface) and going out to the storm barrier. The frames are superimposed in figure 5a. Mainly what happens in the movie is that the event migrates upward toward $t=0$. To remove this dominating effect of vertical translation make another superposition, keeping the hyperbola tops all in the same place. Mathematically, the time t axis is replaced by a so-called *retarded* time axis $t'=t+z/v$, shown in figure 5b. The second, more precise definition of *migration* is the motion of an event in (x, t')-space as z changes. After removing the vertical shift, the residual motion is mainly a shape change. By this definition hyperbola tops, or horizontal layers, don't migrate.

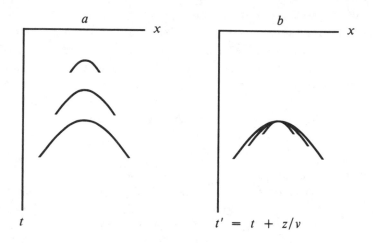

FIG. 1.1-5. Left shows a superposition of the hyperbolas of figure 4. At the right the superposition incorporates a shift, called retardation $t'=t+z/v$, to keep the hyperbola tops together. (Gonzalez)

The hyperbolas in figure 5 really extend to infinity, but the drawing cuts each one off at a time equal $\sqrt{2}$ times its earliest arrival. Thus the hyperbolas shown depict only rays moving within 45° of the vertical. It is good to remember this, that the ratio of first arrival time on a hyperbola to any other arrival time gives the cosine of the angle of propagation. The cutoff on each hyperbola is a ray at 45°. Notice that the end points of the hyperbolas on the drawing can be connected by a straight line. Also, the slope at the end of each hyperbola is the same. For any wavefront, the angle of the wave is $\tan\theta = dx/dz$ in physical space. For any seismic event, the slope $v\,dt/dx$ is $\sin\theta$, as you can see by considering a wavefront intercepting the earth's surface at angle θ. So, energy moving on a straight line in physical (x,z)-space migrates along a straight line in data (x,t)-space. As z increases, the energy of all angles comes together to a focus. The focus is the exploding reflector. It is the gap in the barrier. This third definition of migration is that it is the process that somehow pushes observational data — wave height as a function of x and t — from the beach to the barrier. The third definition stresses not so much the motion itself, but the transformation from the beginning point to the ending point.

To go further, a more general example is needed than the storm barrier example. The barrier example is confined to making Huygens sources only at some particular z. Sources are needed at other depths as well. Then, given a wave-extrapolation process to move data to increasing z values, exploding-reflector images are constructed with

$$Image\ (x,z)\quad =\quad Wave\ (t=0,\ x,\ z\ ) \qquad (2)$$

The fourth definition of migration also incorporates the definition of *diffraction* as the opposite of migration.

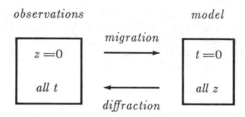

Diffraction is sometimes regarded as the natural process that creates and enlarges hyperboloids. *Migration* is the computer process that does the reverse.

Another aspect of the use of the word *migration* arises in Chapter 3, where the horizontal coordinate can be either shot-to-geophone midpoint y, or offset h. Hyperboloids can be downward continued in both the (y,t)- and the (h,t)-plane. In the (y,t)-plane this is called *migration* or *imaging*, and in the (h,t)-plane it is called *focusing* or *velocity analysis*.

An Impulse in the Data

The Huygens diffraction takes an isolated pulse function (delta function) in (x,z)-space and makes it into a hyperbola in (x,t)-space at $z=0$. The converse is to start from a delta function in (x,t)-space at $z=0$. This converse refers to a seismic survey in which no echoes are recorded except at one particular location, and at that location only one echo is recorded. What earth model is consistent with such observations? As shown in figure 6 this earth must contain a spherical mirror whose center is at the anomalous recording position.

It is unlikely that the processes of nature have created many spherical mirrors inside the earth. But when we look at processed geophysical data, we often see spherical mirrors. Obviously, such input data contains impulses that are not consistent with the wave-propagation theory being explained here.

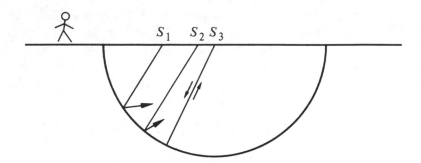

FIG. 1.1-6. When the seismic source S is at the exact center of a semicircular mirror, then, and only then, will an echo return to the geophone at the source. This semicircular reflector is the logical consequence of a dataset where one echo is found at only one place on the earth.

This illustrates why petroleum prospectors study reflection seismic data processing, even though they personally plan to write no processing programs. The raw data is too complex to comprehend. The processed data gives an earth model, but its reliability is difficult to know. You may never plan to build an automobile, but when you drive alone far out into the desert, you should know as much as you can about automobiles.

Hand Migration

Given a seismic event at (x_0, t_0) with a slope $p = dt/dx$, let us determine its position (x_m, t_m) after migration. Consider a planar wavefront at angle θ to the earth's surface traveling a distance dx in a time dt. Assuming a velocity v we have the wave angle in terms of measurable quantities.

$$\sin\theta = \frac{v\, dt}{dx} = p\, v \tag{3}$$

The vertical travel path is less than the angled path by

$$t_m = t_0 \cos\theta = t_0 \sqrt{1 - p^2 v^2} \tag{4a}$$

A travel time t_0 and a horizontal component of velocity $v\sin\theta$ gives the lateral location after migration:

$$x_m = x_0 - t_0 v \sin\theta = x_0 - t_0 p\, v^2 \tag{4b}$$

Consideration of a hyperbola migrating towards its apex shows why (4b) contains a minus sign. Equations (4a) and (4b) are the basic equations for manual migration of reflection seismic data. They tell you where the point migrates, but they do not tell you how the slope p will change.

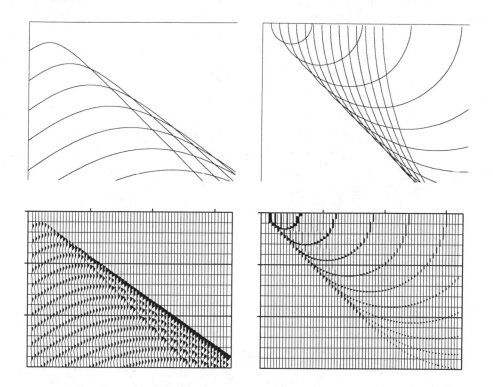

FIG. 1.1-7. Left is a superposition of many hyperbolas. The top of each hyperbola lies along a straight line. That line is like a reflector, but instead of using a continuous line, it is a sequence of points. Constructive interference gives an apparent reflection off to the side.

Right shows a superposition of semicircles. The bottom of each semicircle lies along a line that could be the line of an observed plane wave. Instead the plane wave is broken into point arrivals, each being interpreted as coming from a semicircular mirror. Adding the mirrors yields a more steeply dipping reflector.

Reflector Steepening

Consider a vertical wall, a limiting case of a dipping bed. Its reflections, the asymptotes of a hyperbola, have a nonvertical steepness. This establishes that migration increases the apparent steepness of dipping beds. I use the words *apparent steepness* because it is the slope as seen in the (x, t)-plane that has steepened. Migration really produces its output in z. but z/v is often overlain on t to create a *migrated time section*. When we say a hyperbola migrates to its apex, we are of course thinking of the migrated time section. Let us determine the steepening as a function of angle.

Consider a point $(x_{0+}, t_{0+}) = x_0 + \Delta, t_0 + p \Delta$ neighboring the original point (x_0, t_0). By equation (4), this neighbor migrates to

$$t_{m+} = (t_0 + p \Delta) \sqrt{1 - p^2 v^2} \tag{5a}$$

$$x_{m+} = x_0 + \Delta - (t_0 + p \Delta) p v^2 \tag{5b}$$

Now we compute the stepout p_m of the migrated event

$$p_m = \frac{dt_{m+}}{dx_{m+}} = \frac{dt_{m+}/d \Delta}{dx_{m+}/d \Delta}$$

$$p_m = \frac{p \sqrt{1 - p^2 v^2}}{1 - p^2 v^2} = \frac{p}{\sqrt{1 - p^2 v^2}} = \frac{\tan \theta}{v} \tag{6}$$

So slopes on migrated time sections, like slopes in Cartesian space, imply tangents of angles while slopes on unmigrated time sections imply sines.

It may seem paradoxical that dipping beds change slope on migration whereas flanks of hyperbolas do not change slope during downward continuation. One reason is that migration is downward continuation *plus* imaging (selecting $t=0$). Another reason is that a hyperbola is a special event that comes from a single source at a single depth whereas a dipping bed is a superposition of point sources from different depths. Figure 7 shows how points making up a line reflector diffract to a line reflection, and how points making up a line reflection migrate to a line reflector.

Limitations of the Exploding-Reflector Concept

The exploding-reflector concept is a powerful and fortunate analogy. For people who spend their time working entirely on data interpretation rather than on processing, the exploding-reflector concept is more than a vital crutch. It's the only means of transportation! But for those of us who work on data processing, the exploding-reflector concept has a serious shortcoming. No one has yet figured out how to extend the concept to apply to data recorded at nonzero offset. Furthermore, most data is recorded at rather large offsets. In a modern marine prospecting survey, there is not one hydrophone, but hundreds, which are strung out in a cable towed behind the ship. The recording cable is typically 2-3 kilometers long. Drilling may be about 3 kilometers deep. So in practice the angles are big. Therein lie both new problems and new opportunities, none of which will be considered until Chapter 3.

Furthermore, even at zero offset, the exploding-reflector concept is not quantitatively correct. For the moment, note three obvious failings: figure 8 shows rays that are not predicted by the exploding-reflector model. These rays will be present in a zero-offset section. Lateral velocity variation is required for this situation to exist.

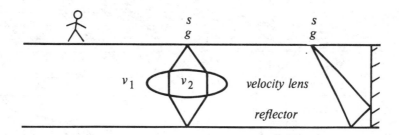

FIG. 1.1-8. Two rays, not predicted by the exploding-reflector model, that would nevertheless be found on a zero-offset section.

Second, the exploding-reflector concept fails with multiple reflections. For a flat sea floor with a two-way travel time t_1, multiple reflections are predicted at times $2t_1$, $3t_1$, $4t_1$, etc. In the exploding-reflector geometry the first multiple goes from reflector to surface, then from surface to reflector, then from reflector to surface, for a total time $3t_1$. Subsequent multiples occur at times $5t_1$, $7t_1$, etc. Clearly the multiple reflections generated on the zero-offset section differ from those of the exploding-reflector model.

The third failing of the exploding-reflector model is where we are able to see waves bounced from both sides of an interface. The exploding-reflector model predicts the waves emitted by both sides have the same polarity. The physics of reflection coefficients says reflections from opposite sides have opposite polarities.

Plate Tectonics Example

Plate tectonic theory says the ocean floors are made of thin plates that are formed at volcanic ridges near the middle of the oceans. These plates move toward trenches in the deepest part of the ocean where they plunge back down into the earth. The best evidence for the theory is the lack of old

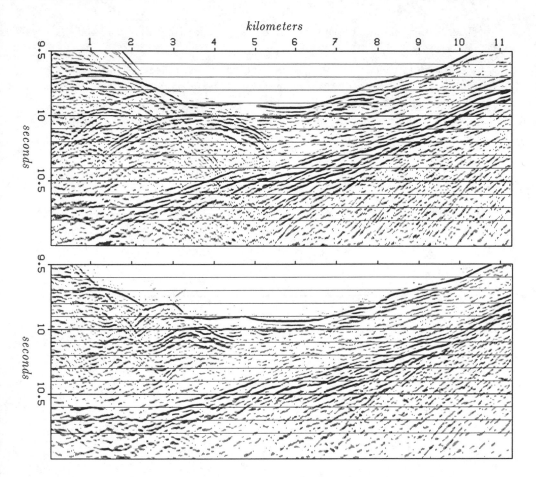

FIG. 1.1-9. Top is 11 kilometers of reflection data from a survey line across the Japan trench (Tokyo University Oceanographic Research Institute). Bottom shows the result of migration processing. (Ottolini)

rocks on the floors of the earth's oceans. Generally, continents are older rocks jostled by the younger moving oceanic plates. The formation of plates by mid-ocean ridge volcanism is readily observed in a variety of ways. Whether the plates really do plunge at the trenches is not so clear observationally. The evidence comes from earthquake locations and from reflection seismology. Figure 9 shows some reflection data from the Japan trench. Two reflections dominate, the sea floor reflection and a deeper layer dipping down to the left. This latter is presumably the top part of a plate that is beginning its descent into the earth. We can examine it for evidence of bending downward, such as tension fractures near the surface. (The topmost layer is soft recent marine sediment loosely attached to the plate).

kilometers

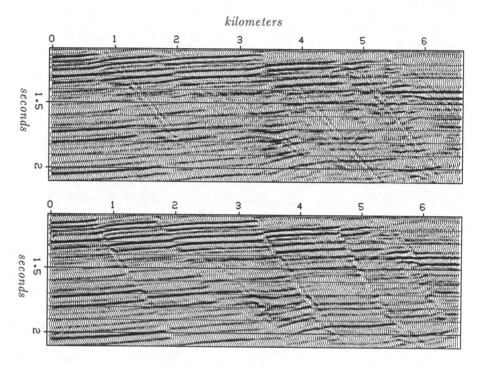

FIG. 1.1-10.   Top is 6.5 kilometers of reflection data from a survey line offshore from the Texas coast of the Gulf of Mexico. Bottom shows the result of migration processing. (Rothman).

Notice that the top of the plot is not zero time. The time axis runs from 9.5 to 11.0 seconds. Before 9.5 sec there are no echoes — we are waiting for the waves to go between the ship and the ocean floor. Hyperbolic reflections around kilometers 1-3 are collapsed by migration to form interesting "blocky" shapes. Look at the sea floor topography near kilometer 8 and the difference between migrated and unmigrated data sections. After migration, the sea floor diffraction hyperbolas move away from the plate echo (kilometer 4). Fractures (especially the one at 6.2 km) are more sharply defined. Finally, if the plate bends downward, it is not apparent from the data given. The bending question really requires a more detailed analysis of lateral variation in seismic velocity.

For an example with petroleum interest, see figure 10, data from offshore Texas. Sediments are dropped where coastal rivers enter the Gulf of Mexico. The added weight causes slumping along steep faults. After a permeable sandstone layer has been identified by drilling, its reflection can be extrapolated up dip to the nearest fault on data like figure 10. The fault is likely to break the continuity of the permeability trapping the upward flowing hydrocarbons. A sandstone at this depth can have a porosity of 25%. Assume a seismic velocity of 2.2 km/sec. Deduce the scale between physical volume and the data in figure 10. Comparing the value of a volume of oil to the size of that same volume on figure 10, you can see the importance of good images.

EXERCISES

1. Prove the Pythagorean theorem, that is, the length of the hypotenuse $v\,t$ of a right triangle is determined by $x^2 + z^2 = v^2 t^2$. Hint:

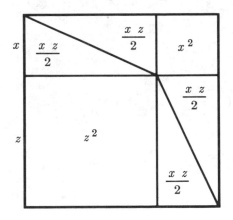 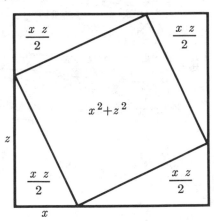

2. Compute propagation angles for the hyperbola flanks in figure 9.

3. Using the result of exercise 2, deduce the plunge angle of the plate.

4. How deep is the Japan trench (water velocity is 1.5 km/sec)?

5. On the Gulf Coast data, which direction is offshore? Why?

1.2 Wave Extrapolation as a 2-D Filter

One of the main ideas in Fourier analysis is that an impulse function (a delta function) can be constructed by the superposition of sinusoids (or complex exponentials). In the study of time series this construction is used for the *impulse response* of a filter. In the study of functions of space, it is used to make a physical point source.

Taking time and space together, Fourier components can be interpreted as monochromatic plane waves. Physical optics (and with it reflection seismology) becomes an extension to filter theory. In this section we learn the mathematical form, in Fourier space, of the Huygens secondary source. It is a two-dimensional (2-D) filter for spatial extrapolation of wavefields.

Rays and Fronts

Figure 1 depicts a ray moving down into the earth at an angle θ from the vertical. Perpendicular to the ray is a wavefront. By elementary geometry the angle between the wavefront and the earth's surface is also θ. The ray increases its length at a speed v. The speed that is observable on the earth's surface is the intercept of the wavefront with the earth's surface. This speed, namely $v/\sin\theta$, is faster than v. Likewise, the speed of the intercept of the wavefront and the vertical axis is $v/\cos\theta$. A mathematical expression for a straight line, like that shown to be the wavefront in figure 1, is

$$z = z_0 - x\tan\theta \tag{1}$$

In this expression z_0 is the intercept between the wavefront and the vertical axis. To make the intercept move downward, replace it by the appropriate velocity times time:

$$z = v\,\frac{t}{\cos\theta} - x\tan\theta \tag{2}$$

Solving for time gives

$$t(x,z) = \frac{z}{v}\cos\theta + \frac{x}{v}\sin\theta \tag{3}$$

Equation (3) tells the time that the wavefront will pass any particular location (x,z). The expression for a shifted waveform of arbitrary shape is $f(t - t_0)$. Using (3) to define the time shift t_0 gives an expression for a

15

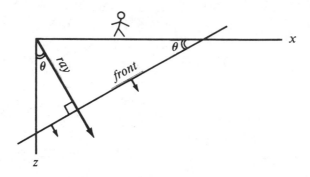

FIG. 1.2-1. Downgoing ray and wavefront.

wavefield that is some waveform moving on a ray.

$$moving \; wavefield \; = \; f\left(t \; - \; \frac{x}{v} \sin\theta - \frac{z}{v}\cos\theta\right) \qquad (4)$$

Waves in Fourier Space

Arbitrary functions can be made from the superposition of sinusoids. Sinusoids and complex exponentials often occur. One reason they occur is that they are the solutions to linear partial differential equations (PDEs) with constant coefficients. The PDEs arise because most laws of physics are expressible as PDEs.

Using Fourier integrals on time functions we encounter the *Fourier kernel* $\exp(-i\omega t)$. Specializing the arbitrary function in equation (4) to be the real part of the function $\exp[-i\omega(t-t_0)]$ gives

$$moving \; cosine \; wave \; = \; \cos\left[\omega\left(\frac{x}{v}\sin\theta + \frac{z}{v}\cos\theta - t\right)\right] \qquad (5)$$

To use Fourier integrals on the space-axis x the spatial angular frequency must be defined. Since we will ultimately encounter many space axes (three for shot, three for geophone, also the midpoint and offset), the convention will be to use a subscript on the letter k to denote the axis being Fourier transformed. So k_x is the angular spatial frequency on the x-axis and $\exp(ik_x x)$ is its Fourier kernel. For each axis and Fourier kernel there is the question of the sign of i. The sign convention used here is the one used in most physics books, namely, the one that agrees with equation (5). Reasons for the choice are given in Section 1.6. With this convention, a wave moves in the *positive* direction along the space axes. Thus the Fourier kernel for (x, z, t)-space will be taken to be

Fourier kernel $=$

$$= e^{i\,k_x\,x}\,e^{i\,k_z\,z}\,e^{-i\omega t} = \exp[i\,(k_x\,x + k_z\,z - \omega t)] \tag{6}$$

Now for the whistles, bells, and trumpets. Equating (5) to the real part of (6), physical angles and velocity are related to Fourier components. These relations should be memorized!

$$(7)$$

Angles and Fourier Components	
$\sin \theta = \dfrac{v\,k_x}{\omega}$	$\cos \theta = \dfrac{v\,k_z}{\omega}$

Equally important is what comes next. Insert the angle definitions into the familiar relation $\sin^2 \theta + \cos^2 \theta = 1$. This gives a most important relationship, known as the *dispersion relation of the scalar wave equation*.

$$k_x{}^2 + k_z{}^2 = \frac{\omega^2}{v^2} \tag{8}$$

We'll encounter *dispersion relations* and the *scalar wave equation* later. The importance of (8) is that it enables us to make the distinction between an arbitrary function and a chaotic function that actually is a wavefield. Take any function $p(t, x, z)$. Fourier transform it to $P(\omega, k_x, k_z)$. Look in the (ω, k_x, k_z)-volume for any nonvanishing values of P. You will have a wavefield if and only if all nonvanishing P have coordinates that satisfy (8). Even better, in practice the (x, t)-dependence at $z=0$ is usually known, but the z-dependence is not. Then the z-dependence is found by assuming P is a wavefield, so the z-dependence is inferred from (8).

Migration Improves Horizontal Resolution

In principle, migration converts hyperbolas to points. In practice, hyperbolas don't collapse to a point, they collapse to a *focus*. A focus has measurable dimensions. Migration is said to be "good" because it increases spatial resolution. It squeezes a large hyperbola down to a tiny focus. To quantitatively describe the improvement of migration, the size of the hyperbola and the size of the focus must be defined. Figure 2 shows various ways of measuring the size of a hyperbola.

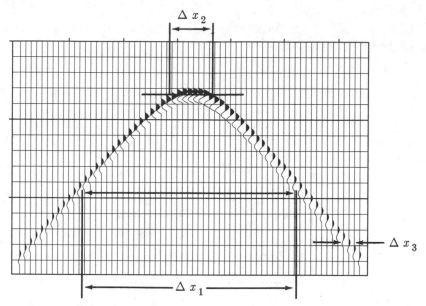

FIG. 1.2-2. Measurements of width parameters of a hyperbola.

The hyperbola carries an impulsive arrival. So the ω-bandwidth of the hyperbola is roughly given by the zero crossings on the time axis of the main energy burst. I'll mention 50 Hz as a typical value, though you could encounter values four times higher or four times lower. Knowledge of a seismic velocity determines depth resolution. I'll suggest 3 km/sec, though once again you could encounter velocities four times greater or four times less. These values imply a seismic wavelength of $v/f = 60$ meters. But the effective seismic wavelength is half the actual wavelength. The half comes from halving the velocity v in exploding reflector calculations, or equivalently, from realizing that the seismic wavelength is divided equally into upgoing and downgoing parts. Resolving power is customarily defined as about half the effective wavelength or about 15 meters. (Whether seismic resolution should be *half* the effective wavelength or a smaller fraction is an issue that involves signal-to-noise considerations outside our present study).

The lateral resolution requires estimates of hyperbola width and focus width. Figure 2 shows three hyperbola widths. The widest, Δx_1, includes about three-quarters of the energy in the hyperbola. Next is the width Δx_2, called the *Fresnel Zone*. It is measured across the hyperbola at the time when the first arrival has just changed polarity. Third is the smallest measurable width, found far out on a flank. This width, Δx_3, is the shortest horizontal wavelength to be found. Resolution is the study of the size of error, and it is not especially useful to be precise about the error in the error. The main idea is that $\Delta x_1 > \Delta x_2 > \Delta x_3$. The bandwidth of the spatial k_x spectrum is roughly $1/\Delta x_3$. How small a focus can migration make? It will be limited by the available bandwidth in the k_x spectrum. The size of the focus will be about the same as Δx_3.

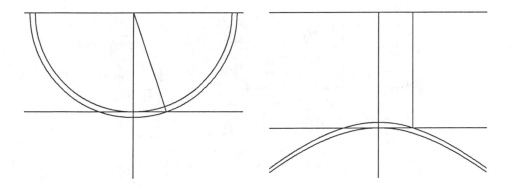

FIG. 1.2-3. Fresnel zone in (x, z)-space (left) and in (x, t)-space (right).

Figure 3 shows the geometry of the Fresnel zone concept. A Fresnel zone is an intercept of a spherical wave with a plane. The intercept is defined when the spherical wave penetrates the plane to a depth of a half wavelength. What is the meaning of the Fresnel width Δx_2? Imagine yourself in Berlin. There is a wall there. You may not go near it. Imagine a hole in the wall. You are shouting to a friend on the opposite side. How does the loudness of the sound depend on the size of the hole ΔX? It is not obvious, but it is well known, both theoretically and experimentally, that holes larger than the Fresnel zone cause little attenuation, but smaller holes restrict the sound in proportion to their size.

Wave propagation is a convolutional filter that smears information from a region Δx_2 along a reflector (or Δx_1 in the subsurface) to a point on the surface. Migration, the reverse of wave propagation, is the deconvolution operation. The final amount of lateral resolution is limited by the spatial bandwidth of the data.

Migration may be called for even where reflectors show no dip. When a well site is to be chosen within an accuracy of less than Δx_2 then the interpreter is looking at subtle changes in amplitude or waveform along the reflector. Migration causes these amplitude and waveform variations to change and to move horizontally along the reflector. The distance moved is about equal the Fresnel zone.

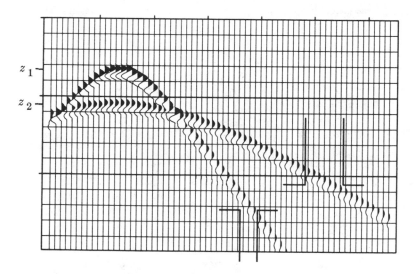

FIG. 1.2-4. Hyperboloids for an earth of velocity increasing with depth. Observable lateral wavelengths get longer with increasing depth. Thus lateral resolving power decreases with depth.

A basic fact of seismology is the resolution limitation caused by the increase with depth of the seismic velocity. What happens is that as the waves get deeper into the earth, their spatial wavelengths get longer because of the increasing velocity. The case of vertical resolution is simply this: longer wavelengths, less resolution. The case of horizontal resolution is similar, but the horizontal wavelength is directly measurable at the earth's surface. Figure 4 demonstrates this. Hyperboloids from shallow and deep scatterers are shown. Shallow hyperbolas have early tops and steep asymptotes. Deep scatterers have late tops and less steep asymptotes. The less steep asymptotes have longer horizontal wavelength. Horizontal wavelengths measured at the surface are unchanged at depth, even though velocity increases with depth. (This implication of Snell's law is shown in Section 1.5). Thus, lateral spatial resolution gets worse with depth. Compounding the above reason for decreasing resolution is the loss of high-frequency energy at late travel time.

Two-Dimensional Fourier Transform

Before going any further, let us review some basic facts about two-dimensional Fourier transformation. A two-dimensional function is represented in a computer as numerical values in a matrix. A one-dimensional Fourier transform in a computer is an operation on a vector. A two-dimensional Fourier transform may be computed by a sequence of one-dimensional Fourier transforms. You may first transform each column vector of the matrix and then transform each row vector of the matrix. Alternately you may first do the rows and later do the columns. This is diagramed as follows:

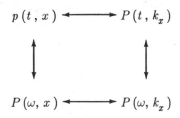

A notational problem on the diagram is that we cannot maintain the usual convention of using a lower-case letter for the domain of physical space and an upper-case letter for the Fourier domain, because that convention cannot include the mixed objects $P(t, k_x)$ and $P(\omega, x)$. Rather than invent some new notation it seems best to let the reader use the context to cope with this notational problem. The arguments of the function must help name the function.

An example of these transformations on typical deep-ocean data is shown in figure 5.

In the deep ocean, sediments are fine-grained and deposit slowly in flat, regular, horizontal beds. The lack of permeable rocks like sandstone severely reduces the potential for petroleum production from the deep ocean. The fine-grained shales overlay irregular, igneous, basement rocks. In the plot of $P(t, k_x)$ the lateral continuity of the sediments is shown by the strong spectrum at low k_x. The igneous rocks show a k_x spectrum extending to such large k_x that the deep data may be somewhat spatially aliased (sampled too coarsely). The plot of $P(\omega, x)$ shows that the data contains no low-frequency energy. At large ω the energy is not dropping off as fast as one might like, which indicates temporal frequency aliasing. This aliasing is also apparent in the plot of $p(t, x)$ in the steplike appearance of the sea-floor

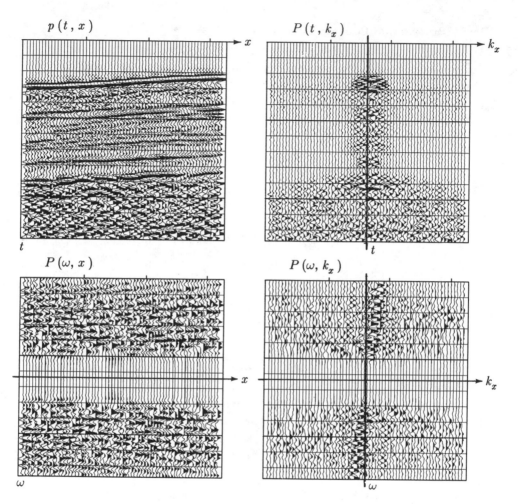

FIG. 1.2-5. A deep-marine dataset $p(t,x)$ from Alaska (U.S. Geological Survey) and the real part of various Fourier transforms of it. Because of the long travel time through the water, the time axis does not begin at $t = 0$.

arrival. The dip of the sea floor shows up in (ω, k_x)-space as the energy crossing the origin at an angle.

Altogether, the two-dimensional Fourier transform of a collection of seismograms involves only twice as much computation as the one-dimensional Fourier transform of each seismogram. This is lucky. Let us write some equations to establish that the asserted procedure does indeed do a two-dimensional Fourier transform. Say first that any function of x and t

may be expressed as a superposition of sinusoidal functions:

$$p(t,x) = \iint e^{-i\omega t + i k_x x} \, P(\omega, k_x) \, d\omega \, dk_x \qquad (9)$$

(Sign convention used in Fourier transformation is explained in Section 1.6). The kernel in this *inverse* Fourier transform has the form of a wave moving in the plus x direction. Likewise, in the *forward* Fourier transform, the signs of both exponentials change, preserving the fact that the kernel is a wave moving positively. The scale factor and the infinite limits are omitted as a matter of convenience. (The limits and scale both differ from the sampled-time computation, so why bother?) The double integration can be nested to show that the temporal transforms are done first (inside):

$$p(t,x) = \int e^{i k_x x} \left[\int e^{-i\omega t} \, P(\omega, k_x) \, d\omega \right] dk_x$$

$$= \int e^{i k_x x} \, P(t, k_x) \, dk_x$$

The quantity in brackets is a Fourier transform over ω done for each and every k_x. Alternately, the nesting could be done with the k_x-integral on the inside. That would imply rows first instead of columns (or vice versa). It is the separability of $\exp(-i\omega t + i k_x x)$ into a product of exponentials that makes the computation this easy and cheap.

The Input-Output Relation

At the heart of the migration process is the operation of downward continuing data. Given the input data on the plane of the earth's surface $z = 0$, we must manufacture the data that could be recorded at depth z. This is most easily done in the Fourier domain. The method will be seen to be simply multiplication by a complex exponential, namely,

$$P(\omega, k_x, z) = P(\omega, k_x, 0) \, e^{i k_z(\omega, k_x) z} \qquad (10)$$

Since the operation is a multiplication in the Fourier domain, it may be described as an engineering diagram.

Filter

input   *output*

$$e^{i \sqrt{\omega^2/v^2 - k_x^2}\, z}$$

$P(\omega, k_x, 0)$ $P(\omega, k_x, z)$

Downward continuation is a product relationship in both the ω-domain and the k_x-domain. What does the filter look like in the time and space domain? It turns out like a cone, that is, it is roughly an impulse function of $x^2 + z^2 - v^2 t^2$. More precisely, it is the Huygens secondary wave source that was exemplified by ocean waves entering a gap through a storm barrier. Adding up the response of multiple gaps in the barrier would be convolution over x. Superposing many incident ocean waves would be convolution over t.

Now let us see why the downward continuation filter has the mathematical form stated. Every point in the (ω, k_x)-plane refers to a sinusoidal plane wave. The variation with depth will also be sinusoidal, namely $\exp(ik_z z)$. The value of k_z for the plane wave is found simply by solving equation (8):

$$k_z = \pm \left(\frac{\omega^2}{v^2} - k_x{}^2 \right)^{1/2} \tag{11a}$$

$$= \pm \frac{\omega}{v} \left\{ 1 - \frac{v^2 k_x{}^2}{\omega^2} \right\}^{1/2} \tag{11b}$$

$$= \pm \frac{\omega}{v} \cos \theta \tag{11c}$$

Choice of the *plus* sign means that $\exp(-i\omega t + i k_z z)$ is a *down*going wave (because the phase will stay constant if z increases as t increases). Choice of a *minus* sign makes the wave *up*coming. The exploding-reflector concept requires upcoming waves, so we nearly always use the minus sign, whether we are migrating or modeling.

The input-output filter, being of the form $e^{i\phi}$, appears to be a phase-shifting filter with *no amplitude scaling*. This bodes well for our plans to deconvolve. It means that signal-to-noise power considerations will be much less relevant for migration than for ordinary filtering.

EXERCISES

1. Suppose that you are able to observe some shear waves at ordinary seismic frequencies. Is the spatial resolution better, equal, or worse than usual? Why?

2. Scan this book for hyperbolic arrivals on field data and measure the Fresnel zone width. Where zero offset recordings are not made, a valid approximation is to measure Δx_2 along a *tilted* line.

3. Explain the horizontal "layering" in figure 1.2-5 in the plot of $P(\omega, x)$. What determines the "layer" separation? What determines the "layer"

slope?

4.   Evolution of a wavefield with time is described by

$$p\,(x\,,z\,,t\,) = \int\!\int \left[P\,(k_x\,,k_z\,,t=0)\, e^{-i\,\omega(k_x\,,\,k_z\,)t} \right] e^{ik_x\,x\,+ik_z\,z}\, dk_x\ dk_z$$

Let $P\,(k_x\,,k_z\,,0)$ be constant, signifying a point source at the origin in $(x\,,z\,)$-space. Let t be very large, meaning that phase $= \phi = [-\omega(k_x\,,k_z\,) + k_x\,(x/t\,) + k_z\,(z/t\,)]t$ in the integration is rapidly alternating with changes in k_x and k_z. Assume that the only significant contribution to the integral comes when the phase is stationary, that is, where $\partial\phi/\partial k_x$ and $\partial\phi/\partial k_z$ both vanish. Where is the energy in $(x\,,z\,,t\,)$-space?

5.   Downward continuation of a wave is expressed by

$$p\,(x\,,z\,,t\,) = \int\!\int \left[P\,(k_x\,,z=0,\,\omega)\, e^{ik_z\,(\omega,\,k_x\,)z} \right] e^{-i\,\omega t\,+ik_x\,x}\, d\omega\ dk_x$$

Let $P\,(k_x\,,0,\,\omega)$ be constant, signifying a point source at the origin in $(x\,,t\,)$-space. Where is the energy in $(x\,,z\,,t\,)$-space?

1.3 Four Wide-Angle Migration Methods

The four methods of migration of reflection seismic data that are described here are all found in modern production environments. As a group they handle wide-angle rays easily. As a group they are used less successfully to deal with lateral velocity variation.

Travel-Time Depth

Conceptually, the output of a migration program is a picture in the $(x\,,z\,)$-plane. In practice the vertical axis is almost never depth z; it is the *vertical travel time τ*. In a constant-velocity earth the time and the depth are related by a simple scale factor. The meaning of the scale factor is that the $(x\,,\tau)$-plane has a vertical exaggeration compared to the $(x\,,z\,)$-plane. In reconnaissance work, the vertical is often exaggerated by about a factor of five. By the time prospects have been sufficiently narrowed for a drill site to be selected, the vertical exaggeration factor in use is likely to be about unity (no exaggeration).

The travel-time depth τ is usually defined to include the time for both the wave going down and the wave coming up. The factor of two thus introduced quickly disappears into the rock velocity. Recall that zero-offset data sections are generally interpreted as exploding-reflector wavefields. To make the correspondence, the rock velocity is cut in half for the wave analysis:

$$\tau = \frac{2z}{v_{true}} = \frac{z}{v_{half}} \tag{1}$$

The first task in interpretation of seismic data is to figure out the approximate numerical value of the vertical exaggeration. It probably won't be printed on the data header because the seismic velocity is not really known. Furthermore, the velocity usually *increases* with depth, which means that the vertical exaggeration *decreases* with depth. For velocity-stratified media, the time-to-depth conversion formula is

$$\tau(z) = \int_0^z \frac{dz}{v(z)} \qquad \text{or} \qquad \frac{d\tau}{dz} = \frac{1}{v} \tag{2}$$

Hyperbola-Summation and Semicircle-Superposition Methods

The methods of hyperbola summation and semicircle superposition are the most comprehensible of all known methods.

Recall the equation for a conic section, that is, a circle in (x, z)-space or a hyperbola in (x, t)-space. Converting to travel-time depth τ

$$x^2 + z^2 = v^2 t^2 \tag{3a}$$

$$\frac{x^2}{v^2} + \tau^2 = t^2 \tag{3b}$$

Figure 1 illustrates the *semicircle-superposition* method. (Both the figure and its caption are from Schneider's classic paper [1971]). Taking the data field to contain a few impulse functions, the output should be a superposition of the appropriate semicircles. Each semicircle denotes the spherical-reflector earth model that would be implied by a dataset with a single pulse. Taking the data field to be one thousand seismograms of one thousand points each, then the output is a superposition of one million semicircles. Since a seismogram has both positive and negative polarities, about half the semicircles will be superposed with negative polarities. The resulting superposition could look like almost anything. Indeed, the semicircles might mutually destroy one another almost everywhere except at one isolated impulse in (x, τ)-

space. Should this happen you might rightly suspect that the input data section in (x, t)-space is a Huygens secondary source, namely, energy concentrated along a hyperbola. This leads us to the *hyperbola-summation* method.

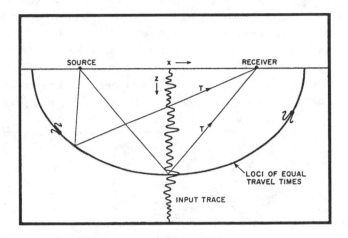

FIG. 1.3-1. The process may be described in numerous ways. Two very simple and equally valid representations are indicated in figures 1 and 2. Shown here is a representation of the process in terms of what happens to a single input trace plotted in depth (time may also be used) midway between its source and receiver. Each amplitude value of this trace is mapped into the subsurface along a curve representing the loci of points for which the travel time from source to reflection point to receiver is constant. If the velocity is constant, these curves are ellipses with source and receiver as foci. The picture produced by this operation is simply a wavefront chart modulated by the trace amplitude information. This clearly is not a useful image in itself, but when the map is composited with similar maps from neighboring traces (and common-depth-point traces of different offsets), useful subsurface images are produced by virtue of constructive and destructive interference between wavefronts in the classical Huygens sense. For example, wavefronts from neighboring traces will all intersect on a diffraction source, adding constructively to produce an image of the diffractor as a high-amplitude blob whose (z, x) resolution is controlled by the pulse bandwidth and the horizontal aperture of the array of neighboring traces composited. For a reflecting surface, on the other hand, wavefronts from adjacent traces are tangent to the surface and produce an image of the reflector by constructive interference of overlapping portions of adjacent wavefronts. In subsurface regions devoid of reflecting and scattering bodies, the wavefronts tend to cancel by random addition. (from Schneider, W. A., 1971 [by permission])

The *hyperbola-summation* method of migration is depicted in figure 2. The idea is to create one point in (x, τ)-space at a time, unlike in the semicircle method, where each point in (x, τ)-space is built up bit by bit as the one million semicircles are stacked together. To create one fixed point in the output (x, τ)-space, imagine a hyperbola, equation (3b), set down with its top on the corresponding position of (x, t)-space. All data values touching the hyperbola are added together to produce a value for the output at the appropriate place in (x, τ)-space. In the same way, all other locations in (x, τ)-space are filled. We can wonder whether the hyperbola-migration method is better or worse than or equivalent to the semicircle method.

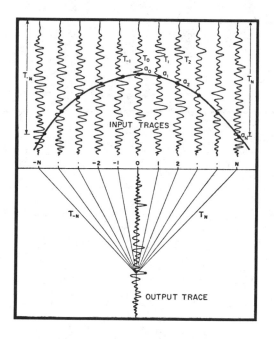

FIG. 1.3-2. A second description of the process is provided here. The process is represented in terms of how an output trace is developed from an ensemble of input traces, shown as CDP-stacked traces in the upper half of the figure. The output in the lower half reflects how each amplitude value at (x, z) is obtained by summing input amplitudes along the travel-time curve shown. This curve defines a diffraction hyperbola. If a diffraction source existed in the subsurface at the output point shown, then a large amplitude would result. The process also works for reflectors since a reflector may be regarded as a continuum of diffracting elements whose individual images merge to produce a smooth continuous boundary. (also from Schneider, 1971)

The opposite of data processing or building models from data is constructing synthetic data from models. With a slight change, the above two processing programs can be converted to modeling programs. Instead of *hyperbola summation* or *semicircle superposition,* you do *hyperbola superposition* or *semicircle summation.* We can also wonder whether the processing programs really are inverse to the modeling programs. Some factors that need to be considered are (1) the angle-dependence of amplitude (the obliquity function) of the Huygens waveform, (2) spherical spreading of energy, and (3) the phase-shift on the Huygens waveform. It turns out that results are reasonably good even when these complicating factors are ignored.

As other methods of migration were developed, the deficiencies of the earlier methods were more clearly understood and found to be largely correctable by careful implementation. One advantage of the later methods is that they implement true all-pass filters. Such migrations preserve the general appearance of the data. This suggests restoration of high frequencies, which tend to be destroyed by hyperbolic integrations. Work with the Kirchhoff diffraction integral by Trorey [1970] and Hilterman [1970] led to forward modeling programs. Eventually (Schneider [1977]) this work suggested quantitative means of bringing hyperbola methods into agreement with other methods, at least for constant velocity. Common terminology nowadays is to refer to any hyperbola or semicircular method as a Kirchhoff method, although, strictly speaking, the Kirchhoff integral applies only in the constant-velocity case.

Spatial Aliasing

Spatial aliasing means insufficient sampling of the data along the space axis. This difficulty is so universal, that all migration methods must consider it.

Data should be sampled at more than two points per wavelength. Otherwise the wave arrival direction becomes ambiguous. Figure 3 shows synthetic data that is sampled with insufficient density along the x-axis. You can see that the problem becomes more acute at high frequencies and steep dips.

There is no generally-accepted, automatic method for migrating spatially aliased data. In such cases, human beings may do better than machines, because of their skill in recognizing true slopes. When the data is adequately sampled, however, computer migration based on the wave equation gives better results than manual methods. Contemporary surveys are usually adequately sampled along the line of the survey, but there is often difficulty in the perpendicular direction.

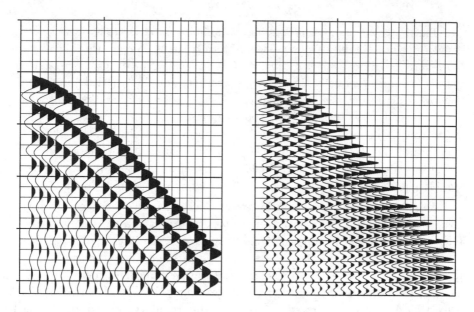

FIG. 1.3-3. Insufficient spatial sampling of synthetic data. To better perceive the ambiguity of arrival angle, view the figures at a grazing angle from the side.

The hyperbola-sum-type methods run the risk of the migration operator itself becoming spatially aliased. This should be avoided by careful implementation. The first thing to realize is that you should be *integrating* along a hyperbolic trajectory. A summation incorporating only one point per trace is a poor approximation. It is better to incorporate more points, as depicted in figure 4. The likelihood of getting an aliased operator increases where the hyperbola is steeply sloped. In production examples an aliased operator often stands out above the sea-floor reflection, where — although the sea floor may be flat — it acquires a noisy precursor due to the steeply flanked hyperbola crossing the sea floor.

The Phase-Shift Method (Gazdag)

The phase-shift method proceeds straightforwardly by extrapolating downward with $\exp(ik_z z)$ and subsequently evaluating the wavefield at $t = 0$ (the reflectors explode at $t = 0$). Of all the wide-angle methods it most easily incorporates depth variation in velocity. Even the phase angle and obliquity function are correctly included, automatically. Unlike Kirchhoff methods, with this method there is no danger of aliasing the operator.

FIG. 1.3-4. For a low-velocity hyperbola, integration will require more than
one point per channel.

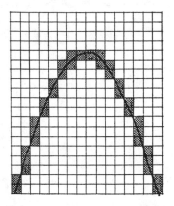

The phase-shift method begins with a two-dimensional Fourier transform
(2D-FT) of the dataset. (Some practical details about 2D-FT are described in
Section 1.7). Then the transformed data values, all in the (ω, k_x)-plane, are
downward continued to a depth Δz by multiplying by

$$e^{i\,k_z\Delta z} \;=\; \exp\left\{-i\,\frac{\omega}{v}\left[1-\left(\frac{v\,k_x}{\omega}\right)^2\right]^{1/2}\Delta z\right\} \tag{4}$$

Ordinarily the time-sample interval $\Delta\tau$ for the output-migrated section is
chosen equal to the time-sample rate of the input data (often 4 milliseconds).
Thus, choosing the depth $\Delta z = v\,\Delta\tau$, the downward-extrapolation operator
for a single time unit is

$$C \;=\; \exp\left\{-i\,\omega\,\Delta\tau\left[1-\left(\frac{v\,k_x}{\omega}\right)^2\right]^{1/2}\right\} \tag{5}$$

Data will be multiplied many times by C, thereby downward continuing it
by many steps of $\Delta\tau$.

Next is the task of imaging. At each depth an inverse Fourier transform
is followed by selection of its value at $t=0$. (Reflectors explode at $t=0$).
Luckily, only the Fourier transform at one point, $t=0$, is needed, so that is
all that need be computed. The computation is especially easy since the value
at $t=0$ is merely a summation of each ω frequency component. (This may
be seen by substituting $t=0$ into the inverse Fourier integral). Finally,
inverse Fourier transform k_x to x. The migration process, computing the
image from the upcoming wave u, may be summarized as follows:

$U(\omega, k_x) \;=\; FT[\,u(t,x)\,]$

For $\tau = \Delta\tau,\ 2\Delta\tau,\ \cdots$, end of time axis on seismogram {

     For all k_x {

         $Image(k_x, \tau) \;=\; 0.$

         For all ω {

             $C \;=\; \exp(-\,i\,\omega\,\Delta\tau\,\sqrt{1 - v^2\,k_x^2/\omega^2}\,)$

             $U(\omega, k_x) \;=\; U(\omega, k_x) * C$

             $Image(k_x, \tau) \;=\; Image(k_x, \tau) \;+\; U(\omega, k_x)$

             }

         }

     $image(x, \tau) \;=\; FT[Image(k_x, \tau)]$

}

Inverse migration (modeling) proceeds in much the same way. Beginning from an upcoming wave that is zero at great depth, the wave is marched upward in steps by multiplication with $\exp(i\,k_z\,\Delta z)$. As each level in the earth is passed, exploding reflectors from that level are added into the upcoming wave. The program for modeling the upcoming wave u is

$Image(k_x, z) \;=\; FT[\,image(x, z)\,]$

For all ω and all k_x

     $U(\omega, k_x) \;=\; 0.$

For all ω {

For all k_x {

For $z = z_{max},\ z_{max}-\Delta z,\ z_{max}-2\Delta z,\ \cdots,\ 0$ {

     $C \;=\; \exp(+\,i\,\Delta z\,\omega\,\sqrt{v^{-2} - k_x^2/\omega^2}\,)$

     $U(\omega, k_x) \;=\; U(\omega, k_x) * C \;+\; Image(k_x, z)$

     } } }

$u(t, x) \;=\; FT[\,U(\omega, k_x)\,]$

The positive sign in the complex exponential is a combination of two negatives, the *up*coming wave and the *up*ward extrapolation. The three loops on ω, k_x, and z are interchangeable. When the velocity v is a constant function of depth the program can be speeded by moving the computation of the complex exponential C out of the inner loop on z.

The velocity is hardly ever known precisely, so although it may be increasing steadily with depth, it is often approximated as constant in layers instead of slowly changing at each of the thousand or so time points on a seismogram. The advantage of this approximation is economy. Once the square root and the sines and cosines in (5) have been computed, the complex multiplier (5) can be reused many times. With a 4-millisecond sample rate and a layer 200 milliseconds thick, the complex multiplier gets used 50 times before it is abandoned.

The Stolt Method

On most computers the Stolt method of migration is the fastest one — by a wide margin. For many applications, this will be its most important attribute. For a constant-velocity earth it incorporates the Huygens wave source exactly correctly. Like the other methods, this migration method can be reversed and made into a modeling program. One drawback, a matter of principle, is that the Stolt method does not handle depth variation in velocity. This drawback is largely offset in practice by an approximate correction that uses an axis-stretching procedure (Section 4.5). A practical problem is the periodicity of all the Fourier transforms. In principle this is no problem at all, since it can be solved by adequately surrounding the data by zeroes.

A single line sketch of the Stolt method is this:

$$p(x,t) \;\rightarrow\; P(k_x,\omega) \;\rightarrow\; P'(k_x, k_z = \sqrt{\omega^2/v^2 - k_x{}^2}) \;\rightarrow\; P(x,z)$$

To see why this works, begin with the input-output relation for down-ward extrapolation of wavefields:

$$P(\omega, k_x, z) \;=\; e^{i\,k_z z}\, P(\omega, k_x, z = 0) \tag{6}$$

Perform a two-dimensional inverse Fourier transform:

$$p(t, x, z) \;=\; \iint e^{i\,k_x x - i\omega t + i\,k_z z}\, P(\omega, k_x, 0)\, d\omega\, dk_x$$

Apply the idea that the image at (x,z) is the exploding-reflector wave at time $t = 0$:

$$Image(x,z) \;=\; \iint e^{i\,k_x x}\, e^{i\,k_z(\omega, k_x) z}\, P(\omega, k_x, 0)\, d\omega\, dk_x \tag{7}$$

Equation (7) gives the final image, but it is in an unattractive form, since it implies that a two-dimensional integration must be done for each and every z-level. The Stolt procedure converts the three-dimensional calculation thus implied by (7) to a single two-dimensional Fourier transform.

So far nothing has been done to specify an *upcoming* wave instead of a downgoing wave. The direction of the wave is defined by the relationship of z and t that is required to keep the phase constant in the expression $\exp(-i\omega t + ik_z z)$. If ω were always positive, then $+k_z$ would always refer to a downgoing wave and $-k_z$ to an upcoming wave. Negative frequencies ω as well as positive frequencies are needed to describe waves that have real (not complex) values. So the proper description for a downgoing wave is that the signs of ω and k_z must be the same. The proper description for an upcoming wave is the reverse. With this clarification the integration variable in (7) will be changed from ω to k_z.

$$\omega = -sgn(k_z)\, v\, \sqrt{k_x^2 + k_z^2} \tag{8a}$$

$$\frac{d\omega}{dk_z} = -sgn(k_z)\, v\, \frac{k_z}{\sqrt{k_x^2 + k_z^2}} \tag{8b}$$

$$\frac{d\omega}{dk_z} = \frac{-v\,|k_z|}{\sqrt{k_x^2 + k_z^2}} \tag{8c}$$

Put (8) into (7), and include also a minus sign so that the integration on k_z goes from minus infinity to plus infinity as was the integration on ω.

$$Image(x,z) = \tag{9}$$

$$= \int\int e^{ik_z z + ik_x x} \left\{ P[\omega(k_x, k_z), k_x, 0] \frac{v\,|k_z|}{\sqrt{k_x^2 + k_z^2}} \right\} dk_z\, dk_x$$

Equation (9) states the result as a two-dimensional inverse Fourier transform. The Stolt migration method is a direct implementation of (9). The steps of the algorithm are

1.   Double Fourier transform data from $p(t,x,0)$ to $P(\omega, k_x, 0)$.

2.   Interpolate P onto a new mesh so that it is a function of k_x and k_z. Multiply P by the scale factor (which has the interpretation $\cos\theta$).

3.   Inverse Fourier transform to (x,z)-space.

Samples of Stolt migration of impulses are shown in figure 5. You can see the expected semicircular smiles. You can also see a semicircular frown hanging from the bottom of each semicircle. The worst frown is on the deepest spike. The semicircular mirrors have centers not only at the earth's surface $z = 0$ but also at the bottom of the model $z = z_{max}$. It is known that these frowns can be suppressed by interpolating more carefully. (Interpolation is the way you convert from a uniform mesh in ω, to a uniform mesh in k_z). Interpolate with say a *sinc* function instead of a linear interpolator. (See Section 4.5). A simpler alternative is to stay away from the bottom of the model, i.e. pad with lots of zeroes.

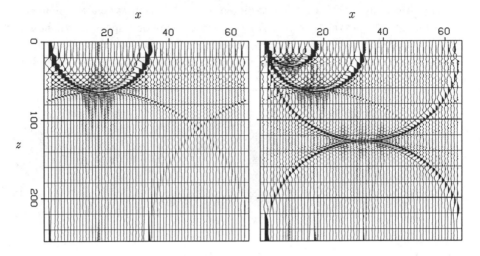

FIG. 1.3-5. Response of Stolt method to data with impulses. Semicircles are seen, along with computation artifacts.

It seems that an extraordinary amount of zero padding is required on the time axis. To keep memory requirements reasonable, the algorithm can be reorganized as described in an exercise. Naturally, the periodicity in x also requires padding the x-axis with zeroes.

Hyperbola Summation Refined into the Kirchhoff Method

Schneider [1978] states the analytic representation for the Huygens secondary wavelet

$$FT^{-1}(e^{i\,k_z\,z}) \;=\; \frac{1}{\pi} \frac{\partial}{\partial z_0} \frac{\text{step}(t - r/v)}{\sqrt{t^2 - r^2/v^2}} \tag{10}$$

where r is the distance $\sqrt{x^2 + (z-z_0)^2}$ between the (exploding reflector) source and the receiver. The function (10) contains a pole and the derivative of a step function. Because of the infinities it really cannot be graphed. But from the mathematical form you immediately recognize that the disturbance concentrates on the expected cone. The derivative of the step function gives a positive impulsive arrival on the cone. The derivative of the inverse square root gives the impulse a tail of negative polarity decaying with a $-3/2$ power. The cosine obliquity arises because the derivative is a z derivative and not an *r* derivative.

Equation (10) states the *two*-dimensional Huygens wavelet, not the *three*-dimensional wavelet (which differs in some minor aspects). Although waves from point sources are mainly spherical, the focusing of bent layers is mainly a two-dimensional focusing, i.e., bent layers are more like cylinders than spheres.

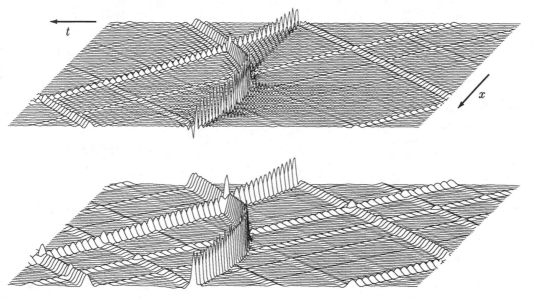

FIG. 1.3-6. The Huygens wavelet (top) and a smoothed time integral (bottom).

You might wonder why anyone would prefer approximations, given the exact inverse transform (10). The difficulty of graphing (10) shows up in practice as a difficulty in convolving it with data. That is why early Kirchhoff migrations were generally recognizable by precursor noise above a flat sea floor. Chapters 2 and 4 are largely devoted to extensions of (10) that are valid with variable velocity and that are better representations on a data mesh.

In the Fourier domain, the Huygens secondary source function is simple and smooth. It is a straightforward matter to evaluate the function on a rectangular mesh and inverse transform with the programs in Section 1.7. Figure 6 shows the result on a 256 × 64 point mesh. (In practice the mesh would be about 1024 × 256 or more, but the coarser mesh used here provides a plot of suitable detail). Because of the difficulty in plotting functions that resemble an impulsive doublet, a second plot of the time integral, (with gentle band limiting) is displayed in the lower part of figure 6.

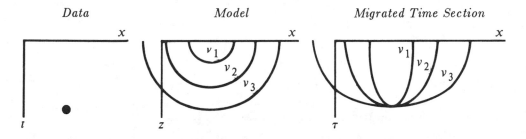

FIG. 1.3-7. Velocity error sensitivity increases with angle up to 90°. Migration of a data impulse as a function of velocity. Three possible choices of constant velocity are shown superposed on one plane.

Sensitivity of Migration to Velocity Error

Figure 7 shows how the migration impulse response depends on velocity. Recall that migrated data is ordinarily displayed as a *time* section. Arbitrary velocity error makes no difference for horizontal bedding.

Different people have different accuracy criteria. A reasonable criterion is that the positioning error of the energy in the semicircles should be less than a half-wavelength. For the energy moving horizontally, the positioning error is simply related to the dominant period ΔT and the travel time T. The ratio $T/\Delta T$ is rarely observed to exceed 100. This 100 seems to be a fundamental observational parameter of reflection seismology in sedimentary rock. (Theoretically, it might be related to the "Q" of sedimentary rock or it may relate to generation of chaotic internal multiple reflections. Larger values than 100 occur when (1) much of the path is in water or (2) at time depths greater than about 4 seconds). Figure 8 compares two nearby migration velocities. The separation of the curves increases with angle. For the separation to be less than a wavelength, for 90° dip the velocity error must be less than one part in 100. For 45° migration velocity error could be larger by $\sqrt{2}$.

FIG. 1.3-8. Timing error of the wrong velocity increases with angle.

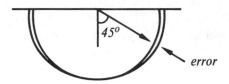

Velocities are rarely known this accurately. So we may question the value of migration at wide angles.

Subjective Comparison and Evaluation of Methods

The three basic methods of migration described in this section are compared subjectively in table 1.

	Hyperbola Sum or Semicircle Superpose	Phase Shift	Stolt
$v(z)$	ray tracing	easily	approximately by stretching
wide angle?	Beware of data alias and operator alias.	Beware of data alias.	Beware of data alias.
correct phase and obliquity?	possible with some effort for const v	easy for any $v(z)$	for const v
wraparound noise?	no	on x, see Section 4.5 to alleviate on t	on x, see Section 4.5 to alleviate on (t, z)
$v(x)$	Production programs have serious pitfalls.	approximately by iteration and interpolation	no known program
side boundaries and irregular spacing	excellent	poor	poor
Speed	slow	average	very fast
memory organization	awkward	good	good

TABLE 1.3-1. Subjective comparison of three wide-angle migration methods.

The perspective of later chapters makes it possible to remark on the quality of the wide-angle methods as a group, and it is useful to do that now. Their greatest weakness is their difficulties with lateral velocity variation. Their greatest strength, wide-angle capability, is reduced by the weakness of other links in the data collection and processing chain, namely:

1. Shot-to-geophone offset angles are often large but ignored. A CDP stack is not a zero-offset section.

2. Why process to the very wide angles seen in the survey line when even tiny angles perpendicular to the line are being ignored?

3. Data is often not sampled densely enough to represent steeply dipping data without aliasing.

4. Accuracy in the knowledge of velocity is seldom enough to justify processing to wide angles.

5. Noise eventually overpowers all echoes and this also implies an angle cutoff. For example, imagine oil reservoirs at a time depth of two seconds, where data recording stopped at four seconds. The implied angle cutoff is at 60°.

EXERCISES

1. The wave modeling program sketch assumes that the exploding reflectors are impulse functions of time. Modify the program sketch for wave modeling to include a source waveform $s(t)$.

2. The migration program sketch allows the velocity to vary with depth. However the program could be speeded considerably when the velocity is a constant function of depth. Show how this could be done.

3. Define the program sketch for the inverse to the Stolt algorithm — that is, create synthetic data from a given model.

4. The Stolt algorithm can be reorganized to reduce the memory requirement of zero padding the time axis. First Fourier transform x to k_x. Then select, from the (t, k_x)-plane of data, vectors of constant k_x. Each vector can be moved into the space of a long vector, then zero padded and interpolated. Sketch the implied program.

5. Given seismic data that is cut off at four seconds, what is the deepest travel time depth from which 80° dips can be observed?

1.4 The Physical Basis

Previous sections have considered the *geometrical* aspects of wave propagation and how they relate to seismic imaging. Now we will consider how the *physical* aspects relate to imaging. The propagation medium has a mass density and compressibility. The waves have a material acceleration vector and a pressure gradient. Static deformation, ground roll, shear, rigidity, dissipation, sedimentary deposition — how are these related to image construction?

Because of the complexity of sedimentary rocks, there is not universal agreement on an appropriate mathematical description. To help you understand the degree to which theory can be used as a guide, I will point to some inconsistencies between theory and current industrial practice.

The Clastic Section

Generally speaking, most petroleum reservoir rocks are sandstones. Sandstones are most often made by the sands that are deposited near the mouths of rivers where the water velocity is no longer sufficient to move them. The sands deposit along the terminus of the sand bars found at the river mouth, often along a slope of 25° or so, as depicted in figure 1. Although the sands are not laid down in flat layers, the process may build a horizontal layer.

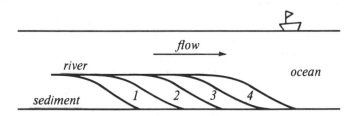

FIG. 1.4-1. Sands (petroleum reservoir rocks) deposit on fairly steep slopes where rivers run into the ocean.

Clays are more fine-grained materials (dirt) that are carried out to deeper water before they settle to make shale. Shale deposits tend to be layered somewhat more horizontally than sandstones. Specific locations of sand deposition change with the passing of storms and seasons, leaving a wood-grain-like appearance in the rock.

The river delta itself is a complicated, ever-changing arrangement of channels and bars, constantly moving along the coastline. At any one time the delta seems to be moving seaward as deposits are left, but later settling, compression, or elevation of sea level can cause it to move landward.

Sand is important because its porosity enables oil to accumulate and its permeability enables the oil to move to a well. Shale is important because it contains the products of former life on earth, and their hydrocarbons. These escape to nearby sands, but often not to the earth's surface, because of covering impermeable shales. The acoustic properties of sands and shales often overlap, though there is a slight tendency for shales to have a lower velocity than sands. Geophysicists on the surface see with seismic wavelengths ($\approx$ 30 meters) the final interbedded three-dimensional mixture of sands and shales.

Mixtures of sands and shales are called *clastic* rocks. The word *clast* means *break*. Clastic rocks are made from broken bits of crystalline rock. Most sedimentary rocks are clastic rocks. Most oil is found in clastic rocks. But much oil is also found in association with carbonates such as limestone. Carbonates are formed in shallow marine environments by marine organisms. Many carbonates (and clastics) contain oil that cannot be extracted because of lack of permeability. Permeability in carbonate rocks arises through several obscure processes. The seismologist knows carbonates as rocks with greater velocity than clastic rocks. Typically, a carbonate has a 20-50% greater velocity than a nearby clastic rock. Clastic rock sometimes contains limestone, in which case it is called *marl*.

Chrono-Stratigraphy

Strange as it may seem, there is not universal agreement about the exact nature of seismic reflections. Physicists tend to think of the reflections as caused by the interface between rock types, as a sand-to-shale contact. The problem with this view is that sands and shales interlace in complex ways, both larger and smaller than the seismic wavelength. Many geologists, particularly a group known as *seismic stratigraphers,* have a different concept. (See *Seismic Stratigraphy — Applications to Hydrocarbon Exploration,* memoir 26 of the American Association of Petroleum Geologists). They have studied thousands of miles of reflection data along with well logs. They believe a reflection marks a constant geological time horizon. They assert that a long, continuous reflector could represent terrigenous deposition on one end and marine deposition on the other end with a variety of rock types in between. Data interpretation based on this assumption is called *chrono-stratigraphy*. The view of the seismic stratigrapher seems reasonable enough for areas that are wholly clastic, but when carbonates and other rocks are present, the physicist's view seems more appropriate. For further details, the book of Sheriff [1980] is recommended.

Nonobservation of Converted Shear Waves

In earthquake seismology and in laboratory measurement there are two clearly observed velocities. The faster velocity is a pressure wave (P-wave), and the slower velocity is a shear wave (S-wave). The shear wave can be polarized with ground motion in a horizontal plane (SH) or in a vertical plane (SV). Theory, field data, and laboratory measurement are in agreement. Successful experimental work with S-waves in the prospecting environment was done by Cherry and Waters [1968] and Erickson, Miller and Waters [1968].

It is remarkable that more than 99% of industrial petroleum prospecting ignores the existence of shear waves. Mathematically the earth is treated as if it were a liquid or a gas. The experimental work with shear waves used special equipment to generate and record vibration perpendicular to the survey line, i.e. SH-waves. The picture of the earth given by these transverse waves is often impaired by the soil layers, but sometimes the SH-wave picture is clear and consistent. Surprisingly, even good SH-wave data is often difficult to relate to the P-wave picture. These experimental studies show that the shear waves typically travel about half the speed of the pressure waves except in the soil layer, where the shear wave speed is often much slower and more variable. Observed shear waves usually have lower frequency than pressure waves. A shear wave with half the frequency and half the velocity of a pressure wave has just the same wavelength and hence the same resolving power as the pressure wave. Indeed, experimental work shows that shear waves do offer us about the same spatial resolution as pressure waves. Most land seismic data shows only the vertical component of motion, and all marine seismic data records the pressure. So in the conventional recording geometry, ideally we should never see SH-waves. More precisely, SH-waves should be small, arising only from the earth's departure from simple stratification.

The puzzling aspect of shear waves in reflection seismology is the failure of petroleum prospectors using the standard operating arrangement to routinely observe P-to-S conversions. Theory predicts that P-waves hitting an interface at an angle should be partially converted to SV-waves.

Furthermore, for the 30°-60° angle reflections that are routinely encountered, these converted waves should have a size comparable to the P-wave.

The routine geometry of recording and processing discriminates somewhat against converted shear waves. But it discriminates against multiple reflections too (in much the same way) and we see multiples all the time. Furthermore the signature of converted waves should resemble that of multiples, but be distinctly different. Converted waves should show up routinely in velocity surveys (Chapter 3). Figure 2 shows a zero-offset section containing some multiple reflections. The multiple reflection is recognizable as a replica of earlier topography. Converted waves would replicate the topography but the time scaling would be in the ratio of about 3/2 instead of exactly 4/2. With a sufficiently complex topography, as in figure 2, the probability is low that the converted wave would be mistaken for another primary reflection.

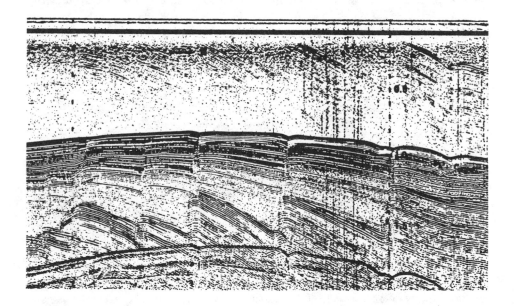

FIG. 1.4-2. A zero-offset section from east Africa with multiple reflections. (Teledyne)

Converted waves should have good diagnostic value in exploration. But the likelihood of seeing converted shear waves in conventional data seems to be so remote that most interpreters have given up looking. Why aren't converted waves seen in conventional data? Some reasons can be offered:

1. In marine data there would have to be an additional conversion to P for the water path.

2. In land data the soil is especially absorptive of shear.

3. Vertical component recorders tend to see P better than S. This is especially so because rays bend toward the vertical in the near surface.

Of all the reasons why converted shear waves should be weaker than pressure waves, none is overwhelming. A wide range of amplitudes are recorded in a wide variety of environments. Data are often displayed with automatic gain control (AGC). Weakened amplitude appears to be insufficient cause for the failure of observation. We should keep looking. Converted waves are certainly more prevalent than our recognition of them. (I have never identified a converted shear wave on conventional recordings).

So although converted shear waves might some day play a significant role in reflection seismology, we now return to the mainstream — how to deal effectively with that which is routinely observable.

Reliability of Reverberation Modeling

The seismological literature contains an abundance of theory to describe seismic waves in layered media. A significant aspect of applied seismology is the general neglect of intrabed reverberation. When a wave reflects from an interface, the strength of the reflected wave is a small fraction, typically less than 10%, of the strength of the incident wave. This reflected wave is the one that is mainly dealt with in this book. However, the reflected wave itself reflects again and again, ad infinitum. For short path geometries, there can be very many of these rays. The question is whether these reverberations can ever amount to enough to make considering them worthwhile. The answer seems to be that although reverberation may be significant, seismologists are rarely able to improve interpretation of reflection survey data with the more complicated theory that is required to incorporate reverberation. A few more details are in Section 5.5.

The situation is somewhat improved when well logs are available, but even then there are serious difficulties. The best possible lateral resolving power, say about 20-50 meters, is obtained after migration. The well log, however, is not a 20-50 meter lateral average of the earth. Next time you see a highway cut through sedimentary rocks, think of the difference between a point and a lateral average over 20-50 meters. In practice, people smooth the well log (vertical smoothing). Too little smoothing gives too much reverberation. Too much smoothing gives no reverberation. The amount of vertical smoothing is an empirically determined parameter, and results are significantly sensitive to it. Vertical averages of the well log may or may not be a satisfactory approximation to the required horizontal average.

Failure of Newtonian Viscosity

Also remarkable is the failure of basic textbook seismology to explain the observed frequency-dependence of the dissipation parameter Q. The simplest theoretical approach to dissipation is to add a strain-rate term to Hooke's stress-strain law. This predicts stronger relative dissipation of high frequencies than of low frequencies. Experimentally, relative dissipation is observed to be roughly constant over many decades of frequency. Other simple Newtonian theories yield polynomial ratios in $-i\omega$ for the stress/strain ratio. These theories contain scale lengths and characteristic frequencies. They do not predict constant Q. The heterogeneity of the rock at all scales seems to be an essential attribute of a successful theory (Section 4.6).

Philosophy of Inverse Problems

Physical processes are often simulated with computers in much the same way they occur in nature. The machine memory is used as a map of physical space, and time evolves in the calculation as it does in the simulated world. A nice thing about solving problems this way is that there is never any question about the uniqueness of the solution. Errors of initial data and model discretization do not tend to have a catastrophic effect. Exploration geophysicists, however, rarely solve problems of this type. Instead of having (x, z)-space in the computer memory and letting t evolve, we usually have (x, t)-space in memory and extrapolate in depth z. This is our business, taking information (data) at the earth's surface and attempting to extrapolate to information at depth. Stable time evolution in nature provides no "existence proof" that our extrapolation goals are reasonable, stable, or even possible.

The time-evolution problems are often called *forward problems* and the depth-extrapolation problems *inverse problems*. In a *forward problem,* such as one with acoustic waves, it is clear what you need and what you can get. You need the density $\rho(x, z)$ and the incompressibility $K(x, z)$, and you need to know the initial source of disturbance. You can get the wavefield everywhere at later times but you usually only want it at the earth's surface for comparison to some data. In the *inverse problem* you have the waves seen at the surface, the source specification, and you would like to determine the material properties $\rho(x, z)$ and $K(x, z)$. What has been learned from experience is that routine observations do not give reasonable estimates of images or maps of ρ and K.

What You Can Get from Reflection Seismology

Luckily, it has been discovered that certain functions of ρ and K can be reliably determined and mapped. The velocity v and the acoustic impedance R are given by the equations

$$v = \sqrt{K/\rho} \tag{1a}$$

$$R = \sqrt{K\rho} \tag{1b}$$

Mathematically, it is a simple job to back-solve equation (1), which gives

$$K = v R \tag{2a}$$

$$\rho = \frac{R}{v} \tag{2b}$$

In practice, the solution (2) has little value because the two parameters v and R are seen through nonoverlapping spectral windows. The acoustic impedance R is seen through the typical 10 to 100 Hz spectral window of good quality reflection data. Since the low frequency part of the spectrum is missing, it is common to say that it is not the impedance which is seen, but the gradient $reflectivity = c(x, z) = \bigtriangledown log(R)$.

The velocity v is seen through a much smaller window. Observation of it involves studying travel time as shot-to-geophone offset varies and will be described in Chapter 3. With this second window it is hard to discern sixteen independent velocity measurements on a 4-second reflection time axis. So this window goes from zero to about 2 Hz, as depicted in figure 3.

Note that there is an information gap from 2-10 Hz. Even presuming that *rock physics* can supply us with a relationship between ρ and K, the gap seriously interferes with the ability of a seismologist to predict a well log before the well is drilled. What seismologists can do somewhat reliably is predict a filtered log.

The observational situation described above has led reflection seismologists to a specialized use of the word *velocity*. To a reflection seismologist, *velocity* means the low spatial frequency part of "real velocity." The high-frequency part of the "real velocity" isn't called *velocity*: it is called *reflectivity*. Density is generally disregarded as being almost unmeasurable by surface reflection seismology.

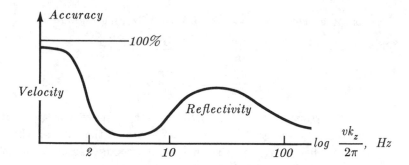

FIG. 1.4-3. Reliability of information obtained from surface seismic measurements.

Mathematical Inverse Problems

In mathematics the solution to an inverse problem has come to mean the "determination" of material properties from wavefields. Often this is achieved with a "convergent sequence." Geophysicists are less precise (or more inclusive) about what they mean by "determination." In Chapters 1-2 of this book reflectors are "determined" by the exploding-reflection concept. In Chapter 3 shot-to-geophone offset is incorporated, and reflectivity $c(x, z)$ and velocity $v(z)$ are "determined" with a buried-experiment concept. In Chapter 5 the concept is developed of suppressing multiple reflections and finding the "true" amplitudes of reflections by having the upcoming wave vanish before the onset of the downgoing wave. Other imaging concepts seem likely to result from future processing schemes. It might be possible to show that some of our "determinations" coincide with those of mathematicians, but such coincidence is not our goal.

Derivation of the Acoustic Wave Equation

The acoustic wave equation describes sound waves in a liquid or gas. Another more complicated set of equations describes elastic waves in solids. Begin with the acoustic case. Define

$$\rho \quad = \quad \text{mass per unit volume of the fluid}$$
$$u \quad = \quad \text{velocity flow of fluid in the } x\text{-direction}$$
$$w \quad = \quad \text{velocity flow of fluid in the } z\text{-direction}$$
$$P \quad = \quad \text{pressure in the fluid}$$

Newton's law of momentum conservation says that a small volume within a gas will accelerate if there is an applied force. The force arises from pressure differences at opposite sides of the small volume. Newton's law says

$$mass \times acceleration \;=\; force \;=\; -\; pressure \; gradient$$

$$\rho \, \frac{\partial u}{\partial t} \;=\; -\, \frac{\partial P}{\partial x} \tag{3a}$$

$$\rho \, \frac{\partial w}{\partial t} \;=\; -\, \frac{\partial P}{\partial z} \tag{3b}$$

The second physical process is energy storage by compression and volume change. If the velocity vector u at $x + \Delta x$ exceeds that at x, then the flow is said to be diverging. In other words, the small volume between x and $x + \Delta x$ is expanding. This expansion must lead to a pressure drop. The amount of the pressure drop is in proportion to a property of the fluid called its *incompressibility* K. In one dimension the equation is

$$pressure \; drop \;=\; (incompressibility) \times (divergence \; of \; velocity)$$

$$-\, \frac{\partial P}{\partial t} \;=\; K \, \frac{\partial u}{\partial x} \tag{4a}$$

In two dimensions it is

$$-\, \frac{\partial P}{\partial t} \;=\; K \left[\frac{\partial u}{\partial x} + \frac{\partial w}{\partial z} \right] \tag{4b}$$

To arrive at the one-dimensional wave equation from (3a) and (4a), first divide (3a) by ρ and take its x-derivative:

$$\frac{\partial}{\partial x} \frac{\partial}{\partial t} u \;=\; -\, \frac{\partial}{\partial x} \frac{1}{\rho} \frac{\partial P}{\partial x} \tag{5}$$

Second, take the time-derivative of (4). In the solid-earth sciences we are fortunate that the material in question does not change during our experiments. This means that K is a constant function of time:

$$\frac{\partial^2 P}{\partial t^2} \;=\; -\, K \, \frac{\partial}{\partial t} \frac{\partial}{\partial x} u \tag{6}$$

Inserting (5) into (6), the one-dimensional scalar wave equation appears.

$$\frac{\partial^2 P}{\partial t^2} \;=\; K \, \frac{\partial}{\partial x} \frac{1}{\rho} \frac{\partial P}{\partial x} \tag{7a}$$

In two space dimensions, the exact, acoustic scalar wave equation is

$$\frac{\partial^2 P}{\partial t^2} \;=\; K \left[\frac{\partial}{\partial x} \frac{1}{\rho} \frac{\partial}{\partial x} + \frac{\partial}{\partial z} \frac{1}{\rho} \frac{\partial}{\partial z} \right] P \tag{7b}$$

You will often see the scalar wave equation in a simplified form, in which it is assumed that ρ is not a function of x and z. Two reasons are often given for this approximation. First, observations are generally unable to determine density, so density may as well be taken as constant. Second, we will soon see that Fourier methods of solution do not work for space variable coefficients. Before examining the validity of this approximation, its consequences will be examined. It immediately reduces (7b) to the usual form of the scalar wave equation:

$$\frac{\partial^2 P}{\partial t^2} = \frac{K}{\rho}\left(\frac{\partial^2}{\partial x^2} + \frac{\partial^2}{\partial z^2}\right)P \qquad (8)$$

To see that this equation is a restatement of the geometrical concepts of previous sections, insert the trial solution

$$P = \exp(-i\omega t + i k_x x + i k_z z) \qquad (9)$$

What is obtained is the *dispersion relation of the two-dimensional scalar wave equation:*

$$\frac{\omega^2}{K/\rho} = k_x^2 + k_z^2 \qquad (10)$$

Earlier (Section 1.2, equation(8)) an equation like (10) was developed by considering only the geometrical behavior of waves. In that development the wave velocity squared was found where K/ρ stands in equation (10). Thus physics and geometry are reconciled by the association

$$v^2 = \frac{K}{\rho} \qquad (11)$$

Last, let us see why Fourier methods fail when the velocity is space variable. Assume that ω, k_x, and k_z are constant functions of space. Substitute (9) into (8) and you get the contradiction that ω, k_x, and k_z must be space variable if the velocity is space variable. Try again assuming space variability, and the resulting equation is still a differential equation, not an algebraic equation like (10).

Evanescence and Ground Roll

Completing the physical derivation of the dispersion relation,

$$k_x^2 + k_z^2 = \frac{\omega^2}{v^2} \qquad (12)$$

we can now have a new respect for it. It carries more meaning than could have been anticipated by the earlier geometrical derivation. The dispersion relation was originally regarded merely as $\sin^2\theta + \cos^2\theta = 1$ where $\sin\theta = v\,k_x/\omega$. There was no meaning in $\sin\theta$ exceeding unity, in other words, in $v\,k_x$ exceeding ω. Now there is. There was a hidden ambiguity in two of the previous migration methods. Since data could be an arbitrary function in the $(t\,,x)$-plane, its Fourier transform could be an arbitrary function in the $(\omega,\,k_x)$-plane. In practice then, there is always energy with an angle sine greater than one. This is depicted in figure 4. What should be done with this energy?

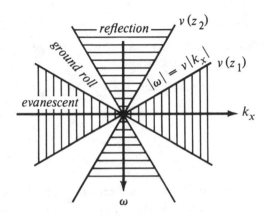

FIG. 1.4-4. The triangle(s) of reflection energy $|\,\omega\,| > v(z)\,|\,k_x\,|$ become narrower with velocity, hence with depth. Ground roll is energy that is propagating at the surface, but evanescent at depth.

When $v\,k_x$ exceeds ω, the familiar downward-extrapolation expression is better rewritten as

$$e^{\pm\,i\,\sqrt{\omega^2/v^2 - k_x^2}\,z} \quad = \quad e^{\pm\sqrt{k_x^2 - \omega^2/v^2}\,z} \tag{13}$$

This says that the depth-dependence of the physical solution is a growing or a damped exponential. These solutions are termed *evanescent* waves. In the most extreme case, $\omega = 0$, k_x is real, and $k_z = \pm i\,k_x$. For elastic waves, that would be the deformation of the ground under a parked airplane. Only if the airplane can move faster than the speed of sound in the earth will a wave be radiated into the earth. If the airplane moves at a subsonic speed the deformation is said to be *quasi-static*.

Perhaps a better physical description is a thought experiment with a sinusoidally corrugated sheet. Such metallic sheet is sometimes used for roofs or garage doorways. The wavelength of the corrugation fixes k_x. Moving such a sheet past your ear at velocity V you would hear a frequency of oscillation equal to $V k_x$, regardless of whether V is larger or smaller than the speed of sound in air. But the sound you hear would get weaker exponentially with distance from the sheet unless it moved very fast, $V > v$, in which case the moving sheet would be radiating sound to great distances. This is why supersonic airplanes use so much fuel.

What should a migration program do with energy that moves slower than the sound speed? Theoretically, such energy should be exponentially damped in the direction going away from the source. The damping in the offending region of (ω, k_x)-space is, quantitatively, extremely rapid. Thus, simple exploding-reflector theory predicts that there should be almost no energy in the data at these low velocities.

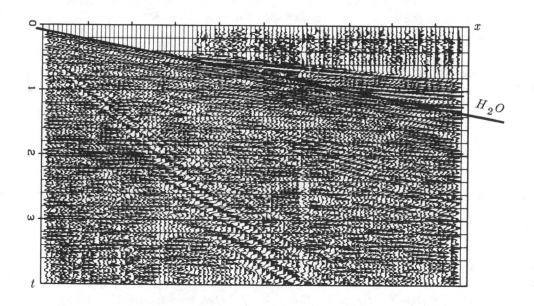

FIG. 1.4-5. Florida shallow marine profile, exhibiting ground roll with frequency dispersion. (Conoco, Yedlin)

The reality is that, instead of tiny amounts of energy in the evanescent region of (ω, k_x)-space, there is often a great deal. This is another breakdown of the exploding-reflector concept. The problem is worst with land data. Waves that are evanescent in deep, fast rocks of interest can be propagating in the low-velocity soil layer. This energy is called *ground roll*. Figure 5 shows an example. Like the surface of the earth, which varies greatly from place to place, the immediate subsurface which controls the ground roll varies substantially. So although figure 5 is a nice example, no example can really be typical. This data is not a zero-offset section. The shot is on the left, and the traces to the right are from geophones at increasing distances from the shot. The straight line drawn onto the data defines a slope equal to the water velocity. Steeper events are ground roll. In this figure there are two types of ground roll, one at about half of water velocity, and a stronger one at about a quarter of water velocity. The later and stronger one shows an interesting feature known as *frequency dispersion*. Viewing the data from the side, you should be able to notice that the high frequencies arrive before the low frequencies.

Ground roll is unwanted noise since its exponential decay effectively prevents it from being influenced by deep objects of interest. In practice, energy in the offending region of (ω, k_x)-space should be attenuated. A mathematical description is to say that the composite mapping from model space to data space and back to model space again is not an identity transformation but an idempotent transformation.

Reflections and the High-Frequency Limit

It is well known that the contact between two different materials can cause acoustic reflections. A material contact is defined to be a place where either K or ρ changes by a spatial step function. In one dimension either $\partial K / \partial x$ or $\partial \rho / \partial x$ or both would be infinite at a point, and we know that either can cause a reflection. So it is perhaps a little surprising that while the density derivative is explicitly found in (7b), the incompressibility derivative is not. This means that dropping the density gradients in (7b) will not eliminate all possible reflections. Dropping the terms will slightly simplify further analysis, however, and since constant density is a reasonable case, the terms are often dropped.

There are also some well-known mathematical circumstances under which the first-order terms may be ignored. Fix your attention on a wave going in any particular direction. Then ω, k_x, and k_z have some prescribed ratio. In the limiting case that frequency goes to infinity, the P_{tt}, P_{xx}, and P_{zz} terms in (8) all tend to the *second* power of infinity. Suppose two media gradually blend into one another so that $\partial \rho / \partial x$ is less than infinity. The terms neglected in going from (7b) to (8) are of the form $\rho_x P_x$ and $\rho_z P_z$. As frequency tends to infinity, these terms only tend to the *first* power of infinity. Thus, in that limit they can be neglected.

These terms are usually included in theoretical seismology where the goal is to calculate synthetic seismograms. But where the goal is to create earth models from seismic field data — as in this book — these terms are generally neglected. Earth imaging is more difficult than calculating synthetic seismograms. But often the reason for neglecting the terms is simply to reduce the clutter. These terms may be neglected for the same reason that equations are often written in two dimensions instead of three: the extension is usually possible but not often required. Further, these terms are often ignored to facilitate Fourier solution techniques. Practical situations might arise for which these terms need to be included. With the finite-difference method (Section 2.2), they are not difficult to include. But any effort to include them in data processing should also take into account other factors of similar significance, such as the assumption that the acoustic equation approximates the elastic world.

EXERCISES

1. Soil is typically saturated with water below a certain depth, which is known as the *water table*. Experience with hammer-seismograph systems shows that seismic velocity typically jumps up to water velocity ($V_{H_2O} = 1500 \ m/s$) at the water table. Say that in a certain location, the ground roll is observed to be greater than the reflected waves, so a decision has been made to bury geophones. The troublesome ground roll is observed to travel at six-tenths the speed of a water wave. How deep must the geophones be buried below the water table to attenuate the ground roll by a factor of ten? Assume the data contains all frequencies from 10 to 100 Hz. (Hints: $\log_e 10 \approx 2$, $2\pi \approx 6$, etc.)

2. Consider the function

$$P(z,t) = P_0 \frac{1}{\sqrt{Y(z)}} e^{\ i\,\omega\,t\ -\ i \int_0^z \omega \sqrt{\frac{\rho(\xi)}{K(\xi)}}\ d\xi} \qquad (1.4E1)$$

where

$$P_0 = \text{constant}$$

$$Y \equiv \frac{1}{\sqrt{\rho(z)\,K(z)}}$$

as a trial solution for the one-dimensional wave equation:

$$\left\{ \frac{\partial^2}{\partial t^2} - \frac{K(z)}{\rho(z)} \frac{\partial^2}{\partial z^2} \right\} P = - \frac{K(z)}{\rho(z)^2} \frac{\partial \rho}{\partial z} \frac{\partial P}{\partial z} \qquad (1.4E2)$$

Substitute the trial solution (E1) into the wave equation (E2). Discuss the trade-off between changes in material properties and the validity of your solution for different wavelengths.

1.5 The Paraxial Wave Equation

The scalar wave equation, unlike Fourier equations, allows arbitrary spatial variation in density and velocity. Because of this you might expect that it would be used directly in the manufacture of migrated sections. But it is used little for migration, and we will first review why this is so. Then we will meet the *paraxial wave equation,* which is the basis for most production migration.

Philosophically, the paraxial wave equation is an intermediary between the simple concepts of rays and plane waves and deeper concepts embodied in the wave equation. (The paraxial wave equation is also called the *single-square-root* equation. In Chapter 2, a specialization of it is called the *parabolic* wave equation). The derivation of the parabolic wave equation does not proceed from simple concepts of classical physics. Its development is more circuitous, like the Schroedinger equation of quantum physics. You must study it for a while to see why it is needed. When I introduced the parabolic wave equation to seismic calculations in 1970, it met with considerable suspicion. Fortunately for you, years of experience have enabled me to do a better job of explaining it, and fortunately for me, its dominance of the industrial scene will give you the interest to persevere.

The paraxial equation will be introduced by means of Fourier methods. Fourier methods are incompatible with space-variable coefficients. Since we want to incorporate spatial variations in velocity, this limitation is ultimately to be avoided, so after getting the paraxial equation in the Fourier domain, ik_z is replaced by $\partial/\partial z$, and ik_x is replaced by $\partial/\partial x$. Then, being in the space domain, the velocity can be space-variable. The result is a partial differential equation often solved by the finite-differencing method. This procedure turns out to be valid, but new students of migration understandably regard it with misgiving. Thus, the final part of this section is a derivation of the paraxial wave equation which makes no use of Fourier methods.

Why the Scalar Wave Equation is Rarely Used for Migration

Life would be simpler if migration could be done with the scalar wave equation instead of the paraxial equation. Indeed, migration can be done with the scalar wave equation, and there are some potential advantages (Hemon [1978], Kosloff and Baysal [1983]). But more than 99% of current industrial migration is done with the paraxial equation.

The main problem with the scalar wave equation is that it will generate unwanted internal multiple reflections. The exploding-reflector concept cannot deal with multiple reflections. Primary reflections can be modeled with only upcoming waves, but multiple reflections involve both up and downgoing paths. The multiple reflections observed in real life are completely different from those predicted by the exploding-reflector concept. For the sea-floor multiple reflection, a sea-floor two-way travel-time depth of t_0 yields sea-floor multiple reflections at times $2t_0$, $3t_0$, $4t_0$, $\cdots$. In the exploding-reflector conceptual model, a sea-floor one-way travel-time depth of t_0 yields sea-floor multiple reflections at times $3t_0$, $5t_0$, $7t_0$, $\cdots$. In building a telescope, microscope, or camera, the designer takes care to suppress backward reflected light because it creates background noise on the image. Likewise, in building a migration program we do not want to have energy moving around that does not contribute to the focused image. The scalar wave equation with space-variable coefficients will generate such energy. This unwanted energy is especially troublesome if it is coherent and migrates to a time when primaries are weak. It is annoying, as the reflection of a bright window seen on a television screen is annoying. So if you were trying to migrate with the scalar wave equation, you would make the velocity as smooth as possible.

Another difficulty of imaging with the scalar wave equation arises with evanescent waves. These are the waves that are exponentially growing or decaying with depth. Nature extrapolates waves forward in time, but we are extrapolating them in depth. Growing exponentials can have tiny sources, even numerical round-off, and because they grow rapidly, some means must be found to suppress them.

A third difficulty of imaging with the scalar wave equation derives from initial conditions. The scalar wave equation has a second depth z-derivative. This means that two boundary conditions are required on the z-axis. Since data is recorded at $z=0$, it seems natural that these boundary conditions should be knowledge of P and $\partial P/\partial z$ at $z=0$. But $\partial P/\partial z$ isn't recorded.

Luckily, in building an imaging device that operates wholly within a computer, we have ideal materials to work with, i.e., reflectionless lenses. Instead of the scalar wave equation of the real world we have the paraxial wave equation.

Fourier Derivation of the Paraxial Wave Equation

Start from the dispersion relation of the scalar wave equation:

$$k_x{}^2 + k_z{}^2 = \frac{\omega^2}{v^2} \tag{1}$$

Take a square root.

$$k_z = \pm \sqrt{\frac{\omega^2}{v^2} - k_x{}^2} \tag{2}$$

The simple act of selecting the minus sign in (2) includes the upcoming waves and eliminates the downgoing waves. Equation (1) is the three-dimensional Fourier transform of the scalar wave equation. Inverse transforming (2) will give us an equation for upcoming (or downgoing) waves only, without the other. Inverse Fourier transformation over a dimension is just a matter of selecting one or more of the following substitutions:

$$\frac{\partial}{\partial t} = -i\,\omega \tag{3a}$$

$$\frac{\partial}{\partial x} = i\,k_x \tag{3b}$$

$$\frac{\partial}{\partial z} = i\,k_z \tag{3c}$$

After inverse transformation over z there is a differential equation in z in which the velocity may be taken to be z-variable. Likewise for x. Any equation resulting from any of the substitutions of (3) into (2) is called a paraxial equation. Chapter 2 of this book goes into great detail about the meaning of these equations. Before beginning this interpretation the paraxial wave equation will be derived without the use of Fourier transformation. Besides giving a clear path to the basic migration equation, this derivation also gives a better understanding of what the equation really does, and how it differs from the scalar wave equation.

Snell Waves

It is natural to begin studies of waves with equations that describe plane waves in a medium of constant velocity. However, in reflection seismic surveys the velocity contrast between shallowest and deepest reflectors ordinarily exceeds a factor of two. Thus depth variation of velocity is almost always included in the analysis of field data. Seismological theory needs to consider waves that are just like plane waves except that they bend to accommodate the velocity stratification $v(z)$. Figure 1 shows this in an idealized geometry: waves radiated from the horizontal flight of a supersonic airplane.

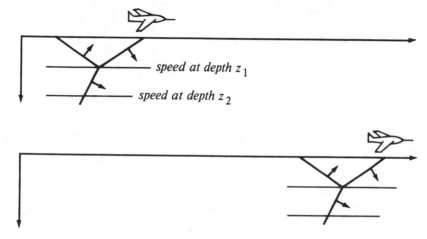

FIG. 1.5-1. Fast airplane radiating a sound wave into the earth. From the figure you can deduce that $\partial t / \partial x$ is the same at depth z_1 as it is at depth z_2. This leads (in isotropic media) to Snell's law.

The airplane flies horizontally at a constant speed. It goes from $x = -\infty$ to $x = +\infty$. Imagine an earth of horizontal plane layers. In this model there is nothing to distinguish any point on the x-axis from any other point on the x-axis. But the seismic velocity varies from layer to layer. There may be reflections, head waves, shear waves, and multiple reflections. Whatever the picture is, it moves along with the airplane. A picture of the wavefronts near the airplane moves along with the airplane. The top of the picture and the bottom of the picture both move laterally at the same speed even if the earth velocity increases with depth. If the top and bottom didn't go at the same speed, the picture would become distorted, contradicting the presumed symmetry of translation. This horizontal speed, or rather its inverse $\partial t / \partial x$, has several names. In practical work it is called the *stepout*. In theoretical work it is called the *ray parameter*. It is very important to note that $\partial t / \partial x$ does not change with depth, even though the seismic velocity does change with depth. In a constant-velocity medium, the angle of a wave does not change with depth. In a stratified medium, $\partial t / \partial x$ does not change with depth.

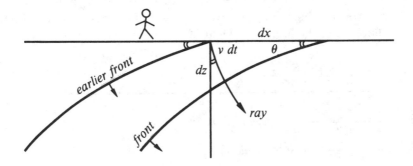

FIG. 1.5-2. Downgoing fronts and rays in stratified medium $v(z)$. The wavefronts are horizontal translations of one another.

Figure 2 illustrates the differential geometry of the wave. The diagram shows that

$$\frac{\partial t}{\partial x} = \frac{\sin \theta}{v} \tag{4a}$$

$$\frac{\partial t}{\partial z} = \frac{\cos \theta}{v} \tag{4b}$$

These two equations define two (inverse) speeds. The first is a horizontal speed, measured along the earth's surface, called the *horizontal phase velocity*. The second is a vertical speed, measurable in a borehole, called the *vertical phase velocity*. Notice that both these speeds *exceed* the velocity v of wave propagation in the medium. Projection of wave *fronts* onto coordinate axes gives speeds larger than v, whereas projection of *rays* onto coordinate axes gives speeds smaller than v. The inverse of the phase velocities is called the *stepout* or the *slowness*.

Snell's law relates the angle of a wave in one layer with the angle in another. The constancy of equation (4a) in depth is really just the statement of Snell's law. Indeed, we have just derived Snell's law. All waves in seismology propagate in a velocity-stratified medium. So they cannot be called plane waves. But we need a name for waves that are near to plane waves. A *Snell wave* will be defined to be the generalization of a plane wave to a stratified medium $v(z)$. A plane wave that happens to enter a medium of depth-variable velocity $v(z)$ gets changed into a Snell wave. While a plane wave has an angle of propagation, a Snell wave has instead a *Snell parameter* $p = \partial t / \partial x$.

It is noteworthy that Snell's parameter $p = \partial t / \partial x$ is directly observable at the surface, whereas neither v nor θ is directly observable. Since $p = \partial t / \partial x$ is not only observable, but constant in depth, it is customary to use it to eliminate θ from equation (4):

$$\frac{\partial t}{\partial x} = \frac{\sin \theta}{v} = p \tag{5a}$$

$$\frac{\partial t}{\partial z} = \frac{\cos \theta}{v} = \left(\frac{1}{v(z)^2} - p^2 \right)^{1/2} \tag{5b}$$

Taking the Snell wave to go through the origin at time zero, an expression for the arrival time of the Snell wave at any other location is given by

$$t(x, z) = \frac{\sin \theta}{v} x + \int_0^z \frac{\cos \theta}{v} dz \tag{6a}$$

$$t(x, z) = p x + \int_0^z \left(\frac{1}{v(z)^2} - p^2 \right)^{1/2} dz \tag{6b}$$

The validity of (6b) is readily checked by computing $\partial t / \partial x$ and $\partial t / \partial z$, then comparing with (5).

An arbitrary waveform $f(t)$ may be carried by the Snell wave. Use (6) to define a delay time t_0 for a delayed wave $f[t - t_0(x, z)]$ at the location (x, z).

$$SnellWave(t, x, z) = f \left\{ t - p x - \int_0^z \left(\frac{1}{v(z)^2} - p^2 \right)^{1/2} dz \right\} \tag{7}$$

Time-Shifting Equations

An important task is to predict the wavefield inside the earth given the waveform at the surface. For a downgoing plane wave this can be done by the time-shifting partial differential equation

$$\frac{\partial P(t, z)}{\partial z} = -\frac{1}{v} \frac{\partial P(t, z)}{\partial t} \tag{8}$$

as may be readily verified by substituting the trial solutions

$$P(t, z) \; = \; f\left(t - \frac{z}{v} \right) \qquad \text{for constant } v \tag{9}$$

or

$$P(t, z) \; = \; f\left(t - \int_0^z \frac{dz}{v(z)} \right) \qquad \text{for } v(z) \tag{10}$$

This also works for nonvertically incident waves with the partial differential equation

$$\frac{\partial P(t, x, z)}{\partial z} \; = \; - \frac{\partial t}{\partial z} \frac{\partial P(t, x, z)}{\partial t} \tag{11}$$

which has the solution

$$P(t, x, z) \; = \; f\left(t - p\, x - \int_0^z \frac{\partial t}{\partial z}\, dz \right) \tag{12}$$

In interpreting (11) and (12) recall that $1/(\partial t / \partial z)$ is the apparent velocity in a borehole. The partial derivative of wavefield $P(t, x, z)$ with respect to depth z is taken at constant x, i.e., the wave is extrapolated down the borehole. The idea that downward extrapolation can be achieved by merely time shifting holds only when a *single* Snell wave is present; that is, the *same* arbitrary time function must be seen at all locations.

Substitution from (5) also enables us to rewrite (11) in the various forms

$$\frac{\partial P(t, x, z)}{\partial z} \; = \; - \frac{\cos \theta}{v} \frac{\partial P(t, x, z)}{\partial t} \tag{13a}$$

$$\frac{\partial P(t, x, z)}{\partial z} \; = \; - \left[\frac{1}{v(z)^2} - p^2 \right]^{1/2} \frac{\partial P(t, x, z)}{\partial t} \tag{13b}$$

$$\frac{\partial P(t, x, z)}{\partial z} \; = \; - \left[\frac{1}{v(z)^2} - \left(\frac{\partial t}{\partial x} \right)^2 \right]^{1/2} \frac{\partial P(t, x, z)}{\partial t} \tag{13c}$$

Equation (13) is a paraxial wave equation. Since $\partial t / \partial x = p$ can be measured along the surface of the earth, it seems that equation (13c), along with an assumed velocity $v(z)$ and some observed data $P(t, x, z=0)$, would enable us to determine $\partial P / \partial z$, which is the necessary first step of downward continuation. But the presumption was that there was only a single Snell wave and not a superposition of several Snell waves. Superposition of different waveforms on different Snell paths would cause different time functions to be seen at different places. Then a mere time shift would not achieve downward continuation. Luckily, a complicated wavefield that is variable from place to place may be decomposed into many Snell waves, each of which can be downward extrapolated with the differential equation (13) or its solution (12). One such decomposition technique is *Fourier analysis*.

Fourier Decomposition

Fourier analyzing the function $f(x, t, z=0)$, seen on the earth's surface, requires the Fourier kernel $\exp(-i\omega t + i k_x x)$. Moving on the earth's surface at an inverse speed of $\partial t / \partial x = k_x / \omega$, the phase of the Fourier kernel, hence the kernel itself, remains constant. Only those sinusoidal components that move at the same speed as the Snell wave can have a nonzero correlation with it. So if the disturbance is a single Snell wave, then all Fourier components vanish except for those that satisfy $p = k_x / \omega$. You should memorize these basic relations:

$$
\frac{\partial t}{\partial x} = \frac{\sin \theta}{v} = p = \frac{k_x}{\omega} \tag{14}
$$

In theoretical seismology a square-root function often appears as a result of using (14) to make a cosine.

Utilization of this Fourier domain interpretation of Snell's parameter p enables us to write the square-root equation (13) in an even more useful form. But first the square-root equation must be reexpressed in the Fourier domain. This is done by replacing the $\partial / \partial t$ operator in (13) by $-i\omega$. The result is

$$
\frac{\partial P(\omega, k_x, z)}{\partial z} = + i\omega \left[\frac{1}{v(z)^2} - \frac{k_x^2}{\omega^2} \right]^{1/2} P(\omega, k_x, z) \tag{15}
$$

At present it is equivalent to specify either the differential equation (15) or its solution (12) with f as the complex exponential:

$$
P(\omega, k_x, z) e^{-i\omega t} = \exp \left\{ i\omega \int_0^z \left[\frac{1}{v(z)^2} - \frac{k_x^2}{\omega^2} \right]^{1/2} dz \right\} \tag{16}
$$

Later, when we consider lateral velocity variation $v(x)$, the solution (16) becomes wrong, whereas the differential equation (13c) is a valid description of any *local* plane-wave behavior. But before going to lateral velocity gradients we should look more carefully at vertical velocity gradients.

Velocity Gradients

Inserting the Snell wavefield expression into the scalar wave equation, we discover that our definition of a Snell wave does not satisfy the scalar wave equation. The discrepancy arises only in the presence of velocity gradients. In other words, if there is a shallow constant velocity v_1 and a deep constant velocity v_2, the equation is satisfied everywhere except where v_1 changes to v_2. Solutions to the scalar wave equation must show amplitude changes across an interface, because of transmission coefficients. Our definition of a Snell wave is a wave of constant amplitude with depth. The paraxial wave equation could be modified to incorporate a transmission coefficient effect. The reason it rarely is modified may be the same reason that density gradients are often ignored. They add clutter to equations while their contribution to better results — namely, more correct amplitudes and possible tiny phase shifts — has marginal utility. Indeed, if they are included, then other deeper questions should also be included, such as the question, why use the acoustic equation instead of various other forms of scalar elastic equations?

Even if the paraxial wave equation were modified to incorporate a transmission coefficient effect, its solution would still fail to satisfy the scalar wave equation because of the absence of the reflected wave. But that is just fine, because it is the paraxial equation, with its reflection-free lenses, that is desired for data processing.

EXERCISES

1. Devise a mathematical expression for a plane wave that is an impulse function of time with a propagation angle of $15°$ from the vertical z-axis in the plus z direction. Express the result in the domain of

    (a) (t, x, z)

    (b) (ω, x, z)

    (c) (ω, k_x, z)

    (d) (ω, p, z)

2. Find an amplitude function $A(z)$ which, when multiplied by f in equation (12), yields an approximate solution to the scalar wave equation for stratified media $v(z)$. For $p = 0$, the solution should reduce to the solution of Exercise 2 in Section 1.4.

1.6 Mastery of 2-D Fourier Techniques

Here is a collection of helpful tips for those of you who will be involved in implementations of migration methods.

Signs and Scales in Fourier Transforms

In Fourier transforming t-, x-, and z-coordinates, a sign convention must be chosen for each coordinate. Electrical engineers have chosen one convention and physicists another. While both have good reasons for their choices, our circumstances more closely resemble the circumstances of physicists, so their convention will be used. For the *inverse* Fourier transformation this is

$$p(t,x,z) \;=\; \iiint e^{-i\omega t + i k_x x + i k_z z}\, P(\omega, k_x, k_z)\, d\omega\, dk_x\, dk_z \qquad (1)$$

For the *forward* Fourier transform, the space variables carry a negative sign and time carries a positive sign. The limits on the integrations and the scale factor in the continuous case differ from the discrete case. We rarely do the transforms analytically in either case. Since the extra notation required for limits and scales usually clutters rather than clarifies a discussion, they will be omitted altogether except when they play a useful role.

The sign convention is more important. Because there are so many space axes (later, midpoint and offset space axes are introduced and transformed as well), it is worthwhile to establish a good sign convention. Someone using the approach of "changing the signs around until it works" is likely to be perplexed by the number of possible permutations. There are good reasons for the sign conventions chosen by physicists, and once the reasons are known, it is easy to remember the conventions.

Waves should, by convention, move in the positive direction on the space axis. This is especially evident on work for which the space axis is a radius. Atoms, like geophysical sources, always radiate from a point to infinity, not the other way around. So our convention will be always to choose waves moving positively on any space axis. In equation (1) this means that the sign on the spatial frequencies must be opposite to the sign on the temporal frequency. This statement applies to both the forward and the inverse transform.

This leaves the choice of whether to use the positive sign for the time axis or the space axes. There are many space axes but only one time axis.

There will be the fewest number of minus signs and the fewest sign changes if the spatial gradient $\partial/\partial x$, $\partial/\partial z$, etc. is chosen to be associated with the positive k-vector, i.e., with ik_x, ik_z, etc. Of course, this leaves the time derivative with $-i\omega$.

This sign convention brings our practice into conflict with the practice of electrical engineers, who rarely work with space axes and naturally enough have chosen to associate $\partial/\partial t$ with $+i\omega$. The only good reason I know to adopt the engineering choice is that we compute with an array processor built and microcoded by engineers who have of course used their own sign convention. It doesn't matter for the programs that transform complex-valued time functions to complex-valued frequency functions, because then the sign convention is under the user's control. But it does make a difference with the program that converts real time functions to complex frequency functions. The way to live in both worlds is to imagine that the frequencies produced by the program do not range from 0 to $+\pi$ as the description says, but from 0 to $-\pi$. Alternately, you could always take the complex conjugate of the transform, which would swap the sign of the ω-axis. With the Stolt algorithm it is common to transform space first. Then the array processor convention turns out to have our notation.

How to Transpose a Big Matrix

It is lucky that very large matrices can easily be transposed. This is what makes wave-equation seismic data processing reasonable on a small minicomputer. By *very large* matrix, I mean one that is too big to fit in a computer's random access memory (RAM). If two copies of the data fit in the RAM, then transposition is simply the copy operation $T(i, j) = M(j, i)$.

The transpose algorithm for very large matrices is simple but tricky. I shall begin, therefore, by describing a card trick. I have in my hands a deck of cards from which I have removed the nines, tens, and face cards. Let a, b, c, and d denote hearts, spades, clubs, and diamonds. Also, I have arranged these cards in the following order (let ace be denoted by one):

$$1a \quad 1b \quad 1c \quad 1d \quad 2a \quad 2b \quad 2c \quad 2d \quad 3a \quad \cdots \quad 8d$$

Now I deal the cards face up alternately, one onto pile A and one onto pile B. You see

$$\text{Pile } A: \quad 1a \quad 1c \quad 2a \quad 2c \quad 3a \quad 3c \quad \cdots \quad 8a \quad 8c$$

$$\text{Pile } B: \quad 1b \quad 1d \quad 2b \quad 2d \quad 3b \quad 3d \quad \cdots \quad 8b \quad 8d$$

Next I place pile A on top of (in front of) pile B, and again deal the cards out alternately onto pile A' and pile B'. You see

> Pile A':   1a   2a   3a $\cdots$ 8a   1b   2b $\cdots$ 8b
>
> Pile B':   1c   2c   3c $\cdots$ 8c   1d   2d $\cdots$ 8d

Now I place pile A' on top of pile B'. We started with all the aces together, the twos together, etc. Now all the hearts are together, the spades together, etc. So you see that in just two deals of the cards, I have transposed the deck. The cards were never spread out all over the table because they never had to be randomly accessed. Transposition was done by making sequential passes over the deck. In principle, this algorithm transposes a matrix requiring four magnetic tapes but almost no core memory.

Now I will try the inverse transpose. Note that it takes me *three* deals of the cards rather than the *two* deals it took for the original transpose. This is because the deck has $2^2 = 4$ suits and $2^3 = 8$ numbers. Actually, there is another algorithm which will allow me to do the inverse transpose in only two passes rather than three. I just do everything backwards. I start with piles A' and B'. Then I create pile A by alternately selecting cards one from pile A' and one from pile B'. Likewise I create pile B. Then I repeat this procedure. The first algorithm is called the *sort* algorithm, and the second is called the *merge* algorithm. With these two algorithms, the matrix transpose of a matrix of size $2^n \times 2^m$ can be done by the lesser of n or m passes over the data.

A variety of generalizations are possible. With four card piles, techniques could be developed for matrices of dimension 4^n. This would decrease the number of passes but increase the required number of tape drives. Likewise, it turns out that arbitrary order can be factored into primes, etc. But this takes us too far afield.

Minimizing the number of passes over the data turns out to maximize the number of tapes. In reality you won't be using real tapes when you are transposing. Instead you will be simulating tape operations on a large disk memory. Then the number of "tapes" you choose to use will be controlled by the ratio of the speed of random transfers to the speed of sequential transfers.

Rocca's 2-D Fourier Transform without Transposing

The most direct method of two-dimensional Fourier transformation in a computer is the repetitive application of a one-dimensional Fourier transform method. The easiest part is the "fast" direction. That is, if the data matrix is stored by columns — as in the Fortran language — then the column transforms are a trivial exercise in the repetitive use of a one-dimensional program. Now for the rows. If the matrix fits in the RAM, then everything is easy: one row at a time can be copied into a vector; the vector can be Fourier transformed and then copied back into the matrix row. The more typical case is that the data doesn't fit in the RAM but does fit in the "virtual" memory. This means that the programmer could write $T(i, j) = M(j, i)$ but the program would run prohibitively slowly because an entire page of vir-. tual memory would be fetched from disk just to find a single number.

Conceptually, an easy way to handle the transformation over the row direction is to transpose the matrix, transform each column, and transpose back. Fabio Rocca suggested a quicker and easier means of Fourier transformation over the row index. The basic Fourier transform program has certain overhead calculations, such as computing or fetching sines and cosines. Ordinarily, these overhead calculations are repeated each time a Fourier transformation is performed. With Rocca's method the overhead calculations are done just once, and all the rows get Fourier transformed. So it is even quicker than the straightforward approach. The method follows.

The data matrix can be regarded as a row vector whose entries are columns. Taking the "fast" index to range down the column, the columns may be transformed by one-dimensional transforms either before or after the row operations are done. To do the row operations, just modify an ordinary one-dimensional Fourier transform program by replacing each scalar add or multiply operation by the same operation on every element in the corresponding column.

The order in which data is accessed makes Rocca's row algorithm efficient in a virtual memory environment. Before the days of virtual memory, we implemented the Rocca row algorithm with reads and writes around the inner loops.

To illustrate Rocca's method, a row Fourier transformation program was written based on the one-dimensional Fourier transformation program found in FGDP. It is included in the next section. That program transforms complex time functions to complex frequency functions. If you should decide to write a real-to-complex Fourier transform, you should beware of the assumption that real and imaginary parts are stored contiguously. This assumption is true for the column index, but not for the row index.

1.7 Sample Programs

The programs in this section generated many of the examples given in this book. They were written for clarity and brevity, and they are excellent for experimental work. Good production programs will be faster (by factors of from 1.01 to about 4). Speed can be gained by taking advantage of various special circumstances. For example, data is real, but these expository programs assume it to be complex.

RATional FORtran = Ratfor

Bare bones Fortran is our most universal computer language. But it is hardly appropriate for expository discussion of algorithms. The ideal expository language is *Ratfor*. Ratfor is *Rat*ional *For*tran, namely, Fortran without the blemishes. Ratfor programs (including the Ratfor preprocessor) are readily converted to Fortran by means of the Ratfor preprocessor. Since the preprocessor is publicly available Ratfor is practically as universal as Fortran.†

You won't really need the preprocessor or any precise definitions if you already know Fortran or almost any other computer language, because then the Ratfor language will be easy to understand. Statements on a line may be separated by ";". Statements may be grouped together with { }. Do loops don't require statement numbers because { } defines the range. Given that "if ( )" is true, the statements in the following { } are done. "Else { }" does what you would expect it to. Indentation is used for readability. Choose your own style. I have overcondensed. Anything following # is a comment. You may omit the braces { } when they contain only one statement. "Break" will cause premature termination of the enclosing { }. "Break 2" escapes from {{ }}. "While ( ) { }" repeats the statements in { } while the condition ( ) is true. "Repeat { } until ( )" is a loop that tests at the bottom. A looping statement more general than "do" is "for(initialize; condition; reinitialize) { }". "Next" causes skipping to the end of any loop and a retrial of the test condition. The Fortran relational operators .gt., .ge., .ne., etc. may be written >, >=, !=, etc. The logical operators .and. and .or. may be written & and |. Anything that doesn't make sense to the Ratfor preprocessor, such as

---

† Kernighan, B.W. and Plauger, P.J., 1976, *Software Tools:* Addison-Wesley Publishing Company.

Fortran input-output, is passed through without change.

Two-Dimensional Fourier Transformation

Two-dimensional Fourier transforms are based on the one-dimensional Fourier transform. An extremely rapid way to compute the one-dimensional Fourier transform exists and is called the *Cooley-Tukey algorithm* or the *Fast Fourier Transform*. Unfortunately it bears little resemblance to the Fourier integral. This method is so fast and effective that you will hardly ever see the transform being done in the obvious way. All functions are taken to be periodic, so physically transient functions must be regarded as functions of very long period. Usually there is the further restriction that the period must be exactly 2^N points long, where N is an integer. To understand this program, you should look at FGDP or any number of electrical engineering books. To write and use two-dimensional Fourier transform programs, it is only necessary to know the one-dimensional definition of inputs and outputs. Figure 1 shows that humans like to have $t=0$ in the middle of the time axis and $\omega=0$ in the middle of the frequency axis, whereas the standard one-dimensional Fourier transform programs place $t=0$ and $\omega=0$ at one end of a vector.

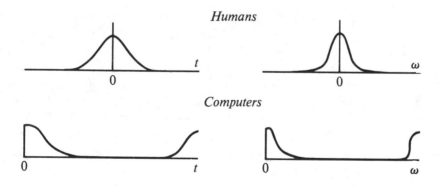

FIG. 1.7-1. Computer storage arrangement in one-dimensional Fourier transform programs.

Take the one-dimensional Fourier transform of an eight-point time function. The zero frequency is output in the first vector element. The Nyquist frequency π, which is the highest frequency representable on a mesh, namely the function $+1, -1, +1, -1, \cdots$, is in the fifth element of the eight point function, after which follow the negative frequencies. The smallest nonzero negative frequency is in the eighth vector element. If there were a ninth element, it would by periodicity be equal to the first element. It is a common beginner's error to find that the output of a migration is not real. The imaginary part should be about 10^{-6} of the real part, as expected for single precision arithmetic. A much larger imaginary part, proportional to $1/N$ where N is the vector length, indicates a programming error.

Below is the test program for the two-dimensional program. The "write" statement is local Fortran, not Ratfor. The function being transformed is something of a low-frequency function on the time axis, and very much a low-frequency function on the space axis.

```
# Test case for two-dimensional Fourier Transformation
integer it,nt,ix,nx;          complex cp(64,64), cwork(64)
open(4,file='plotfile',status='new',access='direct',form='unformatted',recl=1)
nx = 64;        nt = 64;
do it=1,nt
        do ix=1,nx
                cp(it,ix)=0.
cp(16,3)=1.; cp(16,4)=4.; cp(16,5)=6.; cp(16,6)=4.; cp(16,7)=1.
cp(17,3)=1.; cp(17,4)=4.; cp(17,5)=6.; cp(17,6)=4.; cp(17,7)=1.
call ft2d(nt,nx,cp,+1.,+1.,cwork)
write(4,rec=1) ((real(cp(it,ix)),it=1,nt),ix=1,nx)
stop;    end
```

The most basic two-dimensional Fourier transform is shown below.

```
# 2D Fourier transform by using 1D program
subroutine ft2d (n1,n2,cp,sign1,sign2,cwork)
complex  cp(n1,n2),cwork(n2)
integer  n1,n2
real     sign1,sign2
do i2 = 1,n2                        # transform over the fast dimension
        call fork (n1,cp(1,i2),sign1)      # one-dimensional Fourier transform
do i1 = 1,n1 {                      # transform over the slow dimension
        do i2 = 1,n2
                cwork(i2) = cp(i1,i2)
        call fork (n2,cwork,sign2)  # one-dimensional Fourier transform
        do i2 = 1,n2
                cp(i1,i2) = cwork(i2)
        }
return;       end
```

Finally we have the one-dimensional fast Fourier transform program. This one is the Ratfor version of Fortran "fork" found in FGDP on p.12. As usual, lx is a power of 2, the output cx(1) is the zero frequency, cx(lx/2+1) is the so-called Nyquist frequency, and cx(lx) is the smallest negative frequency. The algorithm is short, but tricky, and you should not expect the program to be readable unless you consult other references.

```
# 1D fast Fourier transform
subroutine fork(lx,cx,signi)
complex cx(lx),carg,cexp,cw,ct
j = 1; k = 1;        sc = sqrt(1./lx)
do i = 1,lx {
      if (i<=j) { ct=cx(j)*sc;   cx(j)=cx(i)*sc;   cx(i)=ct }
      m = lx/2
      while (j>m) { j=j-m;    m=m/2;   if (m<1) break }
      j = j+m
      }
repeat {
      istep = 2*k
      do m = 1,k {
            carg = (0.,1.)*(3.14159265*signi*(m-1))/k;      cw = cexp(carg)
            do i = m,lx,istep
                  { ct=cw*cx(i+k);   cx(i+k)=cx(i)-ct;   cx(i)=cx(i)+ct }
            }
      k = istep
      } until(k>=lx)
return;        end
```

Fourier transforms have both real and imaginary parts. Sometimes both are displayed. Often the imaginary part is ignored. This is because most of our time functions vanish before $t = 0$. Thus, their Fourier transforms must satisfy certain conditions, namely, real and imaginary parts must be related by Hilbert transform. Locally, one often looks like cosine, the other like sine. So, seeing the real part, it is often easy to imagine the imaginary part. Figure 2 shows the output of the test program.

Stolt Migration

The Stolt migration program shown next uses linear interpolation to convert the ω-axis to the k_z-axis. The scaling by $dk_z/d\omega$ has little effect, so it was omitted to shorten the program. (Something needed to be saved for the exercises). The test case is to make semicircles from impulses.

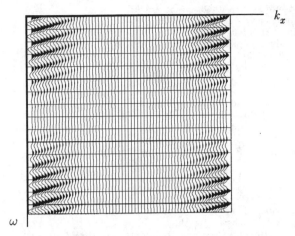

FIG. 1.7-2. Output of two-dimensional Fourier transformation test program.

```
# Test case for Stolt migration.
integer it,nt,ix,nx;   real vdtodx;   complex cp(256,64)
open(4,file='plotfile',status='new',access='direct',form='unformatted',recl=1)
nx = 64;      nt = 256;    vdtodx = 1./4.                # vdtodx = v dt / dx
do it=1,nt
        do ix=1,nx
                cp(it,ix)=0.
cp(32,9)=1.;    cp(64,17)=1.;    cp(128,33)=1.
call stolt(nt,nx,cp,vdtodx)
write(4,rec=1) ((real(cp(it,ix)),it=1,nt),ix=1,nx)
stop;   end
```

```
# Stolt migration subroutine without cosine weight.
subroutine stolt(nt,nx,cp,vdtodx)
integer ikx,nx,nt,nth,iktau,iom
real om,vkx,wl,wh,aktau,pi,pionth,vdtodx
complex cp(nt,nx),cbf(1025)
pi = 3.14159265;   nth=nt/2;   pionth = pi/nth;
call ft2d(nt,nx,cp,1.,-1.,cbf)
do ikx = 1,nx {
      vkx = (ikx-1)*2.*pi*vdtodx/nx
      if ( ikx > nx/2 ) vkx = 2.*pi*vdtodx-vkx        # negative k
      cbf(1) = 0.;         cbf(nt+1)=0.               # cbf = working buffer
      do iom = 1,nt
            cbf(iom) = cp(iom,ikx)              # Omit weighting
      cp(1,ikx)=0.                             # Ignore zero freq
      do iktau = 2,nth+1 {                         # Stretch
            aktau = (iktau-1.01)*pionth
            om = sqrt(aktau*aktau+vkx*vkx);       iom = 1+om/pionth
            if(iom<nth) {
                  wl = iom-om/pionth;       wh = 1.-wl
                  cp(iktau,ikx)    = wl*cbf(iom)    +wh*cbf(iom+1)
                  cp(nt-iktau+2,ikx) = wl*cbf(nt-iom+2)+wh*cbf(nt-iom+1)
                  }
            else
                  cp(iktau,ikx) = 0.
            }
      }
call ft2d(nt,nx,cp,-1.,1.,cbf)
return;       end
```

The output of this test program was shown in Section 1.3. To better illustrate the periodic nature of the solution all but one semicircle was removed and the result plotted with a nonlinear gain. Four identical plots appear side-by-side in figure 3.

Rocca's Row Fourier Transform

Rocca's Fourier transform over rows is somewhat faster than the rudimentary program because the basic overhead is done *once*, while *every* row gets Fourier transformed. But the main advantage of the Rocca method over the rudimentary method is that the data need not be transposed, and the program runs efficiently even in a paged environment.

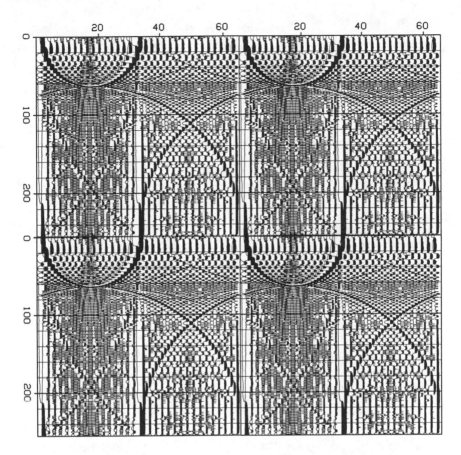

FIG. 1.7-3. Periodicity of the output of the Stolt migration program.

```
# Try Rocca's row Fourier transform.
#     sign2 should be +1. or -1. it is the sign of i.
subroutine rowcc(n1,n2,cx,sign2,scale)
complex cx(n1,n2),cmplx,cw,cdel
do i1 = 1,n1
        do i2 = 1,n2
                cx(i1,i2) = cx(i1,i2)*scale
j = 1
do i = 1,n2 {
        if (i<=j)    call twid1(n1,cx(1,i),cx(1,j))
        m = n2/2
        while (j>m) { j = j-m; m = m/2; if (m<1) break }
        j = j+m }
lstep = 1
repeat {
        istep = 2*lstep;     cw = 1.
        arg = sign2*3.14159265/lstep;  cdel = cmplx(cos(arg),sin(arg))
```

```
        do m = 1,lstep {
                do i = m,n2,istep
                        call twid2(n1,cw,cx(1,i),cx(1,i+lstep))
                cw = cw*cdel
                }
        lstep = istep
        } until(lstep>=n2)
return;         end

subroutine twid1(n,cx,cy)
complex cx(n),cy(n),ct
do i = 1,n { ct = cx(i);    cx(i) = cy(i);    cy(i) = ct }
return;         end

# If you feel like optimizing, this is the place.
subroutine twid2(n,cw,cx,cy)
complex cx(n),cy(n),ctemp,cw
do i = 1,n { ctemp = cw*cy(i);    cy(i) = cx(i)-ctemp;    cx(i) = cx(i)+ctemp }
return;         end
```

EXERCISES

1. Most time functions are real. Their imaginary part is zero. Show that this means that $F(\omega, k)$ can be determined from $F(-\omega, -k)$.

2. Verify by using your computer and plotter that figure 2 is produced by the program given.

3. The real part of the FT plotted in the previous exercise is somewhat difficult to interpret because of the awkward placement of the negative frequencies and wavenumbers. Modify the program so that $F(\omega, k)$ has its origin at the center (33,33) of the plotted grid. Hint: a simple modification of $f(t, x)$ before Fourier transforming is sufficient; recall the "shift theorem." Write $f(t, x)$ and the new, more easily interpreted $F(\omega, k)$. Label axes. (Hale)

4. A point explosion on the earth's surface at time $t=0$ and location $x=32$ provides synthetic observations in the (t, x)-plane shown on the left. On the right is the magnitude of the two-dimensional Fourier transform, (ω, k_x)-plane. The origin is in the upper left corner of each plot. What would these plots look like on an earth of half the velocity? (Toldi)

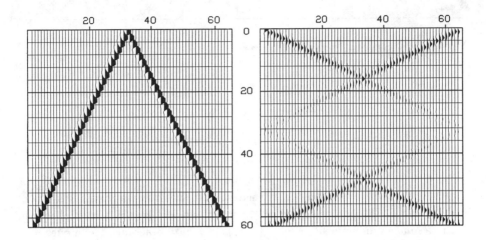

5. Insert the appropriate cosine obliquity function into the Stolt migration program. Test, and verify little difference but some angle-dependent scaling.

6. Write a program for diffraction by the Stolt method. That is, given point scatterers inside the earth, generate the appropriate hyperbolas.

7. If you include the inverse cosine weighting function in a Stolt diffraction program, beware of the pole at the evanescent edge. Is it better to stretch before weighting or after? Why?

8. Interpolation error in the Stolt program may be reduced by reducing the speed of oscillation of $P(\omega)$ with ω. To do this note that $p(t)$ vanishes for negative t. So multiply $P(\omega)$ by $e^{-i\omega\tau}$ before interpolation, and then divide it out after. What is an appropriate value of the constant τ to use in the program?

2

Why Space and Time ?

In the previous chapter we learned how to extrapolate wavefields down into the earth. The process proceeded simply, since it is just a multiplication in the frequency domain by $\exp[ik_z(\omega,k_x)z]$. Finite-difference techniques will be seen to be complicated. They will involve new approximations and new pitfalls. Why should we trouble ourselves to learn them? To begin with, many people find finite-difference methods more comprehensible. In (t,x,z)-space, there are no complex numbers, no complex exponentials, and no "magic" box called FFT.

The situation is analogous to the one encountered in ordinary frequency filtering. Frequency filtering can be done as a product in the frequency domain or a convolution in the time domain. With wave extrapolation there are products in both the temporal frequency ω-domain and the spatial frequency k_x-domain. The new ingredient is the two-dimensional (ω,k_x)-space, which replaces the old one-dimensional ω-space. Our question, why bother with finite differences?, is a two-dimensional form of an old question: After the discovery of the fast Fourier transform, why should anyone bother with time-domain filtering operations?

Our question will be asked many times and under many circumstances. Later we will have the axis of offset between the shot and geophone and the axis of midpoints between them. There again we will need to choose whether to work on these axes with finite differences or to use Fourier transformation. It is not an all-or-nothing proposition: for each axis separately either Fourier transform or convolution (finite difference) must be chosen.

The answer to our question is many-sided, just as geophysical objectives are many-sided. Most of the criteria for answering the question are already familiar from ordinary filter theory. Those electrical engineers and old-time deconvolution experts who have pushed themselves into wave processing have turned out to be delighted by it. They hadn't realized their knowledge had so many applications!

Figure 1 illustrates the differences between Fourier domain calculations and time domain calculations. The figure was calculated on a 256×64 mesh to exacerbate for display the difficulties in either domain. Generally, you notice wraparound noise in the Fourier calculation, and frequency dispersion (Section 4.3) in the time domain calculation. (The "time domain" hyperbola in figure 1 is actually a frequency domain simulation — to wrap the entire hyperbola into view). In this Chapter we will see how to do the time domain calculations. A more detailed comparison of the domains is in Chapter 4.

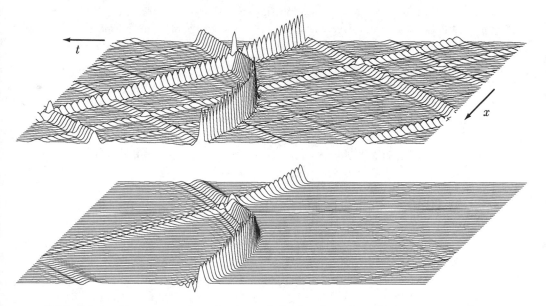

FIG. 2.0-1. Frequency domain hyperbola (top) and time domain hyperbola (bottom).

Even if you always migrate in the frequency domain, it is worth studying time domain methods to help you choose parameters to get a good time domain response. For example both parts of figure 1 were done in the frequency domain, but one simulated the time domain calculation to get a more causal response.

Lateral Variation

In ordinary linear filter theory, a filter can be made time-variable. This is useful in reflection seismology because the frequency content of echoes changes with time. An annoying aspect of time-variable filters is that they cannot be described by a simple product in the frequency domain. So when an application of time-variable filters comes along, the frequency domain is abandoned, or all kinds of contortions are made (stretching the time axis, for example) to try to make things appear time-invariant.

All the same considerations apply to the horizontal space axis x. On space axes, a new concern is the seismic velocity v. If it is space-variable, say $v(x)$, then the operation of extrapolating wavefields upward and downward can no longer be expressed as a product in the k_x-domain. Wave-extrapolation procedures must abandon the spatial frequency domain and go to finite differences. The alternative again is all kinds of contortions (such as stretching the x-axis) to try to make things appear to be space-invariant.

In two or more dimensions, stretching tends to become more difficult and less satisfactory.

A less compelling circumstance of the same type that suggests finite differences rather than Fourier methods is lateral variation in channel location. If geophones somehow have become unevenly separated so that the Δx between channels is not independent of x, then there is a choice of (1) resampling the data at uniform intervals before Fourier analysis, or (2) processing the data directly with finite differences.

Stepout

Much of seismology amounts to measuring time shifts. The word *stepout* denotes a change of travel time with a change in location. Frequency-domain calculations usually conclude with a transform to the time domain to let us see the shifts. An advantage of time-domain computations is that time shifts of wave packets can be measured as the computation proceeds. In the frequency domain it is not difficult to reference one single time point, or to prescribe a shift of the whole time function. But it is not easy to access separate wavelets or wave packets without returning to the time domain.

The upward and downward wavefield extrapolation filter $\exp[i \, k_z(\omega, k_x)z]$ is basically a causal all-pass filter. (Under some circumstances it is anticausal). It moves energy around without amplification or attenuation. I suppose this is why migration filtering is more fun than minimum-phase filtering. Migration filters gather energy from all over and drop it in a good place, whereas minimum-phase filters hardly move things at all — they just scale some frequencies up and others down. Any filter of the form $\exp[i \, \phi(\omega)]$ is an all-pass filter. What are the constraints on the function $\phi(\omega)$ which make the time-domain representation of $\exp(i \, \phi)$ causal?

Causal all-pass filters turn out to have an attractive representation, with Z-transforms as $Z^N \bar{A}(1/Z)/A(Z)$. Those who are familiar with filter theory will realize that the division by $A(Z)$ raises a whole range of new issues: feedback, economy of parameterization, and possible instability. (Section 4.6 covers Z-transforms). These issues will all arise in using finite differences to downward extrapolate wavefields. It is a feedback process. The economy of parameterization is attractive. Taking $A(Z) = 1 + a_1 Z + a_2 Z^2$, the two adjustable coefficients are sufficient to select a frequency and a bandwidth for selective delay. Economy of parameterization also implies economy in application. That is nice. It is also nice having the functional form itself imply causality. On the other hand, the advantages of economy are offset by some dangers. Now we must learn and use some stability theory. $A(Z)$ must be minimum phase.

Being Too Clever in the Frequency Domain

Fourier methods are *global*. That is, the entire dataset must be in hand before processing can begin. Remote errors and truncations can have serious local effects. On the other hand, finite-difference methods are *local*. Data points are directly related only to their neighbors. Remote errors propagate slowly. Let me cite two examples of frequency-domain pitfalls in the field of one-dimensional time-series analysis.

In the frequency domain it is easy to specify sharp cutoff filters, say, a perfectly flat passband between 8 Hz and 80 Hz, zero outside. But such filters cause problems in the time domain. They are necessarily noncausal, giving a response before energy enters the filter. Another ugly aspect is that the time response drops off only inversely with t. Distant echoes that have amplitudes weakened as inverse time squared would get lost in the long filter response of the early echoes.

A more common problem arises with the 60 Hz powerline frequency rejection filters found in much recording equipment. Notch filters are easy to construct in the Z-transform domain. Start with a zero on the unit circle at exactly 60 Hz. That kills the noise but it distorts the passband at other frequencies. So, a tiny distance away, outside the unit circle, place a pole. The separation between the pole and the zero determines the bandwidth of the notch. The pole has the effect of nearly canceling the zero if the pair are seen from a distance. So there is an ideal flat spectrum away from the absorption zone. You record some data with this filter. Late echoes are weaker than early ones, so the plotting program increases the gain with time. After installing your powerline reject filters you discover that they have *increased* the powerline noise instead of decreasing it. Why? The reason is that you tried to be too clever when you put the pole too close to the circle. The exponential gain effectively moved the unit circle away from the zero towards the pole. The pole may end up on the circle! Putting the pole further from the zero gives a broader notch, which is less attractive in the frequency domain, but at least the filter will work sensibly when the gain varies with time.

Zero Padding

When fast Fourier transforms came into use, one of the first applications was convolution. If a filter has more than about fifty coefficients, it may be faster to apply it by multiplication in the frequency domain. The result will be identical to convolution if care has been taken to pad the ends of the data and the filter with enough zeroes. They make invisible the periodic behavior of the discrete Fourier transform. For filtering time functions whose length is typically about one thousand, this is a small price in added memory to pay for the time saved. Seismic sections are often thousands of channels long. For migration, zero padding must simultaneously be done on the space axis and the time axis. There are three places where zeroes may be required, as indicated below:

data	0
0	0

Section 4.5 offers suggestions on how to alleviate the problems of Fourier domain migration techniques.

Looking Ahead

Some problems of the Fourier domain have just been summarized. The problems of the space domain will be shown in this chapter and Chapter 4. Seismic data processing is a multidimensional task, and the different dimensions are often handled in different ways. But if you are sure you are content with the Fourier domain then you can skip much of this chapter and jump directly to Chapter 3, where you can learn about shot-to-geophone offset, stacking, and migration before stack.

2.1 Wave-Extrapolation Equations

A wave-extrapolation equation is an expression for the derivative of a wavefield (usually in the depth z direction). When the wavefield and its derivative are known, extrapolation can proceed by various numerical representations of $P(z + \Delta z) = P(z) + \Delta z \, dP/dz$. So what is really needed is an expression for dP/dz. Two theoretical methods for finding dP/dz are the original *transformation* method and the newer *dispersion-relation* method.

Meet the Parabolic Wave Equation

At the time the parabolic equation was introduced to petroleum prospecting (1969), it was well known that "wave theory doesn't work." At that time, petroleum prospectors analyzed seismic data with rays. The wave equation was not relevant to practical work. Wave equations were for university theoreticians. (Actually, wave theory did work for the surface waves of massive earthquakes, scales 1000 times greater than in exploration). Even for university workers, finite-difference solutions to the wave equation didn't work out very well. Computers being what they were, solutions looked more like "vibrations of a drum head" than like "seismic waves in the earth." The parabolic wave equation was originally introduced to speed finite-difference wave modeling. The following introduction to the parabolic wave equation is via the original transformation method.

The difficulty prior to 1969 came from an inappropriate assumption central to all then-existing seismic wave theory, namely, the horizontal layering assumption. Ray tracing was the only way to escape this assumption, but ray tracing seemed to ignore waveform modeling. In petroleum exploration almost all wave theory further limited itself to vertical incidence. The road to success lay in expanding ambitions from *vertical incidence* to include a small *angular bandwidth* around vertical incidence. This was achieved by abandoning much known, but cumbersome, seismic theory.

A vertically downgoing plane wave is represented mathematically by the equation

$$P(t, x, z) \quad = \quad P_0 \; e^{-i\omega(t - z/v)} \tag{1}$$

In this expression, P_0 is absolutely constant. A small departure from vertical incidence can be modeled by replacing the constant P_0 with something, say, $Q(x, z)$, which is not strictly constant but varies slowly.

$$P(t, x, z) \quad = \quad Q(x, z) \; e^{-i \omega (t - z/v)} \tag{2}$$

Inserting (2) into the scalar wave equation $P_{xx} + P_{zz} = P_{tt}/v^2$ yields

$$\frac{\partial^2}{\partial x^2} \, Q \; + \; \left[\frac{i \omega}{v} + \frac{\partial}{\partial z} \right]^2 Q \; = \; -\frac{\omega^2}{v^2} \, Q$$

$$\frac{\partial^2 Q}{\partial x^2} \; + \; \frac{2 \, i \, \omega}{v} \frac{\partial Q}{\partial z} \; + \; \frac{\partial^2 Q}{\partial z^2} \; = \; 0 \tag{3}$$

The wave equation has been reexpressed in terms of $Q(x, z)$. So far no approximations have been made. To require the wavefield to be near to a plane wave, $Q(x, z)$ must be near to a constant. The appropriate means (which caused some controversy when it was first introduced) is to drop the highest depth derivative of Q, namely, Q_{zz}. This leaves us with the *parabolic wave equation*

$$\frac{\partial Q}{\partial z} \quad = \quad \frac{v}{-2 \, i \, \omega} \frac{\partial^2 Q}{\partial x^2} \tag{4}$$

At the time it was first developed for use in seismology, the most important property of (4) was thought to be this: For a wavefield close to a vertically propagating plane wave, the second x-derivative is small, hence the z-derivative is small. Thus, the finite-difference method should allow a very large Δz and thus be able to treat models more like the earth, and less like a drumhead.

It soon became apparent that the parabolic wave equation was also just what was needed for seismic imaging: it was a wave-extrapolation equation.

It is curious that equation (4) is the Schroedinger equation of quantum mechanics.

This approach, the transformation approach, was and is very useful. But it was soon replaced by the dispersion-equation approach — a way of getting equations to extrapolate waves at wider angles.

Muir Square-Root Expansion

When we use the newer method of finding wave extrapolators, we seek various approximations to a square-root dispersion relation. Then the approximate dispersion relation is inverse transformed into a differential equation. Thanks largely to Francis Muir, the dispersion approach has evolved considerably since the writing of *Fundamentals of Geophysical Data Processing.*

Substitution of the plane wave $\exp(-i\omega t + ik_x x + ik_z z)$ into the two-dimensional scalar wave equation yields the dispersion relation

$$k_z{}^2 + k_x{}^2 \ = \ \frac{\omega^2}{v^2} \tag{5}$$

Solve for k_z selecting the positive square root (thus for the moment selecting *downgoing* waves).

$$k_z \ = \ \frac{\omega}{v} \sqrt{1 - \frac{v^2 k_x{}^2}{\omega^2}} \tag{6a}$$

To inverse transform the z-axis we only need to recognize that ik_z corresponds to $\partial/\partial z$. The resulting expression is a wavefield extrapolator, namely,

$$\frac{\partial P}{\partial z} \ = \ i\,\frac{\omega}{v} \sqrt{1 - \frac{v^2 k_x{}^2}{\omega^2}}\, P \tag{6b}$$

Bringing equation (6b) into the space domain is not simply a matter of substituting a second x derivative for $k_x{}^2$. The problem is the meaning of the square root of a differential operator. The square root of a differential operator is not defined in undergraduate calculus courses and there is no straightforward finite difference representation. The square root becomes meaningful only when the square root is regarded as some kind of truncated series expansion. It will be shown in Section 4.6 that the Taylor series is a poor choice. Francis Muir showed that the original 15° and 45° methods were just truncations of a continued fraction expansion. To see this, let X and R be defined by writing (6a) as

$$k_z \ = \ \frac{\omega}{v} \sqrt{1 - X^2} \ = \ \frac{\omega}{v} R \tag{7}$$

The desired polynomial ratio of order n will be denoted R_n, and it will be determined by the recurrence

$$R_{n+1} \ = \ 1 - \frac{X^2}{1 + R_n} \tag{8}$$

To see what this sequence converges to (if it converges), set $n = \infty$ in (8) and solve

$$R_\infty \ = \ 1 - \frac{X^2}{1 + R_\infty}$$

$$R_\infty (1 + R_\infty) \ = \ 1 + R_\infty - X^2$$

$$R^2 = 1 - X^2 \qquad (9)$$

The square root of (9) gives the required expression (7). Geometrically, (9) says that the cosine squared of the incident angle equals one minus the sine squared. Truncating the expansion leads to angle errors. Actually it is only the low-order terms in the expansion that are ever used. Beginning from $R_0 = 1$ the results in table 1 are found.

$5°$	$R_0 = 1$
$15°$	$R_1 = 1 - \dfrac{X^2}{2}$
$45°$	$R_2 = 1 - \dfrac{X^2}{2 - \dfrac{X^2}{2}}$
$60°$	$R_3 = 1 - \dfrac{X^2}{2 - \dfrac{X^2}{2 - \dfrac{X^2}{2}}}$

TABLE 2.1-1. First four truncations of Muir's continued fraction expansion.

For various historical reasons, the equations in table 1 are often referred to as the $5°$, $15°$, and $45°$ equations, respectively, the names giving a reasonable qualitative (but poor quantitative) guide to the range of angles that are adequately handled. A trade-off between complexity and accuracy frequently dictates choice of the $45°$ equation. It then turns out that a slightly wider range of angles can be accommodated if the recurrence is begun with something like $R_0 = \cos 45°$. Accuracy enthusiasts might even have R_0 a function of velocity, space coordinates, or frequency.

Dispersion Relations

Performing the substitutions of table 1 into equation (7) gives dispersion relationships for comparison to the exact expression (6a). These are shown in table 2.

5°	$k_z = \dfrac{\omega}{v}$
15°	$k_z = \dfrac{\omega}{v} - \dfrac{v\,k_x^{\,2}}{2\,\omega}$
45°	$k_z = \dfrac{\omega}{v} - \dfrac{k_x^{\,2}}{2\,\dfrac{\omega}{v} - \dfrac{v\,k_x^{\,2}}{2\,\omega}}$

TABLE 2.1-2. As displayed in figure 1, the dispersion relations of table 2 tend toward a semicircle.

Depth-Variable Velocity

Identification of ik_z with $\partial/\partial z$ converts the dispersion relations of table 2 into the differential equations of table 3.

The differential equations in table 3 were based on a dispersion relation that in turn was based on an assumption of constant velocity. So you might not expect that the equations have substantial validity or even great utility when the velocity is depth-variable, $v = v(z)$. The actual limitations are better characterized by their inability, by themselves, to describe reflection.

Migration methods based on equation (6b) or on table 3 are called *phase-shift methods.*

Retardation (Frequency Domain)

It is often convenient to arrange the calculation of a wave to remove the effect of overall translation, thereby making the wave appear to "stand still." This subject, wave retardation, will be examined more thoroughly in Section 2.6. Meanwhile, it is easy enough to introduce the time shift t_0 of a vertically propagating wave in a hypothetical medium of velocity $\overline{v}(z)$, namely,

85

$5°$	$\dfrac{\partial P}{\partial z} \;=\; i\left(\dfrac{\omega}{v}\right)P$
$15°$	$\dfrac{\partial P}{\partial z} \;=\; i\left(\dfrac{\omega}{v} - \dfrac{v\,k_x^{\,2}}{2\,\omega}\right)P$
$45°$	$\dfrac{\partial P}{\partial z} \;=\; i\left(\dfrac{\omega}{v} - \dfrac{k_x^{\,2}}{2\,\dfrac{\omega}{v} - \dfrac{v\,k_x^{\,2}}{2\,\omega}}\right)P$

TABLE 2.1-3. Extrapolation equations when velocity depends only on depth.

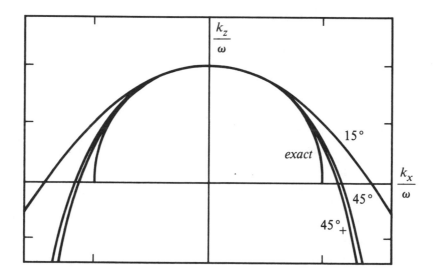

FIG. 2.1-1. Dispersion relation of equations (6a) and table 2. The curve labeled $45°_+$ was constructed with $R_0 = \cos 45°$. It fits exactly at $0°$ and $45°$.

$$t_0 = \int_0^z \frac{dz}{\bar{v}(z)} \tag{10}$$

A time delay t_0 in the time domain corresponds to multiplication by $\exp(i\omega t_0)$ in the ω-domain. Thus, the actual wavefield P is related to the time-shifted wavefield Q by

$$P(z,\omega) = Q(z,\omega) \exp\left[i\omega \int_0^z \frac{dz}{\bar{v}(z)} \right] \tag{11a}$$

(Equation (11) applies in both x- and k_x-space). Differentiating with respect to z gives

$$\frac{\partial P}{\partial z} = \frac{\partial Q}{\partial z} \exp\left[i\omega \int_0^z \frac{dz}{\bar{v}(z)} \right] + Q(z,\omega) \frac{i\omega}{\bar{v}(z)} \exp\left[i\omega \int_0^z \frac{dz}{\bar{v}(z)} \right]$$

or

$$\frac{\partial P}{\partial z} = \exp\left[i\omega \int_0^z \frac{dz}{\bar{v}(z)} \right] \left(\frac{\partial}{\partial z} + \frac{i\omega}{\bar{v}(z)} \right) Q \tag{11b}$$

Next, substitute (11) into table 3 to obtain the retarded equations in table 4.

Lateral Velocity Variation

Having approximated the square root by a polynomial ratio, table 3 or table 4 can be inverse transformed from the horizontal wavenumber domain k_x to the horizontal space domain x by substituting $(ik_x)^2 = \partial^2/\partial x^2$. As before, the result has a wide range of validity for $v = v(x,z)$ even though the derivation would not seem to permit this. Ordinarily $\bar{v}(z)$ will be chosen to be some kind of horizontal average of $v(x,z)$. Permitting $\bar{v}$ to become a function of x generates many new terms. The terms are awkward to implement and ignoring them introduces unknown hazards. So $\bar{v}$ is usually taken to depend on z but not x.

Splitting

The customary numerical solution to the x-domain forms of the equations in tables 3 and 4 is arrived at by splitting. That is, you march forward a small Δz-step alternately with the two extrapolators

$$\frac{\partial Q}{\partial z} = \text{lens term} \tag{12a}$$

$$\frac{\partial Q}{\partial z} = \text{diffraction term} \tag{12b}$$

$5°$	$\dfrac{\partial Q}{\partial z} = zero$	$+ \, i\,\omega \left[\dfrac{1}{v} - \dfrac{1}{\bar{v}(z)} \right] Q$
$15°$	$\dfrac{\partial Q}{\partial z} = -\, i \, \dfrac{v k_x^{\,2}}{2\omega} \, Q$	$+ \, i\,\omega \left[\dfrac{1}{v} - \dfrac{1}{\bar{v}(z)} \right] Q$
$45°$	$\dfrac{\partial Q}{\partial z} = -\, i \, \dfrac{k_x^{\,2}}{2\,\dfrac{\omega}{v} - \dfrac{v k_x^{\,2}}{2\omega}} \, Q$	$+ \, i\,\omega \left[\dfrac{1}{v} - \dfrac{1}{\bar{v}(z)} \right] Q$
general	$\dfrac{\partial Q}{\partial z} = \; diffraction$	$+ \; thin \; lens$

TABLE 2.1-4. Retarded form of phase-shift equations.

Justification of the splitting process is found in Section 2.4. The first equation, called the *lens equation,* is solved analytically:

$$Q(z_2) \; = \; Q(z_1) \exp \left\{ i\,\omega \int_{z_1}^{z_2} \left[\frac{1}{v(x,z)} - \frac{1}{\bar{v}(z)} \right] dz \right\} \tag{13}$$

Migration methods that include the lens equation are called *depth* migrations. The lens equation is often omitted where $v(x)$ is poorly known. Such migrations are called *time* migrations.

Observe that the diffraction parts of tables 3 and 4 are the same. Let us use them and equation (12b) to define a table of diffraction equations. Substitute $\partial/\partial x$ for $i k_x$ and clear $\partial/\partial x$ from the denominators to get table 5.

Time Domain

To put the above equations in the time domain, it is necessary only to get ω into the numerator and then replace $-i\,\omega$ by $\partial/\partial t$. For example, the $15°$, retarded, $v = \bar{v}$ equation from table 5 becomes

$$\frac{\partial^2}{\partial z \, \partial t} \, Q \; = \; \frac{v}{2} \, \frac{\partial^2}{\partial x^2} \, Q \tag{14}$$

5°	$\dfrac{\partial Q}{\partial z} = zero$
15°	$\dfrac{\partial Q}{\partial z} = \dfrac{v(x,z)}{-2\,i\,\omega}\,\dfrac{\partial^2 Q}{\partial x^2}$
45°	$\left\{ 1 - \left(\dfrac{v(x,z)}{-2\,i\,\omega} \right)^2 \dfrac{\partial^2}{\partial x^2} \right\} \dfrac{\partial Q}{\partial z} = \dfrac{v(x,z)}{-2\,i\,\omega}\,\dfrac{\partial^2 Q}{\partial x^2}$

TABLE 2.1-5. Diffraction equations for laterally variable media.

Interpretation of time t for a retarded-time variable Q awaits further clarification in Section 2.6.

Upcoming Waves

All the above equations are for *downgoing* waves. To get equations for *upcoming* waves you need only change the signs of z and $\partial/\partial z$. Letting D denote a *downgoing* wavefield and U an *upcoming* wavefield, equation (14), for example, takes the form

$$\frac{\partial^2}{\partial z\, \partial t}\, D = +\frac{v}{2}\, \frac{\partial^2}{\partial x^2}\, D$$

$$\frac{\partial^2}{\partial z\, \partial t}\, U = -\frac{v}{2}\, \frac{\partial^2}{\partial x^2}\, U$$

TABLE 2.1-6. Time-domain equations for downgoing and upcoming wave diffraction with retardation and the 15° approximation.

Using the exploding-reflector concept, it is the upcoming wave equation that is found in both migration and diffraction programs. The downgoing wave equation is useful for modeling and migration procedures that are more elaborate than those based on the exploding-reflector concept (see Section 5.7).

EXERCISE

1. Consider a tilted straight line tangent to a circle. Use this line to initialize the Muir square-root expansion. State equations and plot them $(-2 \leq X \leq +2)$ for the next two Muir semicircle approximations.

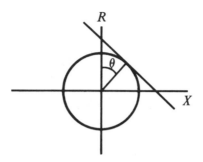

2.2　Finite Differencing

The basic method for solving differential equations in a computer is *finite differencing*. The nicest feature of the method is that it allows analysis of objects of almost any shape, such as earth topography or geological structure. Ordinarily, finite differencing is a straightforward task. The main pitfall is instability. It often happens that a seemingly reasonable approach to a reasonable physical problem leads to wildly oscillatory, divergent calculations. Luckily, there is a fairly small body of important and easily learned tricks that should solve most stability problems.

Of secondary concern are the matters of cost and accuracy. These must be considered together since improved accuracy can be achieved simply by paying the higher price of a more refined computational mesh. Although the methods of the next several pages have not been chosen for their accuracy or efficiency, it turns out that in these areas they are excellent. Indeed, to my knowledge, some cannot be improved on at all, while others can be improved on only in small ways. By "small" I mean an improvement in efficiency of a factor of five or less. Such an improvement is rarely of consequence in research or experimental work; however, its importance in matters of production will justify pursuit of the literature far beyond the succeeding pages.

The Lens Equation

The various wave-extrapolation operators can be split into two parts, a complicated part called the *diffraction* or *migration* part, and an easy part called the *lens* part. The lens equation applies a time shift that is a function of x. The lens equation acquires its name because it acts just like a thin optical lens when a light beam enters on-axis (vertically). Corrections for nonvertical incidence and the thickness of the lens are buried somehow in the diffraction part. The lens equation has an analytical solution, namely, $\exp[i\,\omega t_0(x)]$. It is better to use this analytical solution than to use a finite-difference solution because there are no approximations in it to go bad. The only reason the lens equation is mentioned at all in a chapter on finite differencing is that the companion diffraction equation must be marched forward along with the lens equation, so the analytic solutions are marched along in small steps.

First Derivatives, Explicit Method

The inflation of money q at a 10% rate can be described by the difference equation

$$q_{t+1} - q_t \;=\; .10\, q_t \tag{1a}$$

$$(1.0)\, q_{t+1} + (-1.1)\, q_t \;=\; 0 \tag{1b}$$

This one-dimensional calculation can be reexpressed as a differencing star and a data table. As such it provides a prototype for the organization of calculations with two-dimensional partial-differential equations. Consider

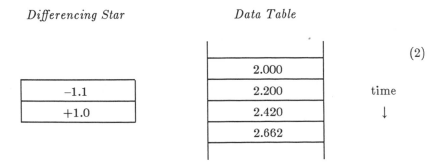

Differencing Star   *Data Table*   (2)

Since the data in the data table satisfy the difference equation (1), the differencing star may be laid anywhere on top of the data table, the numbers in the star may be multiplied by those in the underlying table, and the resulting cross products will sum to zero. On the other hand, if all but one number (the initial condition) in the data table were missing then the rest of the numbers could be filled in, one at a time, by sliding the star along, taking the difference equation to be true, and solving for the unknown data value at each stage.

Less trivial examples utilizing the same differencing star arise when the numerical constant .10 is replaced by a complex number. Such examples exhibit oscillation as well as growth and decay.

First Derivatives, Implicit Method

Let us solve the equation

$$\frac{dq}{dt} = 2\,r\,q \tag{3}$$

by numerical methods. Note that the inflation-of-money equation (1), where $2\,r = .1$, provides an approximation. But then note that in the inflation-of-money equation the expression of dq/dt is centered at $t+1/2$, whereas the expression of q by itself is at time t. There is no reason the q on the right side of equation (3) cannot be averaged at time t with time $t+1$, thus centering the whole equation at $t+1/2$. Specifically, a centered approximation of (3) is

$$q_{t+1} - q_t = 2\,r\,\Delta t \ \frac{q_{t+1} + q_t}{2} \tag{4a}$$

Letting $\alpha = r\,\Delta t$, this becomes

$$(1-\alpha)\,q_{t+1} - (1+\alpha)\,q_t = 0 \tag{4b}$$

which is representable as the difference star

$-1-\alpha$
$+1-\alpha$

t

$\downarrow$

$$\tag{4c}$$

For a fixed Δt this star gives a more accurate solution to the differential equation (3) than does the star for the inflation of money.

Explicit Heat-Flow Equation

The heat-flow equation controls the diffusion of heat. This equation is a prototype for migration. The 15° migration equation is the same equation but the heat conductivity constant is imaginary. (The migration equation is really the Schroedinger equation, which controls the diffusion of probability of atomic particles). Taking σ constant yields

$$\frac{\partial q}{\partial t} = \frac{\sigma}{C}\frac{\partial^2 q}{\partial x^2} \tag{5}$$

Implementing (5) in a computer requires some difference approximations for the partial differentials. The most obvious (but not the only) approach is the basic definition of elementary calculus. For the time derivative, this is

$$\frac{\partial q}{\partial t} \approx \frac{q(t+\Delta t) - q(t)}{\Delta t} \tag{6a}$$

It is convenient to use a subscript notation that allows (6a) to be compacted into

$$\frac{\partial q}{\partial t} \approx \frac{q_{t+1} - q_t}{\Delta t} \tag{6b}$$

In this notation $t+\Delta t$ is abbreviated by $t+1$, a convenience for more complicated equations. The second-derivative formula may be obtained by doing the first derivative twice. This leads to $q_{t+2} - 2q_{t+1} + q_t$. The formula is usually treated more symmetrically by shifting it to $q_{t+1} - 2q_t + q_{t-1}$. These two versions are equivalent as Δt tends to zero, but the more symmetrical arrangement will be more accurate when Δt is not zero. Using superscripts to describe x-dependence gives a finite-difference approximation to the second space derivative:

$$\frac{\partial^2 q}{\partial x^2} \approx \frac{q^{x+1} - 2q^x + q^{x-1}}{\Delta x^2} \tag{7}$$

Inserting the last two equations into the heat-flow equation (and using $=$ to denote $\approx$) gives

$$\frac{q_{t+1}^x - q_t^x}{\Delta t} = \frac{\sigma}{C}\frac{q_t^{x+1} - 2q_t^x + q_t^{x-1}}{(\Delta x)^2} \tag{8}$$

Letting $\alpha = \sigma\,\Delta t/(C\,\Delta x^2)$ (8) can be arranged thus:

$$q_{t+1}^x - q_t^x - \alpha(q_t^{x+1} - 2q_t^x + q_t^{x-1}) = 0 \tag{9}$$

Equation (9) can be interpreted geometrically as a computational star in the (x, t)-plane, as depicted in figure 1. By moving the star around in the data table you will note that it can be positioned so that only one number at a time (the 1) lies over an unknown element in the data table. This enables the computation of subsequent rows beginning from the top. By doing this you are solving the partial-differential equation by the finite-difference method. There are other possible arrangements of initial and side conditions,

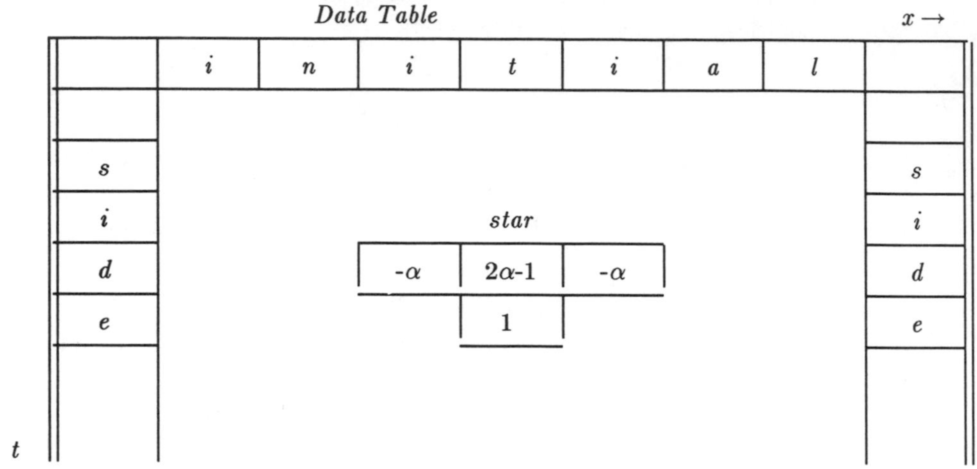

FIG. 2.2-1. Differencing star and table for one-dimensional heat-flow equation.

such as zero-value side conditions. Next is a computer program and a test example.

```
# Explicit heat-flow equation
real q(12), qp(12)
nx=12
do ia=1,2 {                    # stable and unstable cases
        alpha = ia*.3333;       write(6,'(/"alpha =",f4.2)') alpha
        do ix=1,6;   q(ix) = 0.        # Initial temperature step
        do ix=7,12;  q(ix) = 1.
        do it=1,6 {
                write(6,'(20f5.2)') (q(ix),ix=1,nx)
                do ix=2,nx-1
                        qp(ix) = q(ix) + alpha*(q(ix-1)-2.*q(ix)+q(ix+1))
                qp(1) = qp(2);  qp(nx) = qp(nx-1)
                do ix=1,nx
                        q(ix) = qp(ix)
                }
        }
stop;    end
```

alpha = .33

.00	.00	.00	.00	.00	.00	1.00	1.00	1.00	1.00	1.00	1.00
.00	.00	.00	.00	.00	.33	.67	1.00	1.00	1.00	1.00	1.00
.00	.00	.00	.00	.11	.33	.67	.89	1.00	1.00	1.00	1.00
.00	.00	.00	.04	.15	.37	.63	.85	.96	1.00	1.00	1.00
.00	.00	.01	.06	.19	.38	.62	.81	.94	.99	1.00	1.00
.00	.00	.02	.09	.21	.40	.60	.79	.91	.98	1.00	1.00

alpha = .67

.00	.00	.00	.00	.00	.00	1.00	1.00	1.00	1.00	1.00	1.00
.00	.00	.00	.00	.00	.67	.33	1.00	1.00	1.00	1.00	1.00
.00	.00	.00	.00	.44	.00	1.00	.56	1.00	1.00	1.00	1.00
.00	.00	.00	.30	-.15	.96	.04	1.15	.70	1.00	1.00	1.00
.00	.00	.20	-.20	.89	-.39	1.39	.11	1.20	.80	1.00	1.00
.13	.13	-.20	.79	-.69	1.65	-.65	1.69	.21	1.20	.87	.87

The Leapfrog Method

The difficulty with the given program is that it doesn't work for all possible numerical values of α. You can see that when α is too large (when Δx is too small) the solution in the interior region of the data table contains growing oscillations. What is happening is that the low-frequency part of the solution is OK (for a while), but the high-frequency part is diverging. The precise reason the divergence occurs is the subject of some mathematical analysis that will be done in Section 2.8. At wavelengths long compared to Δx or Δt, we expect the difference approximation to agree with the true heat-flow equation, smoothing out irregularities in temperature. At short wavelengths the wild oscillation shows that the difference equation can behave in a way almost opposite to the way the differential equation behaves. The short wavelength discrepancy arises because difference operators become equal to differential operators only at long wavelengths. The divergence of the solution is a fatal problem because the subsequent round-off error will eventually destroy the low frequencies too.

By supposing that the instability arises because the time derivative is centered at a slightly different time $t + 1/2$ than the second x-derivative at time t, we are led to the so-called *leapfrog* method, in which the time derivative is taken as a difference between $t - 1$ and $t + 1$:

$$\frac{\partial q}{\partial t} \approx \frac{q_{t+1} - q_{t-1}}{2\,\Delta t} \tag{10}$$

The resulting leapfrog differencing star is

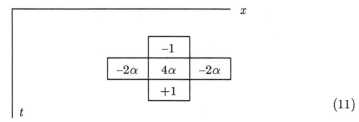

(11)

Here the result is even worse. A later analysis shows that the solution is now divergent for *all* real numerical values of α. Although it was a good idea to center both derivatives in the same place, it turns out that it was a bad idea to express a first derivative over a span of more mesh points. The enlarged operator has two solutions in time instead of just the familiar one. The numerical solution is the sum of the two theoretical solutions, one of which, unfortunately (in this case), grows and oscillates for all real values of α.

To avoid all these problems (and get more accurate answers as well), we now turn to some slightly more complicated solution methods known as *implicit* methods.

The Crank-Nicolson Method

The Crank-Nicolson method solves both the accuracy and the stability problem.

The heat-flow equation (6b) was represented as

$$q_{t+1}^x - q_t^x \;=\; a\left(q_t^{x+1} - 2\,q_t^x + q_t^{x-1}\right)$$

Now, instead of expressing the right-hand side entirely at time t, it will be averaged at t and $t+1$, giving

$$q_{t+1}^x - q_t^x = \frac{a}{2}\left[\left(q_t^{x+1} - 2q_t^x + q_t^{x-1}\right) + \left(q_{t+1}^{x+1} - 2q_{t+1}^x + q_{t+1}^{x-1}\right)\right] \text{(12a)}$$

This is called the Crank-Nicolson method. Letting $\alpha = a/2$, the difference star is

x

$-\alpha$	$2\alpha-1$	$-\alpha$
$-\alpha$	$2\alpha+1$	$-\alpha$

t

$$(12b)$$

When placing this star over the data table, note that, typically, three elements at a time cover unknowns. To say the same thing with equations, move all the $t+1$ terms in (12a) to the left and the t terms to the right, obtaining

$$-\alpha q_{t+1}^{x+1} + (1+2\alpha)q_{t+1}^{x} - \alpha q_{t+1}^{x-1} \;=\; \alpha q_{t}^{x+1} + (1-2\alpha)q_{t}^{x} + \alpha q_{t}^{x-1} \qquad (13a)$$

Taking all the $t+1$ values to be unknown, while all the t values are known the right side of (13a) is known, say, d_t^x, and the left side is a set of simultaneous equations for the unknown q_{t+1}. In other words, (13a) does not give us each q_{t+1}^x *explicitly*. They are given *implicitly* by the solution of simultaneous equations. If the x-axis is limited to five points, these equations are

$$
\begin{bmatrix}
e_{lf} & -\alpha & 0 & 0 & 0 \\
-\alpha & 1+2\alpha & -\alpha & 0 & 0 \\
0 & -\alpha & 1+2\alpha & -\alpha & 0 \\
0 & 0 & -\alpha & 1+2\alpha & -\alpha \\
0 & 0 & 0 & -\alpha & e_{rt}
\end{bmatrix}
\begin{bmatrix}
q_{t+1}^{1} \\
q_{t+1}^{2} \\
q_{t+1}^{3} \\
q_{t+1}^{4} \\
q_{t+1}^{5}
\end{bmatrix}
=
\begin{bmatrix}
d_t^{\,1} \\
d_t^{\,2} \\
d_t^{\,3} \\
d_t^{\,4} \\
d_t^{\,5}
\end{bmatrix}
\qquad (13b)
$$

The values e_{lf} and e_{rt} are adjustable and have to do with the side boundary conditions. The important thing to notice is that the matrix is tridiagonal, that is, except for three central diagonals all the elements of the matrix in (13b) are zero. The solution to such a set of simultaneous equations may be economically obtained. It turns out that the cost is only about twice that of the explicit method given by (9). In fact, this implicit method turns out to be cheaper, since the increased accuracy of (13a) over (9) allows the use of a much larger numerical choice of Δt. A program that demonstrates the stability of the method, even for large Δt, is given next.

A tridiagonal simultaneous equation solving subroutine is used. It is explained subsequently.

```
# Implicit heat-flow equation
real q(12),d(12),e(12),f(12)
nx=12; a = 8.;      write(6,'(/"a =",f4.2)') a;  alpha = .5*a
do ix=1,6;   q(ix) = 0.    # Initial temperature step
do ix=7,12;  q(ix) = 1.
do it=1,4 {
        write(6,'(20f5.2)') (q(ix),ix=1,nx)
        d(1) = 0.;    d(nx) = 0.
        do ix=2,nx-1
                d(ix) = q(ix) + alpha*(q(ix-1)-2.*q(ix)+q(ix+1))
        call rtris(nx,alpha,-alpha,(1.+2.*alpha),-alpha,alpha,d,q,e,f)
        }
stop;   end

# real tridiagonal equation solver
subroutine rtris(n,endl,a,b,c,endr,d,q,e,f)
real q(n),d(n),f(n),e(n),a,b,c,den,endl,endr
e(1) = -a/endl;   f(1) = d(1)/endl
do i = 2,n-1 {
        den = b+c*e(i-1);   e(i) = -a/den;   f(i) = (d(i)-c*f(i-1))/den }
q(n) = (d(n)-c*f(n-1))/(endr+c*e(n-1))
do i = n-1,1,-1
        q(i) = e(i)*q(i+1)+f(i)
return;           end
```

a =8.00

.00	.00	.00	.00	.00	.00	1.00	1.00	1.00	1.00	1.00	1.00
.17	.17	.21	.30	.47	.76	.24	.53	.70	.79	.83	.83
.40	.40	.42	.43	.40	.24	.76	.60	.57	.58	.60	.60
.44	.44	.44	.44	.48	.68	.32	.52	.56	.56	.56	.56

Solving Tridiagonal Simultaneous Equations

Much of the world's computing power gets used up solving tridiagonal simultaneous equations. For reference and completeness the algorithm is included here.

Let the simultaneous equations be written as a difference equation

$$a_j \, q_{j+1} + b_j \, q_j + c_j \, q_{j-1} \; = \; d_j \tag{14}$$

Introduce new unknowns e_j and f_j, along with an equation

$$q_j \; = \; e_j \, q_{j+1} + f_j \tag{15}$$

Write (15) with shifted index:

$$q_{j-1} \; = \; e_{j-1} \, q_j + f_{j-1} \tag{16}$$

Insert (16) into (14):

$$a_j \, q_{j+1} + b_j \, q_j + c_j \, (e_{j-1} \, q_j + f_{j-1}) \; = \; d_j \tag{17}$$

Now rearrange (17) to resemble (15):

$$q_j \;\; = \;\; \frac{-\,a_j}{b_j + c_j \; e_{j-1}} \; q_{j+1} \;\; + \;\; \frac{d_j - c_j \; f_{j-1}}{b_j + c_j \; e_{j-1}} \tag{18}$$

Compare (18) to (15) to see recursions for the new unknowns e_j and f_j:

$$e_j \;\; = \;\; \frac{-\,a_j}{b_j + c_j \; e_{j-1}} \tag{19a}$$

$$f_j \;\; = \;\; \frac{d_j - c_j \; f_{j-1}}{b_j + c_j \; e_{j-1}} \tag{19b}$$

First a boundary condition for the left-hand side must be given. This may involve one or two points. The most general possible end condition is a linear relation like equation (15) at $j=0$, namely, $q_0 = e_0 q_1 + f_0$. Thus, the boundary condition must give us both e_0 and f_0. With e_0 and all the a_j, b_j, c_j, we can use (19a) to compute all the e_j.

On the right-hand boundary we need a boundary condition. The general two-point boundary condition is

$$c_{n-1} \, q_{n-1} + e_{rt} \, q_n \;\; = \;\; d_n \tag{20}$$

Equation (20) includes as special cases the zero-value and zero-slope boundary conditions. Equation (20) can be compared to equation (16) at its end.

$$q_{n-1} \;\; = \;\; e_{n-1} \, q_n + f_{n-1} \tag{21}$$

Both q_n and q_{n-1} are unknown, but in equations (20) and (21) we have two equations, so the solution is easy. The final step is to take the value of q_n and use it in (16) to compute $q_{n-1}, q_{n-2}, q_{n-3}$, etc.

If you wish to squeeze every last ounce of power from your computer, note some facts about this algorithm. (1) The calculation of e_j depends on the *medium* through a_j, b_j, c_j, but it does not depend on the *solution* q_j (even through d_j). This means that it may be possible to save and reuse e_j. (2) In many computers, division is much slower than multiplication. Thus, the divisor in (19a,b) can be inverted once (and perhaps stored for reuse).

The $\partial^3/\partial x^2 \, \partial z$-Derivative

The 45° diffraction equation differs from the 15° equation by the inclusion of a $\partial^3/\partial x^2 \partial z$ -derivative. Luckily this derivative fits on the six-point differencing star

$$\frac{1}{\Delta x^2 \, \Delta z}
\begin{array}{|c|c|c|}
\hline
-1 & 2 & -1 \\
\hline
1 & -2 & 1 \\
\hline
\end{array}
\tag{22}$$

So other than modifying the six coefficients on the star, it adds nothing to the computational cost.

Difficulty in Higher Dimensions

So far we have had no trouble obtaining cheap, safe, and accurate difference methods for solving *partial-differential equations* (PDEs). The implicit method has met all needs. But in space dimensions higher than one, the implicit method becomes prohibitively costly. For the common example of problems in which $\partial^2/\partial x^2$ becomes generalized to $\partial^2/\partial x^2 + \partial^2/\partial y^2$, we will learn the reason why. The simplest case is the heat-flow equation for which the Crank-Nicolson method gave us (13a). Introducing the abbreviation $\delta_{xx} q = q^{x+1} - 2q^x + q^{x-1}$, equation (13a) becomes

$$\left(1 - \alpha \, \delta_{xx}\right) Q_{t+1} = \left(1 + \alpha \, \delta_{xx}\right) Q_t \tag{23}$$

The nested expression on the left represents a tridiagonal matrix. The critical stage is in solving the tridiagonal simultaneous equations for the vector of unknowns Q_{t+1}. Fortunately there is a special algorithm for this solution, and the cost increases only linearly with the size of the matrix. Now turn from the one-dimensional physical space of x to two-dimensional (x, y)-space. Letting α denote the numerical constant in (23), the equation for stepping forward in time is

$$\left[\, 1 - \alpha \left(\delta_{xx} + \delta_{yy}\right)\right] Q_{t+1} = \left[1 + \alpha(\delta_{xx} + \delta_{yy})\right] Q_t \tag{24}$$

The unknowns Q_{t+1} are a two-dimensional function of x and y that can be denoted by a matrix. Next we will interpret the bracketed expression on the left side. It turns out to be a four-dimensional matrix!

To clarify the meaning of this matrix, a mapping from two dimensions to one will be illustrated. Take the temperature Q to be defined on a 4×4 mesh. A natural way of numbering the points on the mesh is

$$
\begin{array}{cccc}
11 & 12 & 13 & 14 \\
21 & 22 & 23 & 24 \\
31 & 32 & 33 & 34 \\
41 & 42 & 43 & 44
\end{array}
\tag{25}$$

For algebraic purposes these sixteen numbers can be mapped into a vector.

There are many ways to do this. A simple way would be to associate the locations in (25) with vector components by the column arrangement

$$
\begin{array}{cccc}
1 & 5 & 9 & 13 \\
2 & 6 & 10 & 14 \\
3 & 7 & 11 & 15 \\
4 & 8 & 12 & 16
\end{array}
\tag{26}
$$

The second difference operator has the following star in the (x, y)-plane:

$$\tag{27}$$

Lay this star down in the (x, y)-plane (26) and move it around. Unfortunately, with just sixteen points, much of what you see is dominated by edges and corners. Try every position of the star that allows the center -4 to overlay one of the sixteen points. Never mind the 1's going off the sides. Start with the -4 in (27) over the 1 in the upper left corner of (26). Observe 1's on the 2 and the 5. Copy the 1's into the top row of table 1 into the second and fifth columns. Then put the -4 in (27) over the 2 in (26). Observe 1's on the 1, 3, and 6. Copy the 1's into the next row of table 1. Then put the -4 over the 3. Observe 1's on the 2, 4, and 7. Continue likewise. The 16×16 square matrix that results is shown in table 1.

Now that table 1 has been constructed we can return to the interpretation of equation (24). The matrix of unknowns Q_{t+1} has been mapped into a sixteen-point column vector, and the bracketed expression multiplying Q_{t+1} can be mapped into a 16×16 matrix. Clearly, the matrix contains zeroes everywhere that table 1 contains dots. It seems fortunate that the table contains many zeroes, and we are led to hope for a rapid solution method for the simultaneous equations. The bad news is that no good method has ever been found. The best methods seem to require effort proportional to N^3, where in this case $N = 4$. Based on our experience in one dimension, those of us who worked on this problem hoped for a method proportional to N^2, which is the cost of an explicit method — essentially the cost of computing the right side of (16). Even all the features of implicit methods do not justify an additional cost of a factor of N. The next best thing is the splitting method.

-4	1	.	.	1	.	.	.	.	.	.	.	.	.	.	.
1	-4	1	.	.	1	.	.	.	.	.	.	.	.	.	.
.	1	-4	1	.	.	1	.	.	.	.	.	.	.	.	.
.	.	1	-4	.	.	.	1	.	.	.	.	.	.	.	.
1	.	.	.	-4	1	.	.	1	.	.	.	.	.	.	.
.	1	.	.	1	-4	1	.	.	1	.	.	.	.	.	.
.	.	1	.	.	1	-4	1	.	.	1	.	.	.	.	.
.	.	.	1	.	.	1	-4	.	.	.	1	.	.	.	.
.	.	.	.	1	.	.	.	-4	1	.	.	1	.	.	.
.	.	.	.	.	1	.	.	1	-4	1	.	.	1	.	.
.	.	.	.	.	.	1	.	.	1	-4	1	.	.	1	.
.	.	.	.	.	.	.	1	.	.	1	-4	.	.	.	1
.	.	.	.	.	.	.	.	1	.	.	.	-4	1	.	.
.	.	.	.	.	.	.	.	.	1	.	.	1	-4	1	.
.	.	.	.	.	.	.	.	.	.	1	.	.	1	-4	1
.	.	.	.	.	.	.	.	.	.	.	1	.	.	1	-4

TABLE 2.2-1. The two-dimensional matrix of coefficients for the Laplacian operator.

EXERCISES

1. Interpret the inflation-of-money equation when the interest rate is the imaginary number $i/10$.

2. Write the 45° diffraction equation in (x, z)-space for fixed ω in the form of (12b).

2.3 Monochromatic Wave Programs

An old professor of education had a monochromatic theme. It was his only theme and the topic of his every lecture. It was this:

People learn by solving problems. Solving problems is the only way people learn, etc., etc., etc.......

All he ever did was lecture; he never assigned any problems.

Your first problems relate to the computer program in figure 1. As it stands it will produce a movie (three-dimensional matrix) of waves propagating through a focus. The whole process from compilation through computation to finally viewing the film loop takes about a minute (when you are the only user on the computer).

Analysis of Film Loop Program

For a film loop to make sense to a viewer, the subject of the movie must be periodic, and organized so that the last frame leads naturally into the first. In the movie created by the program in figure 1, there is a parameter *lambda* that controls the basic repetition rate of wave pulses fired onto the screen from the top. When a wavelet travels one-quarter of the way down the frame, another is sent in. This is defined by the line

$$\text{lambda} \;=\; \text{nz} * \text{dz} \,/4 \;=\; \frac{N_z \,\Delta z}{4}$$

The pulses are a superposition of sinusoids of nw frequencies, namely, $\Delta\omega$, $2\,\Delta\omega$, ..., nw $\Delta\omega$. The lowest frequency dw $= \Delta\omega$ has a wavelength inverse to *lambda*. Thus the definition

$$\text{dw} \;=\; \text{v} * \text{pi2} \,/\, \text{lambda} = \frac{2\,\pi\,v}{\lambda}$$

Finally, the time duration of the film loop must equal the period of the lowest-frequency sinusoid

$$N_t \,\Delta t \;=\; \frac{2\,\pi}{\Delta\omega}$$

This latter equation defines the time interval on the line

$$\text{dt} \;=\; \text{pi2} \,/\, (\,\text{nt} * \text{dw}\,)$$

The differential equation solved by the program is

$$\frac{\partial P}{\partial z} \;=\; \frac{i\,\omega}{v\,(x\,,\,z)}\;P \;+\; \frac{v}{-\,i\,\omega\,2}\;\frac{\partial^2 P}{\partial x^2} \tag{1}$$

For each Δz-step the calculation is done in two stages. The first stage is to solve

$$\frac{\partial Q}{\partial z} \;=\; \frac{v}{-\,i\,\omega\,2}\;\frac{\partial^2 Q}{\partial x^2} \tag{2}$$

Using the Crank-Nicolson differencing method this becomes

$$\frac{q^x_{z+1} - q^x_z}{\Delta z} \;=$$

$$=\; \frac{v}{-\,i\,\omega\,2}\left[\; \frac{q_z^{x+1} - 2\,q_z^x + q_z^{x-1}}{2\,\Delta x^2} \;+\; \frac{q_{z+1}^{x+1} - 2\,q_{z+1}^x + q_{z+1}^{x-1}}{2\,\Delta x^2} \;\right]$$

Absorb all the constants into one and define

$$\alpha \;=\; \frac{v\,\Delta z}{-\,i\,\omega\,4\,\Delta x^2} \tag{3}$$

getting

$$q^x_{z+1} - q^x_z \;=\; \alpha\left[\,(\,q_z^{x+1} - 2\,q_z^x + q_z^{x-1}) + (\,q_{z+1}^{x+1} - 2\,q_{z+1}^x + q_{z+1}^{x-1}\,)\,\right]$$

Bring the unknowns to the left:

$$-\alpha q_{z+1}^{x+1} + (1+2\alpha)q^x_{z+1} - \alpha q_{z+1}^{x-1} \;=\; \alpha q_z^{x+1} + (1-2\alpha)q_z^x + \alpha q_z^{x-1} \tag{4}$$

The second stage is to solve the equation

$$\frac{\partial Q}{\partial z} \;=\; \frac{i\,\omega}{v}\;Q \tag{5}$$

analytically by

$$Q\,(z+\Delta z) \;=\; Q\,(z)\,e^{\,i\,\frac{\omega}{v}\,\Delta z} \tag{6}$$

The program closely follows the notation of equations (3), (4), and (6).

To make a wave pulse, some frequency components are added together. In this program, only two frequencies nw=2 were used. If you try a single frequency nw=1 several things become less clear. Waves reflected at side boundaries (see especially exercise 2) look more like *standing waves*. If you try more frequencies, the program will take longer, but you might like the movie better, because the quiet zones between the pulses will get longer and quieter. Frequency components can be weighted differently.

```
# Wave field extrapolation program
implicit undefined (a-z)
complex cd(48),ce(48),cf(48),q(48),aa,a,b,c,cshift
real p(96,48,12),phase,pi2,dx,dz,v,z0,x0,dt,dw,lambda,w,wov,x
integer ix,nx,iz,nz,iw,nw,it,nt
open(3,file='plot30',status='new',access='direct',form='unformatted',recl=1)

nt=12; nx=48; nz=96; dx=2; dz=1; pi2=2.*3.141592
v=1; lambda=nz*dz/4; dw=v*pi2/lambda; dt=pi2/(nt*dw); nw=2

do iz=1,nz;  do ix=1,nx;  do it=1,nt  { p(iz,ix,it) = 0. }
do iw = 1,nw {                          # superimpose nw frequencies
       w =  iw*dw;      wov = w/v        # frequency / velocity
       x0 = nx*dx/3; z0 = nz*dz/3
       do ix = 1,nx {
            x = ix*dx-x0;                # initial conditions for a
            phase = -wov*sqrt(z0**2+x**2) # collapsing spherical wave
            q(ix) = cexp(cmplx(0.,phase))
            }
       aa = dz/(4.*(0.,-1.)*wov*dx**2)        # tridiagonal matrix coefficients
       a = -aa;      b = 1.+2.*aa;    c = -aa
       do iz = 1,nz {                          # extrapolation in depth
            do ix = 2,nx-1        # diffraction term
                 cd(ix) = aa*q(ix+1) + (1.-2.*aa)*q(ix) + aa*q(ix-1)
            cd(1) = 0.;    cd(nx) = 0.
            call ctris(nx,-a,a,b,c,-c,cd,q,ce,cf)
                      # "ctris" solves complex tridiagonal equations
                      # i.e. "rtris" with complex variables
            cshift = cexp(cmplx(0.,wov*dz))
            do ix = 1,nx                       # shifting term
                 q(ix) = q(ix) * cshift
            do it=1,nt {                        # evolution in time
                 cshift = cexp(cmplx(0.,-w*it*dt))
                 do ix = 1,nx
                      p(iz,ix,it) = p(iz,ix,it)+q(ix)*cshift
                 }
            }
       }
write(3,rec=1) (((p(iz,ix,it),iz=1,nz),ix=1,nx),it=1,nt)
stop;          end
```

FIG. 2.3-1. Computer program to make a movie of a sum of monochromatic waves. (Lynn, Gonzalez, JFC, Hale)

Phase Shift

Theory predicts that in two dimensions waves going through a focus suffer a 90° phase shift. You should be able to notice that a symmetrical waveform is incident on the focus, but an asymmetrical waveform emerges. (This is best seen in figure 6, but is clearer in a movie). In migrations, waves go just *to* a focus, not *through* it. So the migration impulse response in two dimensions carries a 45° phase shift. Even though real life is three dimensional, the two dimensional response is appropriate for migrating seismic lines where focusing is presumed to arise from cylindrical, not spherical, reflectors.

Lateral Velocity Variation

Lateral velocity variation $v = v(x)$ has not been included in the program, but it is not difficult to install. It enters in two places. It enters first in equation (6). If the data is such that k_x is small enough to be neglectable, then equation (6) is the only place it is needed. Second, it enters in the tridiagonal coefficients. The so-called thin-lens approximation of optics seems to amount to including the equation (6) part only.

Side-Boundary Analysis

In geophysics, we usually wish the side-boundary question would go away. The only real reason for side boundaries is that either our survey or our processing activity is necessarily limited in extent. Given that side boundaries are inevitable, we must think about them. The program of figure 1 included zero-slope boundary conditions. This type of boundary treatment resulted from taking

$$d(1) \;=\; 0. \qquad ; \qquad d(nx) \;=\; 0.$$

and in the call to "ctris" taking

$$endl \;=\; -a \qquad ; \qquad endr \;=\; -c$$

A quick way to get zero-value side-boundary conditions is to take

$$endl \;=\; endr \;=\; 10^{30} \;\approx\; \infty$$

The above approach is slightly wasteful of computer memory, because the end zero is stored, and the zero slope is explicitly visible as two identical traces. This waste is avoided in Dave Hale's coding of the boundary conditions as given, but not derived, below:

```
q0 = bl * q(1);        qnxp1 = br * q(nx)

cd(1)   = aa * q(2)    + ( 1. - 2. * aa ) * q(1) + aa * q0
cd(nx)  = aa * q(nx-1) + ( 1. - 2. * aa ) * q(nx) + aa * qnxp1

endl = c * bl + b;
endr = a * br + b

call ctris(nx,endl,a,b,c,endr,cd,q,ce,cf)
```

Note that $bl = br = 0$ for zero-value boundaries, and $bl = br = 1$ for zero-slope boundaries. Absorbing side boundaries, derived in Section 4.4, are obtained by letting bl and br be complex.

Variations on the Film Loop Program

Keep a record of your progress through these exercises. It will be helpful when preparing for the final exam. And several years hence you will be able to refresh your memory.

Get a three-ring notebook. Cut all plots and program listings to 8-*1/2* by 11 size and three-hole punch them. If algebraic analysis is required, do it on the same size paper. Avoid leaving important bits of analysis on scraps of paper. Either keep this material with your lecture notes or maintain it as a laboratory notebook, filing consistently by date.

For each of these exercises, hand in a program listing and a plot of the first frame.

1.   Specify program changes that give an initial plane wave propagating downward at an angle of 15° to the right of vertical.

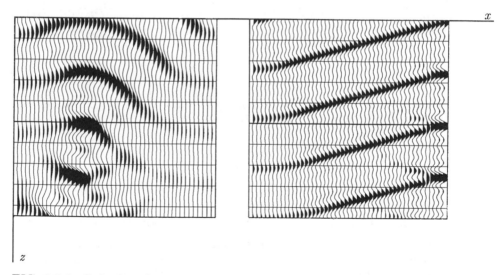

FIG. 2.3-2. Left, first frame of movie generated by figure 1. Right, solution to exercise 1. (Li Zhiming).

2. Given that the domain of computation is $0 < x <= $ xmax and $0 < z \leq$ zmax, how would you modify the initial conditions at $z=0$ to simulate a point source at $(x, z) = $ (xmax/3, –zmax/2)? Try it.

3. Modify the program so that zero-slope side boundaries are replaced by zero-value side boundaries.

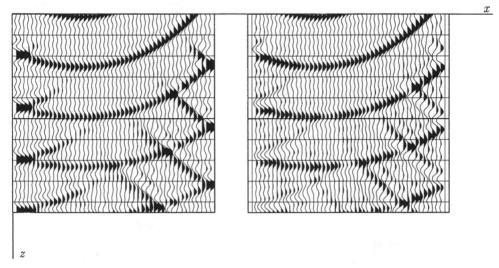

FIG. 2.3-3.   Left, exercise 2, expanding spherical wave. Right, exercise 3, zero-value side boundaries. (Li Zhiming).

4. Incorporate the 45° term, ∂_{xxz}, for the collapsing spherical wave. Use zero-slope sides. Compare your result with the 15° result obtained via the program in figure 1. Mark an X at the theoretical focus location.

5. Make changes to the program to include a thin-lens term with a lateral velocity change of 40% across the frame produced by a constant slowness gradient. Identify other parts of the program which are affected by lateral velocity variation. You need not make these other changes. Why are they expected to be small?

6. Observe and describe various computational artifacts by testing the program using a point source at $(x, z) = $ (xmax/2,0). Such a source is rich in the high spatial frequencies for which difference equations do not mimic their differential counterparts.

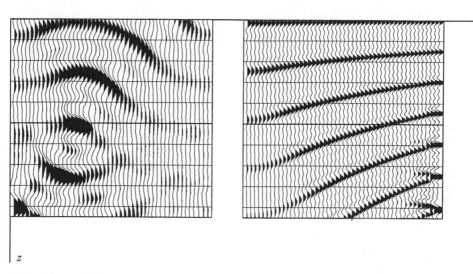

FIG. 2.3-4. Left, exercise 4, 45° term. Right, exercise 5, lateral velocity variation. (Li Zhiming).

7.   Section 4.4 explains how to absorb energy at the side boundaries. Make the necessary changes to the program.

8.   The accuracy of the x-derivative may be improved by a technique that is analyzed later in Section 4.3. Briefly, instead of representing $k_x^2 \, \Delta x^2$ by the tridiagonal matrix **T** with $(-1, 2, -1)$ on the main diagonal, you use **T**/(**I**−**T**/6). Modify the extrapolation analysis by multiplying through by the denominator. Make the necessary changes to the 45° collapsing wave program.

Migration Program in the $(\omega,\mathbf{x},\mathbf{z})$-Domain (Kjartansson, Jacobs)

The migration program is similar to the film loop program. But there are some differences. The film loop program has "do loops" nested four deep. It produces results for many values of t. Migration requires a value only at $t = 0$. So one loop is saved, which means that for the same amount of computer time, the space volume can be increased. Unfortunately, loss of a loop seems also to mean loss of a movie. With ω-domain migration, it seems that the only interesting thing to view is the input and the output.

The input for this process will probably be field data, unlike for the film loop movie, so there will not be an analytic representation in the ω-domain.

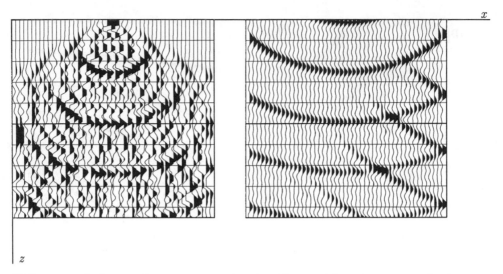

FIG. 2.3-5. Left, exercise 6, computational artifacts with point source. Right, exercise 7, absorbing side. (Li Zhiming).

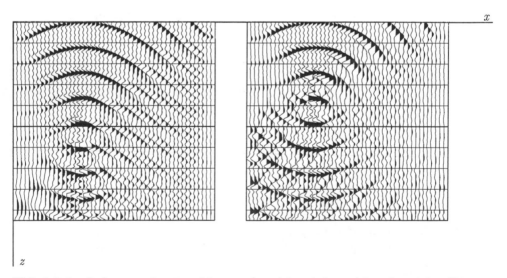

FIG. 2.3-6. Left, exercise 8, without 1/6 trick; right, with 1/6 trick. (Li Zhiming).

The input will be in the time domain and will have to be Fourier transformed. The beginning of the program in figure 6 defines some pulses to simulate field data. The pulses are broadened impulses and should migrate to approximate semicircles. Exact impulses were not used because the departure of difference operators from differential operators would make a noisy mess.

Next the program Fourier transforms the pseudodata from the time domain into the ω-frequency domain.

Then comes the downward continuation of each frequency. This is a loop on depth z and on frequency ω. Either of these loops may be on the inside. The choice can be made for machine-dependent efficiency.

```
# Migration in the (omega,x,z)-domain
real q(48,64),pi2,alpha,dt,dtau,dw
complex cq(48,64),cd(48),ce(48),cf(48),aa,a,b,c,cshift
integer ix,nx,iz,nz,iw,nw,it,nt
open(4,file='plot36',status='new',access='direct',form='unformatted',recl=1)

nt = 64;   nz = nt;    nx = 48;      pi2=2.*3.141592
dt=1.;   dtau=1.;    dw=pi2/(dt*nt);    nw=nt/2;
alpha = .25                                  # alpha = v*v*dtau/(4*dx*dx)
do iz=1,nz;   do ix=1,nx;   { q(ix,iz) = 0.;    cq(ix,iz)=0. }
do it=nt/3,nt,nt/4
        do ix=1,4                            # Broadened impulse source
               { cq(ix,it) = (5.-ix);    cq(ix,it+1) = (5.-ix) }
call rowcc(nx,nt,cq,+1.,+1.)                 # F.T. over time.
do iz = 1,nz {                               # iz and iw loops interchangeable
do iw = 2,nw {                               # iz and iw loops interchangeable
        aa = - alpha /( (0.,-1.)*(iw-1)*dw )
        a = -aa;      b = 1.+2.*aa;      c = -aa
        do ix = 2,nx-1
               cd(ix) = aa*cq(ix+1,iw) + (1.-2.*aa)*cq(ix,iw) + aa*cq(ix-1,iw)
        cd(1) = 0.;      cd(nx) = 0.
        call ctris(nx,-a,a,b,c,-c,cd,cq(1,iw),ce,cf)
        cshift = cexp(cmplx(0.,-(iw-1)*dw*dtau))
        do ix=1,nx
               cq(ix,iw) = cq(ix,iw) * cshift
        do ix = 1,nx
               q(ix,iz) = q(ix,iz)+cq(ix,iw)       # q(t=0) = Σ Q(ω)
        }}
write(4,rec=1) ((q(ix,iz),iz=1,nz),ix=1,nx)
stop;      end
```

FIG. 2.3-7.   Migration program in the (ω, x, z)-domain.

For migration an equation for upcoming waves is required, unlike the downgoing wave equation required for the film loop program. Change the sign of the z-axis in equation (1). This affects the sign of aa and the sign of the phase of *cshift*.

Another difference with the film loop program is that the input now has a time axis whereas the output is still a depth axis. It is customary and convenient to reorganize the calculation to plot travel-time depth, instead of depth, making the vertical axes on both input and output the same. Using $\tau = z/v$, equivalently $d\tau/dz = 1/v$, the chain rule gives

$$\frac{\partial}{\partial z} \;=\; \frac{\partial \tau}{\partial z}\frac{\partial}{\partial \tau} \;=\; \frac{1}{v}\frac{\partial}{\partial \tau} \tag{7}$$

Substitution into (1) gives

$$\frac{\partial P}{\partial \tau} \;=\; -i\,\omega\,P \;-\; \frac{v^2}{-i\,\omega\,2}\frac{\partial^2 P}{\partial x^2} \tag{8}$$

In the program, the time sample size $dt = \Delta t$ and the travel-time depth sample $dtau = \Delta \tau$ are taken to be unity, so the maximum frequency is the Nyquist. Notice that the frequency loop covers only the positive frequency axis. The negative frequencies serve only to keep the time function real, a task that is more easily done by simply taking the real part.

The output of the program is shown in figure 8. Mainly, you see semicircle approximations. There are also some artifacts at late time that may be ω-domain wraparounds. The input pulses were apparently sufficiently broadbanded in dip that the figure provides a preview of the fact, to be proved later, that the actual semicircle approximation is an ellipse going through the origin.

Notice that the waveform of the original pulses was a symmetric function of time, whereas the semicircles exhibit a waveform that is neither symmetric nor antisymmetric, but is a 45° phase-shifted pulse. Waves from a point in a three-dimensional world would have a phase shift of 90°. Waves from a two-dimensional exploding reflector in a three-dimensional world have the 45° phase shift.

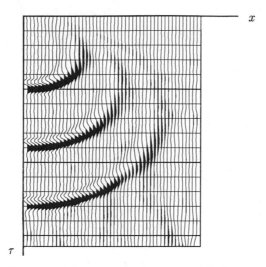

FIG. 2.3-8. Output of figure 7 program: semicircle approximations.

2.4 Splitting and Full Separation

Two processes, **A** and **B,** which ordinarily act simultaneously, may or may not be interconnected. The case where they are independent is called *full separation.* In this case it is often useful, for thought and for computation, to imagine process **A** going to completion before process **B** is begun. Where the processes are interconnected it is possible to allow **A** to run for a short while, then switch to **B,** and continue in alternation. This alternation approach is called *splitting.*

The Heat-Flow Equation

The diffraction or migration equation could be called the "wavefront healing" equation. It smooths back together any lateral breaks in the wavefront that may have been caused by initial conditions or by the lens term. The 15° migration equation has the same mathematical form as the heat-flow equation. But the heat-flow equation has all real numbers, and its physical behavior is more comprehensible. This makes it a worthwhile detour. A two-sentence derivation of it follows. (1) The heat flow H_x in the x-

direction equals the negative of the gradient $-\partial/\partial x$ of temperature T times the heat conductivity σ. (2) The decrease of temperature $-\partial T/\partial t$ is proportional to the divergence of the heat flow $\partial H_x/\partial x$ divided by the heat storage capacity C of the material. Combining these, extending from one dimension to two, taking σ constant and $C=1$, gives the equation

$$\frac{\partial T}{\partial t} \;=\; \sigma \left[\frac{\partial^2}{\partial x^2} + \frac{\partial^2}{\partial y^2} \right] T \tag{1}$$

Splitting

The splitting method for numerically solving the heat-flow equation is to replace the two-dimensional heat-flow equation by two one-dimensional equations, each of which is used on alternate time steps:

$$\frac{\partial T}{\partial t} \;=\; 2\,\sigma\,\frac{\partial^2 T}{\partial x^2} \qquad \textit{(all y)} \tag{2a}$$

$$\frac{\partial T}{\partial t} \;=\; 2\,\sigma\,\frac{\partial^2 T}{\partial y^2} \qquad \textit{(all x)} \tag{2b}$$

In equation (2a) the heat conductivity σ has been doubled for flow in the x-direction and zeroed for flow in the y-direction. The reverse applies in equation (2b). At odd moments in time heat flows according to (2a) and at even moments in time it flows according to (2b). This solution by alternation between (2a) and (2b) can be proved mathematically to converge to the solution to (1) with errors of the order of Δt. Hence the error goes to zero as Δt goes to zero. The motivation for splitting is the infeasibility of higher-dimensional implicit methods (end of Section 2.2).

Full Separation

Splitting can turn out to be much more accurate than might be imagined. In many cases there is *no* loss of accuracy. Then the method can be taken to an extreme limit. Think about a radical approach to equations (2a) and (2b) in which, instead of alternating back and forth between them at alternate time steps, what is done is to march (2a) through all time steps. Then this intermediate result is used as an initial condition for (2b), which is marched through all time steps to produce a final result. It might seem surprising that this radical method can produce the correct solution to equation (1). But if σ is a constant function of x and y, it does. The process is depicted in figure 1 for an impulsive initial disturbance. A differential equation like (1) is said to be *fully separable* when the correct solution is obtainable by the radical method. It should not be too surprising that full separation works when σ is a constant, because then Fourier transformation may be used, and the two-dimensional solution $\exp[-\sigma\,(k_x{}^2 + k_y{}^2)t\,]$ equals the succession of one-dimensional solutions $\exp(-\sigma\,k_x{}^2t\,)\ \exp(-\sigma\,k_y{}^2t\,)$. It turns out, and will later be shown, that the condition required for applicability of full separation is that $\sigma\,\partial^2/\partial x^2$ should commute with $\sigma\,\partial^2/\partial y^2$, that is, the order of differentiation should be irrelevant. Technically there is also a boundary-condition requirement, but it creates no difficulty when the disturbance dies out before reaching a boundary.

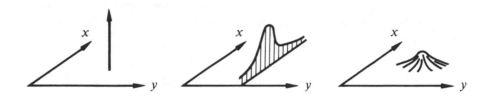

FIG. 2.4-1. Temperature distribution in the (x,y)-plane beginning from a delta function (left). After heat is allowed to flow in the x-direction but not in the y-direction the heat is located in a "wall" (center). Finally allowing heat to flow for the same amount of time in the y-direction but not the x-direction gives the same symmetrical Gaussian result that would have been found if the heat had moved in x- and y-directions simultaneously (right).

Surprisingly, no notice is made of full separability in many textbooks on numerical solutions. Perhaps this is because the total number of additions and multiplications is the same whether a solution is found by splitting or by full separation. But as a practical matter, the cost of solving large problems does not mount up simply according to the number of multiplications. When the data base does not fit entirely into the random-access memory, as is almost the definition of a *large problem,* then each step of the splitting method demands that the data base be transposed, say, from (x,y) storage order to (y,x) storage order. Transposing requires no multiplications, but in many environments transposing would be by far the most costly part of the whole computation. So if transposing cannot be avoided, at least it should be reduced to a practical minimum.

There are circumstances which dictate a middle road between splitting and separation — for example, if σ were a slowly variable function of x or y. Then you might find that although $\sigma\,\partial^2/\partial x^2$ does not strictly commute with $\sigma\,\partial^2/\partial y^2$, it comes close enough that a number of time steps may be made with (2a) before you transpose the data and switch over to (2b). Circumstances like this one but with more geophysical interest arise with the wave-extrapolation equation that is considered next. The significance in seismology of the splitting and full separation concepts was first recognized by Brown [1983].

Application to Lateral Velocity Variation

A circumstance in which the degree of noncommutativity of two differential operators has a simple physical meaning and an obviously significant geophysical application is the so-called monochromatic 15° wave-extrapolation equation in inhomogeneous media. Taking $v \approx \overline{v}$ this equation is

$$\frac{\partial U}{\partial z} \;=\; \left\{ \frac{i\,\omega}{\overline{v}(z)} + i\,\omega \left[\frac{1}{v(x\,,z)} - \frac{1}{\overline{v}(z)} \right] - \frac{\overline{v}(z)}{2\,i\,\omega}\,\frac{\partial^2}{\partial x^2} \right\} U \quad (3)$$

$$=\; (retardation \,+\, thin\ lens \,+\, diffraction)\ U$$

Inspection of (3) shows that the retardation term commutes with the thin-lens term and with the free-space diffraction term. But the thin-lens term and the diffraction term do not commute with one another. In practice it seems best to split, doing the thin-lens part analytically and the diffraction part by the Crank-Nicolson method. Then stability is assured because the stability of each separate problem is known. Also, the accuracy of the analytic solution is an attractive feature. Now the question is, to what degree do these two terms commute?

The problem is just that of focusing a slide projector. Adjusting the focus knob amounts to repositioning the thin-lens term in comparison to the free-space diffraction term. There is a small range of knob positions over which no one can notice any difference, and a larger range over which the people in the back row are not disturbed by misfocus. Much geophysical data processing amounts to downward extrapolation of data. The lateral variation of velocity occurring in the lens term is known only to a limited accuracy. The application could be to determine $v(x)$ by the extrapolation procedure.

For long lateral spatial wavelengths the terms commute. Then diffraction may proceed in ignorance of the lateral variation in v. At shorter wavelengths the diffraction and lensing effects must be interspersed. So the real issue is not merely computational convenience but the interplay between data accuracy and the possible range for velocity in the underlying model.

Application to 3-D Downward Continuation

The operator for migration of zero-offset reflection seismic data in three dimensions is expandable to second order by Taylor series expansion to the so-called 15° approximation

$$\left[\frac{(-i\,\omega)^2}{v^2} - \frac{\partial^2}{\partial x^2} - \frac{\partial^2}{\partial y^2}\right]^{1/2} \approx \frac{-i\,\omega}{v} - \frac{v}{-2i\,\omega}\frac{\partial^2}{\partial x^2} - \frac{v}{-2i\,\omega}\frac{\partial^2}{\partial y^2} \tag{4}$$

The most common case is when v is slowly variable or independent of x and y. Then the conditions of full separation do apply. This is good news because it means that we can use ordinary 2-D wave-extrapolation programs for 3-D, doing the in-line data and the out-of-line data in either order. The bad news comes when we try for more accuracy. Keeping more terms in the Taylor series expansion soon brings in the cross term $\partial^4/\partial x^2 \partial y^2$. Such a term allows neither full separation nor splitting. Fortunately, present-day marine data-acquisition techniques are sufficiently crude in the out-of-line direction that there is little justification for out-of-line processing beyond the 15° equation. Francis Muir had the good idea of representing the square root as

$$\left[\frac{(-i\,\omega)^2}{v^2} - \frac{\partial^2}{\partial x^2} - \frac{\partial^2}{\partial y^2}\right]^{1/2} \approx$$

$$\approx \left[\frac{(-i\,\omega)^2}{v^2} - \frac{\partial^2}{\partial x^2}\right]^{1/2} - \frac{v}{-2i\,\omega}\frac{\partial^2}{\partial y^2} \tag{5}$$

There may be justification for better approximations with land data. Fourier transformation of at least one of the two space axes will solve the computational problem. This should be a good approach when the medium velocity does not vary laterally so rapidly as to invalidate application of Fourier transformation.

Separability of 3-D Migration (the Jakubowicz Justification)

In an operations environment, 3-D is much harder to cope with than 2-D. Therefore, it may be expedient to suppose that 3-D migration can be achieved merely by application of 2-D migration twice, once in the x-direction and once in the y-direction. The previous section would lead you to believe that such an expedient process would result in a significant degradation of accuracy. In fact, the situation is *much* better than might be supposed. It has been shown by Jakubowicz and Levin [1983] that, wonder of wonders, for a constant-velocity medium, the expedient process is exact.

The explanation is this: migration consists of more than downward continuation. It also involves imaging, that is, the selection of data at $t=0$. In principle, downward continuation is first completed, for both the x and the y directions. After that, the imaging condition is applied. In the expedient process there are four steps: downward continuation in x, imaging, downward continuation in y, and finally a second imaging. Why it is that the expedient procedure gives the correct result seems something of a puzzle, but the validity of the result is easy to demonstrate.

First note that substitution of (6) into (7) gives (8) where

$$t_1^2 \;=\; t_0^2 + (x-x_0)^2/v^2 \tag{6}$$

$$t^2 \;=\; t_1^2 + (y-y_0)^2/v^2 \tag{7}$$

$$t^2 \;=\; t_0^2 + (x-x_0)^2/v^2 + (y-y_0)^2/v^2 \tag{8}$$

Equation (8) represents travel time to an arbitrary point scatterer. For a 2-D survey recorded along the y-axis, i.e., at constant x, equation (7) is the travel-time curve. In-line hyperbolas cannot be distinguished from sideswipe hyperbolas. 2-D migration with equation (7) brings the energy up to t_1. Subsequently migrating the other direction with equation (6) brings the energy up the rest of the way to t_0. This is the same result as the one given by the more costly 3-D procedure migrating with (8).

The Jakubowicz justification is somewhat more mathematical, but may be paraphrased as follows. First note that substitution of (9) into (10) gives (11) where

$$k_\tau^2 \;=\; \omega^2 - v^2\,k_x^2 \tag{9}$$

$$k_z^2 \;=\; \frac{k_\tau^2}{v^2} - k_y^2 \tag{10}$$

$$k_z^2 \;=\; \frac{\omega^2}{v^2} - k_x^2 - k_y^2 \tag{11}$$

Two-dimensional Stolt migration over x may be regarded as a transformation from travel-time depth t to a pseudodepth τ by use of equation (9). The second two-dimensional migration over y may be regarded as a transformation from pseudodepth τ to true depth z by use of equation (10). The composite is the same as equation (11), which depicts 3-D migration.

The validity of the Jakubowicz result goes somewhat beyond its proof. Our two-dimensional geophysicist may be migrating other offsets besides zero offset. (In Chapter 3 nonzero-offset data is migrated). If a good job is done, all the reflected energy moves up to the apex of the zero-offset hyperbola. Then the cross-plane migration can handle it if it can handle zero offset. So offset is not a problem. But can a good job be done of bringing all the energy up to the apex of the zero-offset hyperbola?

Difficulty arises when the velocity of the earth is depth-dependent, as it usually is. Then the Jakubowicz proof fails, and so does the expedient 3-D method. With a 2-D survey you have the problem that the sideswipe planes require a different migration velocity than the vertical plane. Rays propagating to the side take longer to reach the high-velocity media deep in the earth. So sideswipes usually require a lower migration velocity. If you really want to do three-dimensional migration with $v(z)$, you should forget about separation and do it the hard way. Since we know how to transpose (Section 1.6), the hard way really isn't much harder.

Separability in Shot-Geophone Space

Reflection seismic data gathering is done on the earth's surface. One can imagine the appearance of the data that would result if the data were generated and recorded at depth, that is, with deeply buried shots and geophones. Such buried data could be synthesized from surface data by first downward extrapolating the geophones, then using the reciprocal principle to interchange sources and receivers, and finally downward extrapolating the surface shots (now the receivers). A second, equivalent approach would be to march downward in steps, alternating between shots and geophones. This latter approach is developed in Chapter 3, but the result is simply stated by the equation

$$\frac{\partial U}{\partial z} \;=\; \left\{ \left[\frac{(-i\omega)^2}{v(s)^2} - \frac{\partial^2}{\partial s^2}\right]^{1/2} + \left[\frac{(-i\omega)^2}{v(g)^2} - \frac{\partial^2}{\partial g^2}\right]^{1/2} \right\} U \qquad (12)$$

The equivalence of the two approaches has a mathematical consequence. The shot coordinate s and the geophone coordinate g are independent variables, so the two square-root operators commute. Thus the same solution is obtained by splitting as by full separation.

Validity of the Splitting and Full-Separation Concepts

When Fourier transformation is possible, extrapolation operators are complex numbers like $e^{ik_z z}$. With complex numbers a and b there is never any question that $ab = ba$. Then both splitting and full separation are always valid, but the proof will be given only for a more general arrangement.

Suppose Fourier transformation has not been done, or could not be done because of some spatial variation of material properties. Then extrapolation operators are built up by combinations of the finite-differencing operators described in previous sections. Let **A** and **B** denote two such operators.

For example, $\mathbf{A}$ could be a matrix containing the second x differencing operator. Seen as matrices, the boundary conditions of a differential operator are incorporated in the corners of the matrix. The bottom line is whether $\mathbf{A\,B} = \mathbf{B\,A}$, so the question clearly involves the boundary conditions as well as the differential operators.

Extrapolation forward a short distance can be done with the operator $(\mathbf{I}+\mathbf{A}\,\Delta z)$. In two-dimensional problems $\mathbf{A}$ was seen to be a four-dimensional matrix. For convenience the terms of the four-dimensional matrix can be arranged into a super-large, ordinary two-dimensional matrix. Implicit finite-differencing calculations gave extrapolation operators like $(\mathbf{I}+\mathbf{A}\,\Delta z)/(\mathbf{I}-\mathbf{A}\,\Delta z)$. Let $\mathbf{p}$ denote a vector where components of the vector designate the wavefield at various locations. As has been seen, the locations need not be constrained to the x-axis but could also be distributed throughout the (x,y)-plane. Numerical analysis gives us a matrix operator, say $\mathbf{A}$, which enables us to project forward, say,

$$\mathbf{p}(z+\Delta z) \;=\; \mathbf{A}_1\,\mathbf{p}(z)$$

The subscript on $\mathbf{A}$ denotes the fact that the operator may change with z. To get a step further the operator is applied again, say,

$$\mathbf{p}(z+2\,\Delta z) \;=\; \mathbf{A}_2\,[\mathbf{A}_1\,\mathbf{p}(z)]$$

From an operational point of view the matrix $\mathbf{A}$ is never squared, but from an analytical point of view, it really is squared.

$$\mathbf{A}_2\,[\mathbf{A}_1\,\mathbf{p}(z)] \;=\; (\mathbf{A}_2\,\mathbf{A}_1)\,\mathbf{p}(z)$$

To march some distance down the z-axis we apply the operator many times. Take an interval $z_1 - z_0$, to be divided into N subintervals. Since there are N intervals, an error proportional to $1/N$ in each subinterval would accumulate to an unacceptable level by the time z_1 was reached. On the other hand, an error proportional to $1/N^2$ could only accumulate to a total error proportional to $1/N$. Such an error would disappear as the number of subintervals increased.

To prove the validity of splitting, we take $\Delta z = (z_1 - z_0)/N$. Observe that the operator $\mathbf{I}+(\mathbf{A}+\mathbf{B})\Delta z$ differs from the operator $(\mathbf{I}+\mathbf{A}\,\Delta z)(\mathbf{I}+\mathbf{B}\,\Delta z)$ by something in proportion to Δz^2 or $1/N^2$. So in the limit of a very large number of subintervals, the error disappears.

It is much easier to establish the validity of the full-separation concept. Commutativity is whether or not $\mathbf{A}=\mathbf{B\,A}$. Commutativity is always true for scalars. With finite differencing the question is whether the two matrices commute. Taking $\mathbf{A}$ and $\mathbf{B}$ to be differential operators,

commutativity is defined with the help of the family of all possible wavefields
P. Then $\mathbf{A}$ and $\mathbf{B}$ are commutative if $\mathbf{A}\,\mathbf{B}\,P = \mathbf{B}\,\mathbf{A}\,P$.

The operator representing $\partial P/\partial z$ will be taken to be $\mathbf{A}+\mathbf{B}$. The simplest numerical integration scheme using the splitting method is

$$P(z_0 + \Delta z) \;=\; (\mathbf{I} + \mathbf{A}\,\Delta z)\,(\mathbf{I} + \mathbf{B}\,\Delta z)\,P(z_0) \qquad (13)$$

Applying (13) in many stages gives a product of many operators. The operators $\mathbf{A}$ and $\mathbf{B}$ are subscripted with j to denote the possibility that they change with z.

$$P(z_1) \;=\; \prod_{j=1}^{N} \left[(\mathbf{I} + \mathbf{A}_j\,\Delta z)(\mathbf{I} + \mathbf{B}_j\,\Delta z)\right] P(z_0) \qquad (14)$$

As soon as $\mathbf{A}$ and $\mathbf{B}$ are assumed to be commutative, the factors in (14) may be rearranged at will. For example, the $\mathbf{A}$ operator could be applied in its entirety before the $\mathbf{B}$ operator is applied:

$$P(z_1) \;=\; \left[\prod_{j=1}^{N} (\mathbf{I} + \mathbf{B}\,\Delta z)\right] \left[\prod_{j=1}^{N} (\mathbf{I} + \mathbf{A}\,\Delta z)\right] P(z_0) \qquad (15)$$

Thus the full-separation concept is seen to depend on the commutativity of operators.

EXERCISES

1. With a splitting method, Ma Zaitian (Ma [1981]) showed how very wide-angle representations may be implemented with successive applications of an equation like a 45° equation. This avoids the band matrix solving inherent in the high-order Muir expansion. Specifically, one chooses coefficients a_j and b_j, in the square-root fitting function

$$i\,k_z \;=\; \sum_{j=1}^{n-1} \frac{k_x^{\,2}\,b_j}{-i\,\omega + a_j\,ik_x}$$

    The general n^{th}-order case is somewhat complicated, so your job is simply to find a_1, a_2, b_1, and b_2, to make the fitting function match the 45° equation.

2. Migrate a two dimensional data set with velocity v_1. Then migrate the migrated data set with a velocity v_2. Rocca pointed out that this double migration simulates a migration with a third velocity v_3. Using a method of deduction similar to the Jakubowicz deduction equations (9), (10), and (11) find v_3 in terms of v_1 and v_2.

3.   Consider migration of zero-offset data $P(x, y, t)$ recorded in an area of the earth's surface plane. Assume a computer with a random access memory (RAM) large enough to hold several planes (any orientation) from the data volume. (The entire volume resides in slow memory devices). Define a migration algorithm by means of a program sketch (such as in Section 1.3). Your method should allow velocity to vary with depth.

2.5 Recursive Dip Filters

Recursive filtering is a form of filtering where the output of the filter is fed back as an input. This can achieve a long impulse response for a tiny computational effort. It is particularly useful in computing a running mean. A running mean could be implemented as a low-pass filter in the frequency domain, but it is generally much better to avoid transform space. Physical space is cheaper, it allows for variable coefficients, and it permits a more flexible treatment of boundaries. Geophysical datasets are rarely stationary over long distances in either time or space, so recursive filtering is particularly helpful in statistical estimation.

The purpose of most filters is to make possible the observation of important weak events that are obscured by strong events. *One*-dimensional filters can do this only by the selection or rejection of *frequency* components. In *two* dimensions, a different criterion is possible, namely, selection by *dip*.

Dip filtering is a process of long-standing interest in geophysics (Embree, Burg, and Backus [1963]). Steep dips are often ground-roll noise. Horizontal dips can also be noise. For example, weak fault diffractions carry valuable information, but they may often be invisible because of the dominating presence of flat layers.

To do an ordinary dip-filtering operation ("pie slice"), you simply transform data into (ω, k)-space, multiply by any desired function of k/ω, and transform back. Pie-slice filters thus offer complete control over the filter response in k/ω dip space. While the recursive dip filters are not controlled so easily, they do meet the same general needs as pie-slice filters and offer the additional advantages of

1. time- and space-variability
2. causality
3. ease of implementation
4. orders of magnitude more economic than (ω, k)-implementation

The causality property offers an interesting opportunity during data recording. Water-velocity rejection filters could be built into the recording apparatus of a modern high-density marine cable.

Definition of a Recursive Dip Filter

Let P denote raw data and Q denote filtered data. When seismic data is quasimonochromatic, dip filtering can be achieved with spatial frequency filters. The table below shows filters with an adjustable cutoff parameter α.

Dip Filters for Monochromatic Data $(\omega \approx Const\,)$	
Low Pass	*High Pass*
$Q = \dfrac{\alpha}{\alpha + k^2}\, P$	$Q = \dfrac{k^2}{\alpha + k^2}\, P$

To apply these filters in the space domain it is necessary only to interpret k^2 as the tridiagonal matrix $\mathbf{T}$ with $(-1, 2, -1)$ on the main diagonal. Specifically, for the low-pass filter it is necessary to solve a tridiagonal set of simultaneous equations like

$$(\alpha\, \mathbf{I} + \mathbf{T})\, \mathbf{q} \;=\; \alpha\, \mathbf{p} \tag{1}$$

in which $\mathbf{q}$ and $\mathbf{p}$ are column vectors whose elements denote different places on the x-axis. Previously, this was done while solving the heat-flow equation. To make the filter space-variable, the parameter α can be taken to depend on x so that $\alpha\,\mathbf{I}$ is replaced by an arbitrary diagonal matrix. It doesn't matter whether $\mathbf{p}$ and $\mathbf{q}$ are represented in the ω-domain or the t-domain.

Turn your attention from narrow-band data to data with a somewhat broader spectrum and consider

Dip Filters for Moderate Bandwidth Data ($\Delta \omega$)	
Low Pass	*High Pass*
$Q = \dfrac{\alpha}{\alpha + \dfrac{k^2}{-i\omega}} P$	$Q = \dfrac{\dfrac{k^2}{-i\omega}}{\alpha + \dfrac{k^2}{-i\omega}} P$

Naturally these filters can be applied to data of any bandwidth. However the filters are appropriately termed "dip filters" only over a modest bandwidth.

To understand these filters look in the (ω, k)-plane at contours of constant k^2/ω, i.e. $\omega \approx k^2$. Such contours, examples of which are shown in figure 1, are curves of constant attenuation and constant phase shift. The low-pass filter has no phase shift in the pass zone, but there is time differentiation in the attenuation zone. This is apparent from the defining equation. The high-pass filter has no phase shift in the flat pass zone, but there is time integration in the attenuating zone.

An interesting feature of these dip filters is that the low-pass and the high-pass filters constitute a pair of filters which sum to unity. So nothing is lost if a dataset is partitioned by them in two. The high-passed part could be added to the low-passed part to recover the original dataset. Alternately, once the low-pass output is computed, it is much easier to compute the high-pass output, because it is just the input minus the low-pass.

Recursive-Dip-Filter Implementation

Implementation of the moderate bandwidth dip filters is, again, a straightforward matter. For example, clearing fractions, the low-pass filter becomes

$$(-i\omega\alpha \mathbf{I} + \mathbf{T})\mathbf{Q} = -i\omega\alpha\mathbf{P} \tag{2}$$

The main trick is to realize that the differentiation implied by $-i\omega$ is performed in a Crank-Nicolson sense. That is, terms not differentiated are averaged over adjacent values.

$$\mathbf{I}\frac{\alpha}{\Delta t}\left(\mathbf{q}_{t+1} - \mathbf{q}_t\right) + \mathbf{T}\frac{\mathbf{q}_{t+1} + \mathbf{q}_t}{2} = \frac{\alpha}{\Delta t}\left(\mathbf{p}_{t+1} - \mathbf{p}_t\right) \tag{3}$$

Gathering the unknowns to the left gives

$$\left[\frac{\alpha}{\Delta t}\mathbf{I} + \frac{1}{2}\mathbf{T}\right]\mathbf{q}_{t+1} = \left[\frac{\alpha}{\Delta t}\mathbf{I} - \frac{1}{2}\mathbf{T}\right]\mathbf{q}_t + \frac{\alpha}{\Delta t}\left(\mathbf{p}_{t+1} - \mathbf{p}_t\right) \tag{4}$$

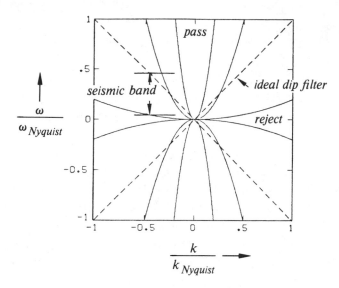

FIG. 2.5-1. Constant-attenuation contours of dip filters. Over the seismic frequency band these parabolas may be satisfactory approximations to the dashed straight line. Pass/reject zones are indicated for the low-pass filter. (Hale)

Equation (4) is a tridiagonal system of simultaneous equations for the un-knowns q_{t+1}. The system may be solved recursively for successive values of t.

The parameter α determines the filter cutoff. It can be chosen to be any function of time and space. However, if the function is to vary extremely rapidly, then it may be necessary to incorporate some of the stability analysis that is developed in a later chapter for use with wave equations.

Side Boundaries

Usually geophysicists wish that there were no boundaries on the sides, or that they were infinitely far away. There are two kinds of side conditions to think about, those in x, and those in k.

Often the side conditions on x are best approximated by zero-slope side conditions. It is possible to use more general side conditions because we have previously learned to solve any tridiagonal system of equations.

The side conditions in k-space relate to the steepest dips. A way to han-dle these dips is to use $\mathbf{T}/(\mathbf{I}-\beta\mathbf{T})$ to represent k^2. This introduces another adjustable parameter β, which must be kept less than $1/4$. Details are studied in Section 4.3.

Slicing Pies

Naturally we may prefer true dip filters, that is, functions of k/ω instead of the functions of k^2/ω described above. But it can be shown that replacing k^2/ω in the above expressions by k^2/ω^2 gives recursions that are unstable.

Sharper pie slices (filters which are more strictly a rectangle function of k/ω), may be defined through a variety of approximation methods described by Hale and Claerbout [1983]. Generally, $|k|$ can be expanded in a power series in $\partial^2/\partial x^2$. If the approximation to $|k|$ is ensured positive, you can expect stability of the recursion that represents $|k|/i\omega$.

More simply, you might be willing to Fourier transform time or space, but not both. In the remaining dimension (the one not transformed) the required operation is a highpass or lowpass filter. This is readily implemented by a variety of techniques, such as the Butterworth filter.

Higher Dimensionality

It is natural to think of a recursive three-dimensional low-pass dip filter as the functional form

$$\frac{\alpha}{\alpha + \dfrac{k_x^{\,2} + k_y^{\,2}}{-i\,\omega}} \tag{5}$$

This, however, leads to an infeasible Crank-Nicolson situation. Multidimensional low-pass filtering *is* possible with

$$\left(\frac{\alpha_x}{\alpha_x + \dfrac{k_x^{\,2}}{-i\,\omega}}\right)\left(\frac{\alpha_y}{\alpha_y + \dfrac{k_y^{\,2}}{-i\,\omega}}\right) \tag{6}$$

2.6 Retarded Coordinates

To examine running horses it may be best to jump on a horse. Likewise, to examine moving waves, it may be better to move along with them. So to describe waves moving downward into the earth we might abandon (x, z)-coordinates in favor of moving (x, z')-coordinates, where $z' = z + t\,v$.

An alternative to the moving coordinate system is to define *retarded coordinates* (x, z, t') where $t' = t - z/v$. The classical example of retarded coordinates is solar time. Time seems to stand still on an airplane that moves westward at the speed of the sun.

The migration process resembles the simulation of wave propagation in either a moving coordinate frame or a retarded coordinate frame. Retarded coordinates are much more popular than moving coordinates. Here is the reason: In solid-earth geophysics, velocity may depend on both x and z, but the earth doesn't change with time t during our seismic observations. In a moving coordinate system the velocity could depend on all three variables, thus unnecessarily increasing the complexity of the calculations. Fourier transformation is a popular means of solving the wave equation, but it loses most of its utility when the coefficients are nonconstant.

Definition of Independent Variables

The specific definition of retarded coordinates is a matter of convenience. Often the retardation is based on hypothetical rays moving straight down with velocity $\overline{v}(z)$. The definition of these coordinates has utility even in problems in which the earth velocity varies laterally, say $v(x, z)$, even though there may be no rays going exactly straight down. In principle, any coordinate system may be used to describe any circumstance, but the utility of the retarded coordinate system generally declines as the family of rays defining it departs more and more from the actual rays.

Despite the simple case at hand it is worthwhile to be somewhat formal and precise. Define the retarded coordinate system (t', x', z') in terms of ordinary Cartesian coordinates (t, x, z) by the set of equations

$$t' = t'(t, x, z) = t - \int_0^z \frac{dz}{\bar{v}(z)} \tag{1a}$$

$$x' = x'(t, x, z) = x \tag{1b}$$

$$z' = z'(t, x, z) = z \tag{1c}$$

The purpose of the integral is to accumulate the travel time from the surface to depth z. The reasons to define (x', z') when it is just set equal to (x, z) are, first, to avoid confusion during partial differentiation and, second, to prepare for later work in which the family of rays is more general.

Definition of Dependent Variables

There are two kinds of *dependent* variables, those that characterize the medium and those that characterize the waves. The medium is characterized by its velocity v and its reflectivity c. The waves are characterized by using U for an upcoming wave, D for a downgoing wave, P for the pressure, and Q for a modulated form of pressure. Let us say $P(t, x, z)$ is the mathematical function to find pressure, given (t, x, z); and $P'(t', x', z')$ is the mathematical function given (t', x', z'). The statement that the two mathematical functions P and P' both refer to the same physical variable is this:

$$P(t, x, z) = P'[t'(t, x, z), x'(t, x, z), z'(t, x, z)] \tag{2}$$
$$P(t, x, z) = P'(t', x', z')$$

Obviously there are analogous expressions for the other dependent variables and medium parameters like velocity $v(x, z)$.

The Chain Rule and the High Frequency Limit

The familiar partial-differential equations of physics come to us in (t, x, z)-space. The chain rule for partial differentiation will convert the partial derivatives to (t', x', z')-space. For example, differentiating (2) with respect to z gives

$$\frac{\partial P}{\partial z} = \frac{\partial P'}{\partial t'}\frac{\partial t'}{\partial z} + \frac{\partial P'}{\partial x'}\frac{\partial x'}{\partial z} + \frac{\partial P'}{\partial z'}\frac{\partial z'}{\partial z} \tag{3a}$$

Using (1) to evaluate the coordinate derivatives gives

$$\frac{\partial P}{\partial z} = -\frac{1}{\bar{v}}\frac{\partial P'}{\partial t'} + \frac{\partial P'}{\partial z'} \tag{3b}$$

There is nothing special about the variable P in (3). We could as well write

$$\frac{\partial}{\partial z} = -\frac{1}{v}\frac{\partial}{\partial t'} + \frac{\partial}{\partial z'} \tag{4}$$

where the left side is for operation on functions that depend on (t, x, z) and the right side is for functions of (t', x', z'). Differentiating twice gives

$$\frac{\partial^2}{\partial z^2} = \left(\frac{-1}{v}\frac{\partial}{\partial t'} + \frac{\partial}{\partial z'}\right)\left(\frac{-1}{v}\frac{\partial}{\partial t'} + \frac{\partial}{\partial z'}\right) \tag{5}$$

Using the fact that the velocity is always time-independent results in

$$\frac{\partial^2}{\partial z^2} = \frac{1}{v^2}\frac{\partial^2}{\partial t'^2} - \frac{2}{v}\frac{\partial^2}{\partial t'\partial z'} + \frac{\partial^2}{\partial z'^2} + \left[\frac{1}{v^2}\frac{d\bar{v}}{dz'}\right]\frac{\partial}{\partial t'} \tag{6}$$

Except for the rightmost term with the square brackets it could be said that "squaring" the operator (4) gives the second derivative. This last term is almost always neglected in data processing. The reason is that its effect is similar to the effect of other first-derivative terms with material gradients for coefficients. Such terms, as described in Section 1.5, cause amplitudes to be more carefully computed. If the last term in (6) is to be included, then it would seem that all such terms should be included, from the beginning.

Fourier Transforms in Retarded Coordinates

Given a pressure field $P(t, x, z)$, we may Fourier transform it with respect to any or all of its independent variables (t, x, z). Likewise, if the pressure field is specified in retarded coordinates, we may Fourier transform with respect to (t', x', z'). Since the Fourier dual of (t, x, z) is (ω, k_x, k_z), it seems appropriate for the dual of (t', x', z') to be (ω', k'_x, k'_z). Now the question is, how are (ω', k'_x, k'_z) related to the familiar (ω, k_x, k_z)? The answer is contained in the chain rule for partial differentiation. Any expression like

$$\frac{\partial}{\partial z} = -\frac{1}{v}\frac{\partial}{\partial t'} + \frac{\partial}{\partial z'} \tag{4}$$

on Fourier transformation says

$$i\,k_z = -\frac{-i\,\omega'}{v} + i\,k'_z \tag{7}$$

Computing all the other derivatives, we have the transformation

$$\omega = \omega' \tag{8a}$$

$$k_x = k'_x \tag{8b}$$

$$k_z = k'_z + \frac{\omega'}{v} \tag{8c}$$

Recall the dispersion relation for the scalar wave equation:

$$\frac{\omega^2}{v^2} = k_x^2 + k_z^2 \tag{9}$$

Performing the substitutions from (8) into (9) we have the expression of the scalar wave equation in retarded time, namely,

$$\left\{ \frac{\omega'}{v} \right\}^2 = k_x'^2 + \left\{ k_z' + \frac{\omega'}{v} \right\}^2 \tag{10}$$

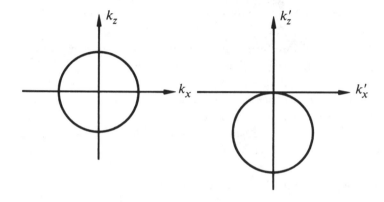

FIG. 2.6-1. Dispersion relation of the wave equation in usual coordinates (left) and retarded time coordinates (right).

These two dispersion relations are plotted in figure 1 for the retardation velocity chosen equal to the medium velocity.

Figure 1 graphically illustrates that retardation can reduce the cost of finite-difference calculations. Waves going straight down are near the top of the dispersion curve (circle). The effect of retardation is to shift the circle's top down to the origin. Discretizing the x- and z-axes will cause spatial frequency aliasing on them. The larger the frequency ω, the larger the circle. Clearly the top of the shifted circle is further from folding. Alternately, Δz may be increased (for the sake of economy) before k_z' exceeds the Nyquist frequency $\pi/\Delta z$.

Interpretation of the Modulated Pressure Variable Q

Earlier a variable Q was defined from the pressure P by the equation

$$P(\omega) \;=\; Q(\omega)\exp\left[\,i\,\omega\!\int_0^z \frac{dz}{\overline{v}(z)}\,\right] \tag{11}$$

The right side is a product of two functions of ω . At constant velocity (11) is expressed as

$$P(\omega) \;=\; Q(\omega)\,e^{\,i\,\omega z/v} \;=\; Q(\omega)\,e^{\,i\,\omega t_0} \tag{12}$$

In the time domain $e^{\,i\,\omega t_0}$ becomes a delta function $\delta(t - t_0)$. Equation (12) is a product in the frequency domain, so in the time domain it is the convolution

$$
\begin{aligned}
p(t) \;&=\; q(t) * \delta(t - z/v) \\
&=\; q(t - z/v) \\
&=\; q(t')
\end{aligned}
\tag{13}
$$

This confirms that the definition of a dependent variable Q is equivalent to introducing retarded time t' .

Einstein's Special Relativity Theory

There is no known application of Einstein's theory of special relativity to seismic imaging. But some of the mathematical methods are related, and now is the appropriate time to take a peek at this famous theory.

In 1887 the Michelson-Morley interferometer experiment established with high accuracy that light travels in all directions at the same speed, day and night, winter and summer. We have seen that the dispersion relation of the scalar wave equation is a circle centered at the origin, meaning that waves go the same speed in all directions. But if the coordinate system is moving with respect to the medium, then the dispersion relation loses directional symmetry. For light propagating in the vacuum of outer space, there seems to be no natural reference coordinate system. If the earth is presumed to be at rest in the summer, then by winter, the earth is moving around the sun in the opposite direction. The summer coordinates relate to the winter coordinates by something like $x' = x - 2\,v_{earth}\,t$. While analysis of the Michelson-Morley experiment shows that such motion should have a measurable asymmetry, measurements show that the predicted asymmetry is absent. Why? One theory is the "ether" theory. Ether is a presumed substance that explains the paradox of the Michelson-Morley experiment. It is presumed to be of minuscule density and viscosity, allowing us to imagine that it is somehow dragged around the earth in such a way that earthbound experimenters are always moving at the same speed as it is. Other measurements, however, also contradict the presumption of ether. Just as wind refracts atmospheric sound waves, ether should cause a measurable refraction of starlight, but this is not observed.

Einstein's explanation of the experiments is based on a mathematical fact that you can easily verify. Let a coordinate frame be defined by

$$z' = z - v \frac{t}{\sqrt{1 - v^2/c^2}} \tag{14a}$$

$$x' = x \tag{14b}$$

$$t' = \frac{t - \frac{v}{c^2} z}{\sqrt{1 - v^2/c^2}} \tag{14c}$$

The amazing thing about this transformation, which you can easily prove, is that it converts the equation $P_{xx} + P_{zz} = c^{-2}P_{tt}$ to the equation $P_{x'x'} + P_{z'z'} = c^{-2}P_{t't'}$. The transformed wave equation is independent of velocity v which is what led Einstein to his surprising conclusions.

2.7 Finite Differencing in (t, x, z)-Space

Much, if not most, production migration work is done in (t, x, z)-space. To avoid being overwhelmed by the complexity of this three-dimensional space, we will first look at migration in (z, t)-space for fixed k_x.

Migration in (z, t)-Space

Migration and data synthesis may be envisioned in (z', t')-space on the following table, which contains the upcoming wave U:

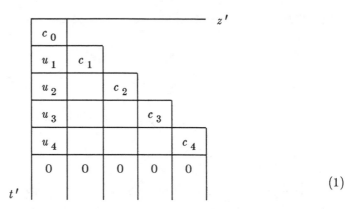

$$(1)$$

In this table the observed upcoming wave at the earth's surface $z' = 0$ is denoted by u_t. The migrated section, denoted by c_t, is depicted along the diagonal because the imaging condition of exploding reflectors at time $t = 0$ is represented in retarded space as

$$z' = z \tag{2a}$$

$$t' = t + z/v \qquad (+ \text{ for } up) \tag{2b}$$

$$0 = t = t' - z'/v \tag{3}$$

The best-focused migration need not fall on the 45° line as depicted in (1); it might be on any line or curve as determined by the earth velocity. This curve forms the basis for velocity determination (Section 3.5). You couldn't determine velocity this way in the frequency-domain.

From Section 2.1, the equation for upcoming waves U in retarded coordinates (t', x', z') is

$$\frac{\partial^2 U}{\partial z' \, \partial t'} = -\frac{v}{2} \frac{\partial^2 U}{\partial x'^2} \tag{4}$$

Next, Fourier transform the x-axis. This assumes that v is a constant function of x and that the x-dependence of U is the sinusoidal function $\exp(ik_x x)$. Thus,

$$0 = \left\{ \frac{v}{2} k_x^2 - \frac{\partial^2}{\partial z' \, \partial t'} \right\} U \tag{5}$$

Now this partial-differential equation will be discretized with respect to t' and z'. Matrix notation will be used, but the notation does not refer to matrix algebra. Instead the matrices refer to differencing stars that may be placed on the (t', z')-plane of (1). Let $*$ denote convolution in (z, t)-space. A succession of derivatives is really a convolution, so the concept of

$\partial/\partial z \; \partial/\partial t = \partial^2/\partial z \, \partial t$ is expressed by is expressed by

$$[-1 \;\; +1] \; * \; \begin{bmatrix} -1 \\ 1 \end{bmatrix} = \begin{bmatrix} 1 & -1 \\ -1 & 1 \end{bmatrix} \tag{6}$$

Thus, the differenced form of (5) is

$$0 = \left\{ \frac{v}{2} \frac{\Delta z' \, \Delta t'}{4} \; k_x^{\,2} \begin{bmatrix} 1 & 1 \\ 1 & 1 \end{bmatrix} - \begin{bmatrix} 1 & -1 \\ -1 & 1 \end{bmatrix} \right\} * U \tag{7}$$

The $1/4$ enters in because the average of U is taken over four places on the mesh.

The sum of the two operators always has $|\, b \,| \geq |\, s \,|$ in the form

$$0 = \begin{bmatrix} s & b \\ b & s \end{bmatrix} * U \tag{8}$$

Now the differencing star in (8) will be used to fill table (1) with values for U.

Given the three values of U in the boxes, a missing one, M, may be determined by either of the implied two operations

or   (9a,b)

It turns out that because $|\, b \,| \geq |\, s \,|$, the implied filling operations by

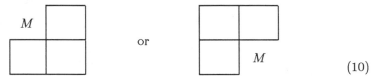

or   (10)

are unstable. It is obvious that there would be a zero-divide problem if s were equal to 0, and it is not difficult to do the stability analysis that shows that (10) causes exponential growth of small disturbances.

It is a worthwhile exercise to make the zero-dip assumption $(k_x = 0)$ and use the numerical values in the operator of (8) to fill in the elements of table (1). It will be found that the values of u_t move laterally in z across the table with no change, predicting, as the table should, that $c_t = u_t$. Slow change in z suggests that we have oversampled the z-axis. In practice, effort is saved by sampling the z-axis with fewer points than are used to sample the t-axis.

(t, x, z)-Space, 15° Diffraction Program

The easiest way to understand 15° migration in (t, x, z)-space is to refer to the (z, t)-space migration. Instead of a scalar function $U(k_x)$, we use **u**, a vector whose components u_j measure pressure at $x = j \, \Delta x$. Think of k_x^2 as a tridiagonal matrix, call it **T**, with $(-1, 2, -1)$ on the main diagonal. Note that k_x^2 is positive, and that **T** is a positive definite matrix with a positive element on its main diagonal. Take equation (7) and use α to denote the left constant. This gives

$$0 = \left\{ \alpha \, \mathbf{T} \begin{bmatrix} \mathbf{I} & \mathbf{I} \\ \mathbf{I} & \mathbf{I} \end{bmatrix} - \begin{bmatrix} \mathbf{I} & -\mathbf{I} \\ -\mathbf{I} & \mathbf{I} \end{bmatrix} \right\} * \mathbf{u} = \begin{bmatrix} \alpha\mathbf{T}-\mathbf{I} & \alpha\mathbf{T}+\mathbf{I} \\ \alpha\mathbf{T}+\mathbf{I} & \alpha\mathbf{T}-\mathbf{I} \end{bmatrix} * \mathbf{u} \qquad (11)$$

Consider a modeling program. It begins down inside the earth with the differencing star (9b). Solving (11) for the unknown $\mathbf{u}_{t'+1, z'}$ and dropping all primes yields

$$(\alpha\mathbf{T}+\mathbf{I}) \, \mathbf{u}_{t+1, z} = - \left[(\alpha\mathbf{T}+\mathbf{I}) \, \mathbf{u}_{t, z+1} + (\alpha\mathbf{T}-\mathbf{I})(\mathbf{u}_{t, z} + \mathbf{u}_{t+1, z+1}) \right] \qquad (12)$$

First, evaluate the expression on the right. The left side is a tridiagonal system to be solved for the unknown $\mathbf{u}_{t+1, z}$. Allowable sequences in which (12) may be applied are dictated by the differencing star (9b).

Heeding the earlier remark that with waves of modest dip, the z'-axis need not be sampled so densely as the t'-axis, we do a computation that skips alternate levels of z'. The specific order chosen in the computer program in figure 1 is indicated by the numbers in the following table:

				z'
c_0	c_0			
5	c_1			
6	c_2	c_2		
7	2	c_3		
8	3	c_4	c_4	
9	4	1	c_5	

t' $\qquad\qquad\qquad\qquad\qquad\qquad\qquad\qquad\qquad\qquad$ (13)

An inescapable practical problem shown in the table when the number of points in t'-space is not exactly equal that in z'-space is that the earth image must be *interpolated* along a diagonal on the mesh. The crude interpolation in (13) illustrates the assumption that the wave field changes rapidly in t' but slowly in z', i.e. the small-angle assumption.

```
# Time Domain 15-degree Diffraction Movie
# Star:          w=p(t ,z)          y=p(t ,z+1)
# Star:          u=p(t+1,z)         v=p(t+1,z+1)
real p(36,96),u(36),w(36),v(36),y(36),e(36),f(36),d(36),z(96),alfa,beta
integer ix,nx,iz,nz,it,nt,kbyte
nx = 36;   nz = 96;   nt = 96;     kbyte=1
alfa = .125                 # v*dz*dt/(8*dx*dx)
beta = .140                 # accurate x derivative parameter; simplest case b=0.
open(3,file='plot40',status='new',access='direct',form='unformatted',recl=1)
do iz=1,nz;  do ix=1,nx;  p(ix,iz) = 0.         # clear space
do iz=nz/5,nz,nz/4                    # Set up initial model
         do it=1,15                   # of 4 band limited
             do ix=1,4                # "point" scatterers.
                 p(ix,it+iz) = (5.-it)*(8-it)*exp(-.1*(it-8)**2)
apb = alfa+beta; amb = alfa-beta          # tridiagonal coefficients
diag = 1.+2.*amb;  offdi = -amb
do iz=nz,2,-2 {                                # Climb up in steps of 2 z-levels
         do i=1,nz;  z(i)=0.;  z(iz)=1.    # Pointer to current z-level
         write(3,rec=kbyte) (z(i),i=1,nz),((p(ix,i),i=1,nz),ix=1,nx)
         kbyte = kbyte + nx*nz*4 + nz*4
         do ix=1,nx
             { u(ix) = p(ix,iz-1);          v(ix) = u(ix) }
         do it=iz,nt {
                 do ix=1,nx        #update the differencing star
                     { w(ix) = u(ix);  y(ix) = v(ix);  v(ix) = p(ix,it) }
                 dd = (1.-apb)*(v(1)+w(1))+apb*(v(2)+w(2))
                 d(1) = dd-diag*y(1)-offdi*(y(1)+y(2))
                 do ix=2,nx-1 {
                         dd    = (1.-2.*apb)*(v(ix)+w(ix))
                         dd    = dd + apb*(v(ix-1)+w(ix-1)+v(ix+1)+w(ix+1))
                         d(ix) = dd-diag*y(ix)-offdi*(y(ix-1)+y(ix+1))    }
                 dd = (1.-apb)*(v(nx)+w(nx))+apb*(v(nx-1)+w(nx-1))
                 d(nx) = dd-diag*y(nx)-offdi*(y(nx)+y(nx-1))
                 call rtris(nx,diag+offdi,offdi,diag,offdi,diag+offdi,d,u,e,f)
                 do ix=1,nx
                         p(ix,it) = u(ix)
                 }
         }
do i=1,nz;  z(i)=0.;  z(1)=1.
write(3,rec=kbyte) (z(i),i=1,nz),((p(ix,i),i=1,nz),ix=1,nx)
stop;     end
```

FIG. 2.7-1.   Time-domain diffraction movie program. (Clayton, Gonzalez, JFC, Hale)

Figure 2 shows the last frame in the movie produced by the test program. Exercise 1 suggests minor changes to the program of figure 1 to convert it from diffraction to migration. As modified, the program is essentially the original wave equation migration program introduced by Johnson and Claerbout [1971] and Doherty and Claerbout [1972].

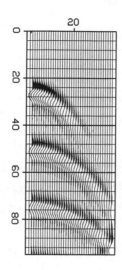

FIG. 2.7-2. Diffractions in the last frame of the downward-continuation movie.

You Can't Time Shift in the Time Domain.

You might wish to do migration in (x, z, t)-space with lateral velocity variation. Then the thin-lens stage would be implemented by time shifting instead of by multiplying by $\exp\{i\omega\,[v(x, z)^{-1}- \bar{v}(z)^{-1}]\Delta z\}$. Time shifting is a delightfully easy operation when what is needed is to shift data by an integral number of sample units. Repetitive time shifting by a fractional number of digital units, however, is a nightmare. Multipoint interpolation operators are required. Even then, pulses tend to disperse. So the lens term is probably best left in the frequency domain.

(t, x, z)-Space, 45° Equation

The 45° migration is a little harder than the 15° migration because the operator in the time domain is higher order, but the methods are similar to those of the 15° equation and the recursive dip filter. The straightforward approach is just to write down the differencing stars. When I did this kind of work I found it easiest to use the Z-transform approach where $1/(-i\omega\Delta t)$ is represented by the bilinear transform $1\!/\!2(1+Z)/(1-Z)$. There are various ways to keep the algebra bearable. One way is to bring all powers of Z to the numerator and then collect powers of Z. Another way, called the integrated approach, is to keep $1/(1-Z)$ with some of the terms. Terms including $1/(1-Z)$ are represented in the computer by buffers that contain

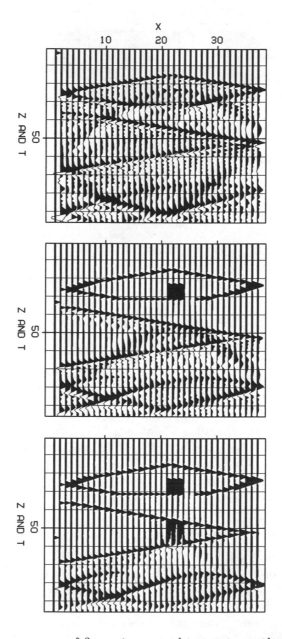

FIG. 2.7-3. The program of figure 1 was used to create synthetic data from a more complicated model. The three frames depict three depths. The leftmost channel of each frame depicts the depth of the frame. (Hale)

the sum from infinite time to time t. The Z-transform approach is developed in Section 4.6. Its real advantage is that it systematizes the stability analysis.

EXERCISES

1. Alter the program given in figure 1 so that it does migration. The delta-function inputs should turn into approximate semicircles.

2. Perform major surgery on the program in figure 1 so that it becomes a low-pass dip filter.

3. Consider a 45° migration program in the space of (z, t, k_x). Find the coefficients in a 6-point differencing star, three points in time and two points in depth. For simplicity, take $v = 1$, $\Delta t = 1$, and $\Delta z = 1$. Suppose this analysis were transformed into the x-domain ($\Delta x = 1$) by replacing k_x^2 with $\mathbf{T}$. What set of tridiagonal equations would have to be solved?

2.8 Introduction to Stability

Experience shows that as soon as you undertake an application that departs significantly from textbook situations, stability becomes a greater concern than accuracy. Stability, or its absence, determines whether the goal is achievable at all, whereas accuracy merely determines the price of achieving it. Here we will look at the stability of the heat-flow equation with real and with imaginary heat conductivity. Since the latter case corresponds to seismic migration, these two cases provide a useful background for stability analysis.

Most stability analysis is based on Fourier transformation. More simply, single sinusoidal or complex exponential trial solutions are examined. If a method becomes unstable for any frequency, then it will be unstable for any realistic case, because realistic functions are just combinations of all frequencies. Begin with the sinusoidal function

$$P(x) = P_0 \, e^{i k x} \tag{1}$$

The second derivative is

$$\frac{\partial^2 P}{\partial x^2} = -k^2 P \tag{2}$$

An expression analogous to the second difference operator defines $\hat{k}$:

$$\frac{\delta^2 P}{\delta x^2} = \frac{P(x + \Delta x) - 2P(x) + P(x - \Delta x)}{\Delta x^2} \tag{3a}$$

$$= -\hat{k}^2 P \tag{3b}$$

Ideally $\hat{k}$ should equal k. Inserting the complex exponential (1) into (3a) gives an expression for $\hat{k}$:

$$-\hat{k}^2 P = \frac{P_0}{\Delta x^2} \left[e^{i\,k(x+\Delta x)} - 2\,e^{i\,k\,x} + e^{i\,k(x-\Delta x)} \right] \tag{4a}$$

$$(\hat{k}\,\Delta x)^2 = 2\,[1 - \cos(k\,\Delta x)] \tag{4b}$$

It is a straightforward matter to make plots of (4b) or its square root. The square root of (4b), through the half-angle trig identity, is

$$\hat{k}\,\Delta x = 2\,\sin\frac{k\,\Delta x}{2} \tag{4c}$$

Series expansion shows that $\hat{k}$ matches k well at low spatial frequencies. At the Nyquist frequency, defined by $k\,\Delta x = \pi$, the value of $\hat{k}\,\Delta x = 2$ is a poor approximation to π. As with any Fourier transform on the discrete domain, $\hat{k}$ is a periodic function of k above the Nyquist frequency. Although k ranges from minus infinity to plus infinity, $\hat{k}^2$ is compressed into the range zero to four. The limits to the range are important since instability often starts at one end of the range.

Explicit Heat-Flow Equation

Begin with the heat-flow equation and Fourier transform over space. Thus $\partial^2/\partial x^2$ becomes simply $-k^2$, and

$$\frac{\partial q}{\partial t} = -\frac{\sigma}{c}\,k^2\,q \tag{5}$$

Finite differencing explicitly over time gives an equation that is identical in form to the inflation-of-money equation:

$$\frac{q_{t+1} - q_t}{\Delta t} = -\frac{\sigma}{c}\,k^2\,q_t \tag{6a}$$

$$q_{t+1} = \left(1 - \frac{\sigma\,\Delta t}{c}\,k^2 \right) q_t \tag{6b}$$

For stability, the magnitude of q_{t+1} should be less than or equal to the magnitude of q_t. This requires the factor in parentheses to have a magnitude less than or equal to unity. The dangerous case is when the factor is more negative than -1. There is instability when $k^2 > 2c/(\sigma \Delta t)$. This means that the high frequencies are diverging with time. The explicit finite differencing on the time axis has caused disaster for short wavelengths on the space axis. Surprisingly, this disaster can be recouped by differencing the space axis coarsely enough! The second space derivative in the Fourier transform domain is $-k^2$. When the x-axis is discretized it becomes $-\hat{k}^2$. So, to discretize (5) and (6), just replace k by $\hat{k}$. Equation (4c) shows that $\hat{k}^2$ has an upper limit of $\hat{k}^2 = 4/\Delta x^2$ at the Nyquist frequency $k\,\Delta x = \pi$. Finally, the factor in (6b) will be less than unity and there will be stability if

$$\hat{k}^2 \;=\; \frac{4}{\Delta x^2} \;\leq\; \frac{2c}{\sigma \Delta t} \tag{7}$$

Evidently instability can be averted by a sufficiently dense sampling of time compared to space. Such a solution becomes unbearably costly, however, when the heat conductivity $\sigma(x)$ takes on a wide range of values. For problems in one space dimension, there is an easy escape in implicit methods. For problems in higher-dimensional spaces, explicit methods must be used.

Explicit 15° Migration Equation

We saw in Section 2.1 that the retarded 15° wave-extrapolation equation is like the heat-flow equation with the exception that the heat conductivity σ must be replaced by the purely imaginary number i. The amplification factor (the magnitude of the factor in parentheses in equation (6b)) is now the square root of the sum squared of real and imaginary parts. Since the real part is already one, the amplification factor exceeds unity for all nonzero values of k^2. The resulting instability is manifested by the growth of dipping plane waves. The more dip, the faster the growth. Furthermore, discretizing the x-axis does not solve the problem.

Implicit Equations

Recall that the inflation-of-money equation

$$q_{t+1} - q_t \;=\; r\,q_t \tag{8}$$

is a simple explicit finite differencing of the differential equation $dq/dt \approx q$. And recall that a better approximation to the differential equation is given by the Crank-Nicolson form

$$q_{t+1} - q_t \;=\; r \, \frac{q_{t+1} + q_t}{2} \tag{9a}$$

that may be rearranged to

$$\left(1 - \frac{r}{2}\right) q_{t+1} \;=\; \left(1 + \frac{r}{2}\right) q_t \tag{9b}$$

or

$$\frac{q_{t+1}}{q_t} \;=\; \frac{1 + r/2}{1 - r/2} \tag{9c}$$

The amplification factor (9c) has magnitude less than unity for all negative r values, even r equal to minus infinity. Recall that the heat-flow equation corresponds to

$$r \;=\; -\frac{\sigma \, \Delta t}{c} \, k^2 \tag{10}$$

where k is the spatial frequency. Since (9c) is good for all negative r, the heat-flow equation, implicitly time-differenced, is good for all spatial frequencies k. The heat-flow equation is stable whether or not the space axis is discretized (then $k \to \hat{k}$) and regardless of the sizes of Δt and Δx. Furthermore, the $15°$ wave-extrapolation equation is also unconditionally stable. This follows from letting r in (9c) be purely imaginary: the amplification factor (9c) then takes the form of some complex number $1+r/2$ divided by its complex conjugate. Expressing the complex number in polar form, it becomes clear that such a number has a magnitude exactly equal to unity. Again there is unconditional stability.

At this point it seems right to add a historical footnote. When finite-difference migration was first introduced many objections were raised on the basis that the theoretical assumptions were unfamiliar. Despite these objections finite-difference migration quickly became popular. I think the reason for its popularity was that, compared to other methods of the time, it was a gentle operation on the data. More specifically, since (9c) is of exactly unit magnitude, the output has the same (ω, k)-spectrum as the input. There may be a wider lesson to be learned from this experience: any process acting on data should do as little to the data as possible.

Leapfrog Equations

The leapfrog method of finite differencing, it will be recalled, requires expressing the time derivative over two time steps. This keeps the centers of the differencing operators in the same place. For the heat-flow equation Fourier-transformed over space,

$$\frac{q_{t+1} - q_{t-1}}{2\,\Delta t} = -\frac{\sigma}{c}\,k^2\,q_t \tag{11}$$

It is a bit of a nuisance to analyze this equation because it covers times $t-1$, t, and $t+1$ and requires slightly more difficult analytical techniques. Therefore, it seems worthwhile to state the results first. The result for the heat-flow equation is that the solution always diverges. The result for the wave-extrapolation equation is much more useful: there is stability provided certain mesh-size restrictions are satisfied, namely, Δz must be less than some factor times Δx^2. This result is not exciting in one space dimension (where implicit methods seem ideal), but in higher-dimensional space, such as in the so-called 3-D prospecting surveys, we may be thankful to have the leapfrog method.

The best way to analyze equations like (11) which range over three or more time levels is to use Z-transform filter analysis. Converted to a Z-transform filter problem, the question posed by (11) becomes whether the filter has zeroes inside (or outside) the unit circle. Z-transform stability analysis is described in Section 4.6. Such analysis is necessary for all possible numerical values of k^2. Its result is that there is always trouble if k^2 ranges from zero to infinity. But with the wave-extrapolation equation, instability can be avoided with certain mesh-size restrictions, because $(\hat{k}\,\Delta x)^2$ lies between zero and four.

Tridiagonal Equation Solver

The tridiagonal algorithm is stable for all positive definite matrices. If you have any problems with the tridiagonal solver, you should question the validity of your problem formulation. What is there about your application that seems to demand division by zero?

3

Offset, Another Dimension

Earlier chapters have assumed that the shot and the geophone are located in the same place. The reality is that there is often as much as a 3-km horizontal separation between them. The 3-km offset is comparable to the depth of many petroleum reservoirs.

Offset is another *dimension* in the analysis of data. At the time of writing, this dimension is often represented in field operations by about 48 channels. No one seems to believe, however, that 48 channels is enough. Recording systems with as many as 1024 channels are coming into use.

The offset dimension adds three important aspects to reflection seismology. First, it enables us to routinely measure the velocity of seismic waves in rocks. This velocity has been assumed to be known in the previous chapters of this book. Second, it gives us data redundancy: it gives independent measurements of quantities that should be the same. Superposition of the measurements (stacking) offers the potential for signal enhancement by destructive interference of noise. Third (a disadvantage), since the offset is nonzero, procedures for migration take on another element of complexity. By the end of this chapter we will be trying to deal with three confusing subjects at the same time — dip, offset, and lateral velocity variation.

Theoretically it seems that offset should offer us the possibility of identifying rocks by observing the reflection coefficient as a function of angle, both for P waves and for P-to-S converted waves. The reality seems to be that neither measurement can be made reliably, if at all. See Section 1.4 for a fuller discussion of converted waves, an interesting subject for research, with a large potential for practical rewards. See also Ostrander [1984] and Tatham and Stoffa [1976]. The reasons for the difficulty in measurement, and the resolution of the difficulty, are, however, not the goal of this book. This goal is instead to enable us to deal effectively with that which is routinely observable.

Stacking Diagrams

First, define the midpoint y between the shot and geophone, and define h to be half the horizontal offset between the shot and geophone:

$$y = \frac{g + s}{2} \tag{1a}$$

$$h = \frac{g - s}{2} \tag{1b}$$

The reason for using *half* the offset in the equations is to simplify and symmetrize many later equations. Offset is defined with $g - s$ rather than with $s - g$ so that positive offset means waves moving in the positive x direction. In the marine case, this means the ship is presumed to sail negatively along the x-axis. In reality the ship may go either way, and shot points may either increase or decrease as the survey proceeds. In some situations you can clarify matters by setting the field observer's shot-point numbers to negative values.

Data is defined experimentally in the space of (s, g). Equation (1) represents a change of coordinates to the space of (y, h). Midpoint-offset coordinates are especially useful for interpretation and data processing. Since the data is also a function of the travel time t, the full dataset lies in a volume. Because it is so difficult to make a satisfactory display of such a volume, what is customarily done is to display slices. The names of slices vary slightly from one company to the next. The following names seem to be well known and clearly understood:

$(y,\ h = 0,\ t)$	zero-offset section
$(y,\ h = h_{min},\ t)$	near-trace section
$(y,\ h = const,\ t)$	constant-offset section
$(y,\ h = h_{max},\ t)$	far-trace section
$(y = const,\ h,\ t)$	common-midpoint gather
$(s = const,\ g,\ t)$	field profile (or common-shot gather)
$(s,\ g = const,\ t)$	common-geophone gather
$(s,\ g,\ t = const)$	time slice
$(h,\ y,\ t = const)$	time slice

A diagram of slice names is in figure 1. Figure 2 shows three slices from the data volume. The first mode of display is "engineering drawing mode." The second mode of display is on the faces of a cube. But notice that although the data is displayed on the surface of a cube, the slices themselves are taken from the interior of the cube. The intersections of slices across one another are shown by dark lines.

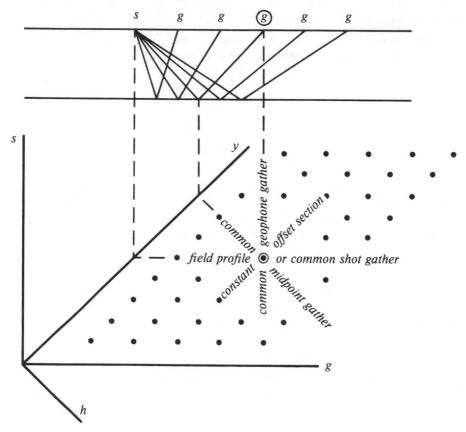

FIG. 3.0-1. Top shows field recording of marine seismograms from a shot at location *s* to geophones at locations labeled *g*. There is a horizontal reflecting layer to aid interpretation. The lower diagram is called a stacking diagram. (It is *not* a perspective drawing). Each dot in this plane depicts a possible seismogram. Think of time running out from the plane. The center geophone above (circled) records the seismogram (circled) that may be found in various geophysical displays. Labels in the diagram below give common names for the displays.

A common-depth-point (CDP) gather is defined by the industry and by common usage to be the same thing as a common-midpoint (CMP) gather. But in this book a distinction will be made. A CDP gather will be considered to be a CMP gather with its time axis stretched according to some velocity model, say,

$$(y = const, \ h, \ \sqrt{t^2 - 4h^2/v^2}) \qquad \text{common-depth-point gather}$$

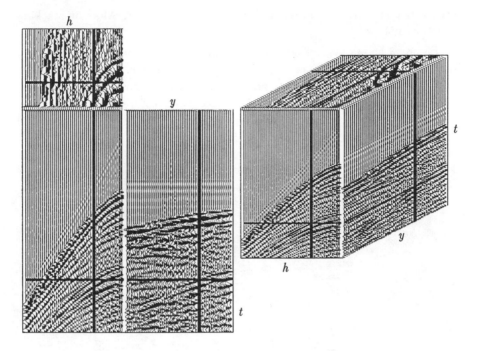

FIG. 3.0-2. Slices from a cube of data from the Grand Banks. Left is "engineering drawing" mode. At the right slices from within the cube are shown as faces on the cube. (Data from Amoco. Display via Rick Ottolini's movie program).

This offset-dependent stretching makes the time axis of the gather become more like a *depth* axis, thus providing the D in CDP. The stretching is called *normal moveout correction* (NMO). Notice that as the velocity goes to infinity, the amount of stretching goes to zero.

In industrial practice the data is not routinely displayed as a function of offset. Instead, each CDP gather is summed over offset. The resulting sum is a single trace. Such a trace can be constructed at each midpoint. The collection of such traces, a function of midpoint and time, is called a CDP stack. Roughly speaking, a CDP stack is like a zero-offset section, but it has a less noisy appearance.

The construction of a CDP stack requires that a numerical choice be made for the moveout-correction velocity. This choice is called the *stacking velocity*. The stacking velocity may be simply someone's guess of the earth's velocity. Or the guess may be improved by stacking with some trial velocities to see which gives the strongest and least noisy CDP stack. More on stacking in Section 3.5.

Figure 3 shows typical land and marine profiles (common-shot gathers). The land data has geophones on both sides of the source. The arrangement shown is called an *uneven split spread*. The energy source was a vibrator. The marine data happens to nicely illustrate two or three head waves (see Sections 3.5 and 5.2). The marine energy source was an air gun. These field profiles were each recorded with about 120 geophones.

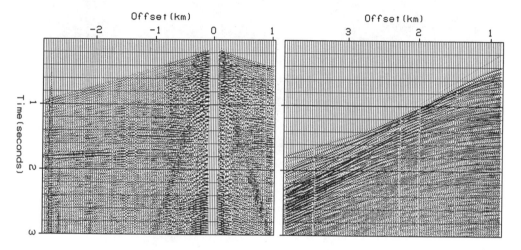

FIG. 3.0-3. Field profiles. Left is a land profile from West Texas. Right is a marine profile off the Aleutian Islands. (Western Geophysical).

What is "Poor Quality" Data?

Vast regions of the world have good petroleum potential but are hard to explore because of the difficulty of obtaining good quality reflection seismic data. The reasons are often unknown. What is "poor quality" data? From an experimental view, almost all seismic data is good in the sense that it is repeatable. The real problem is that the data makes no sense.

Take as an earth model a random arrangement of point reflectors. Its migrated zero-offset section should look random too. Given the repeatability that is experienced in data collection, data with a random appearance implies a random jumble of reflectors. With only zero-offset data little else can be deduced. But with the full range of offsets at our disposal, a more thoughtful analysis can be tried. This chapter provides some of the required techniques.

An interesting model of the earth is a random jumble of point scatterers in a constant-velocity medium. The data would be a random function of time and a random function of the horizontal location of the shot-geophone midpoint. But after suitable processing, for each midpoint, the data should be a perfectly hyperbolic function of shot-geophone offset. This would determine the earth velocity exactly, even if the random scatterers were distributed in three dimensions, and the survey were only along a surface line.

This particular model could fail to explain the "poor quality" data. In that case other models could be tried. The effects of random velocity variations in the near surface or the effects of multiple reflections could be analyzed. Noise in seismology can usually be regarded as a failure of analysis rather than as something polluting the data. It is the offset dimension that gives us the redundancy we need to try to figure out what is really happening.

Texture of Horizontal Bedding, Marine Data

Gravity is a strong force for the stratification of rocks, and in many places in the world rocks are laid down in horizontal beds. Yet even in the most ideal environment the bedding is not mirror smooth; it has some *texture*. We begin the study of offset with synthetic data that mimics the most ideal environment. Such an environment is almost certainly marine, where sedimentary deposition can be slow and uniform. The wave velocity will be taken to be constant, and all rays will reflect as from horizontally lying mirrors. Mathematically, *texture* is introduced by allowing the reflection coefficients of the beds to be laterally variable. The lateral variation is presumed to be a random function, though not necessarily with a white spectrum. Let us examine the appearance of the resulting field data.

Randomness is introduced into the earth with a random function of midpoint y and depth z. This randomness is impressed on some geological "layer cake" function of depth z. For every point in (y, z)-space, a hyperbola of the appropriate random amplitude must be superposed in the space of offset h and travel time t.

What does the final data space look like? This question has little meaning until we decide how the three-dimensional data volume will be presented to the eye. Let us view the data much as it is recorded in the field. For each shot point we see a frame in which the vertical axis is the travel time and the horizontal axis is the distance from the ship down the towed hydrophone cable. The next shot point gives us another frame. Repetition gives us a movie. And what does the movie show?

A single frame shows hyperbolas with imposed texture. The movie shows the texture moving along each hyperbola to increasing offsets. (I find that no

```
# Synthetic marine data tape movie generation
integer kbyte,it,nt,ih,nh,is,ns,iz,nz,it0,iy
real p(512),b(512),refl(25,16),z(25),geol(25),random
open(3,file="plot",status='new',access='direct',form='unformatted',recl=1)
nt = 512;        nh = 48;          ns = 10;          nz = 25;kbyte = 1
do iz=1,nz                              # Reflector depth
     z(iz) =  nt*random()    # random() is on the interval (0.,1.)
do iz=1,nz                              # Reflector strength with depth.
     geol(iz) = 2.*random()-1.
do is = 1,ns                         # Give texture to the Geology
     do iz = 1,nz
          refl(iz,is) = (1.+random())*geol(iz)
do it = 1,nt                            # Prepare a wavelet
     b(it) = exp(-it*.08)*sin(.5*it-.5)
do is = ns,1,-1 {                    # Shots. Run backwards.
     do ih = 1,nh {                  # down cable h = (g-s)/2
          iy = (is-1)+(ih-1)          # y = midpoint
          iy = 1 + (iy-ns*(iy/ns)) # periodic with midpoint
          do it = 1,nt
               p(it) = 0.
          do iz = 1,nz {             # Add in a hyperbola for each layer
               it0 = sqrt( z(iz)**2 + 100.*(ih-1)**2 )
               do it = 1,nt-it0 {      # Add in the wavelet
                    p(it+it0) = p(it+it0) + refl(iz,iy)*b(it)
                    }
               }
          write(3,rec=kbyte) (p(it),it=1,nt);       kbyte = kbyte+nt*4
          }
     }
stop;           end
```

FIG. 3.0-4. Computer program to make synthetic field tapes in an ideal marine environment.

sequence of still pictures can give the impression that the movie gives). Really the ship is moving; the texture of the earth is remaining stationary under it. This is truly what most marine data looks like, and the computer program of figure 4 simulates it. Comparing the simulated data to real marine-data movies, I am impressed by the large amount of random lateral variation required in the simulated data to achieve resemblance to field data. The randomness seems too great to represent lithologic variation. Apparently it is the result of something not modeled. Perhaps it results from our incomplete understanding of the mechanism of reflection from the quasi-random earth. Or perhaps it is an effect of the partial focusing of waves sometime after they reflect from minor topographic irregularities. A full explanation awaits more research.

Texture of Land Data: Near-Surface Problems

Reflection seismic data recorded on land frequently displays randomness because of the irregularity of the soil layer. Often it is so disruptive that the seismic energy sources are deeply buried (at much cost). The geophones are too many for burial. For most land reflection data, the texture caused by these near-surface irregularities exceeds the texture resulting from the reflecting layers.

To clarify our thinking, an ideal mathematical model will be proposed. Let the reflecting layers be flat with no texture. Let the geophones suffer random time delays of several time points. Time delays of this type are called *statics*. Let the shots have random strengths. For this movie, let the data frames be *common-midpoint gathers,* that is, let each frame show data in (h, t)-space at a fixed midpoint y. Successive frames will show successive midpoints. The study of figure 1 should convince you that the travel-time irregularities associated with the geophones should move leftward, while the amplitude irregularities associated with the shots should move rightward. In real life, both amplitude and time anomalies are associated with both shots and geophones.

EXERCISES

1. Note that figure 1 is drawn for a shot interval Δs equal to half the geophone interval Δg. Redraw figure 1 for $\Delta s = \Delta g$. Common-midpoint gathers now come in two types. Suggest two possible definitions for "near-offset section."

2. Modify the program of figure 4 to produce a movie of synthetic *midpoint* gathers with random shot amplitudes and random geophone time delays. Observing this movie you will note the perceptual problem of being able to see the leftward motion along with the rightward motion. Try to adjust anomaly strengths so that both left-moving and right-moving patterns are visible.

    Your mind will often see only one, blocking out the other, similar to the way you perceive a 3-D cube, from a 2-D projection of its edges.

3. Define recursive dip filters to pass and reject the various textures of shot, geophone, and midpoint.

3.1 Absorption and a Little Focusing

Sometimes the earth strata lie horizontally with little irregularity. There we may hope to ignore the effects of migration. Seismic rays should fit a simple model with large reflection angles occurring at wide offsets. Such data should be ideal for the measurement of reflection coefficient as a function of angle, or for the measurement of the earth acoustic absorptivity $1/Q$. In his doctoral dissertation, Einar Kjartansson reported such a study. The results were so instructive that the study will be thoroughly reviewed here. I don't know to what extent the Grand Isle gas field (Pan [1983]) typifies the rest of the earth, but it is an excellent place to begin learning about the meaning of shot-geophone offset.

The Grand Isle Gas Field: A Classic Bright Spot

The dataset Kjartansson studied was a seismic line across the Grand Isle gas field, off the shore of Louisiana, and was supplied by the Gulf Oil Company. The data contain several classic "bright spots" (strong reflections) on some rather flat undisturbed bedding. Of interest are the lateral variations in amplitude on reflections at a time depth of about 2.3 seconds. (See figure 3). It is widely believed that such bright spots arise from shallow gas-bearing sands.

Theory predicts that reflection coefficient should be a function of angle. For an anomalous physical situation like gas-saturated sands, the function should be distinctive. Evidence should be found on common-midpoint gathers like those shown in figure 1. Looking at any one of these gathers you will note that the reflection strength versus offset seems to be a smooth, sensibly behaved function, apparently quite measurable. Using layered media theory, however, it was determined that only the most improbably bizarre medium could exhibit such strong variation of reflection coefficient with angle, particularly at small angles of incidence. (The reflection angle of the energy arriving at wide offset at time 2.5 seconds is not a large angle. Assuming constant velocity, $\arccos(2.3/2.6) = 28°$). Compounding the puzzle, each common-midpoint gather shows a *different* smooth, sensibly behaved, measurable function. Furthermore, these midpoints are near one another, ten shot points spanning a horizontal distance of 820 feet.

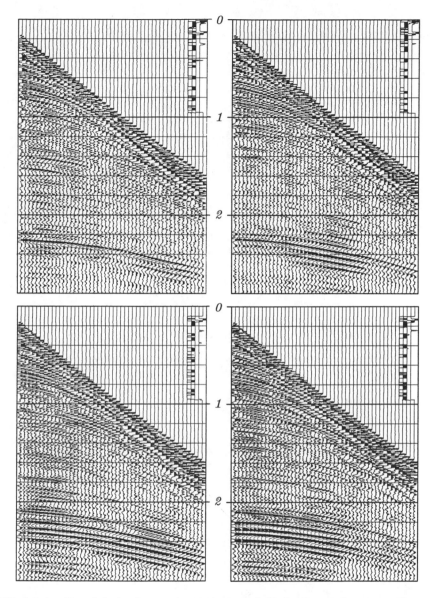

FIG. 3.1-1. Top left is shot point 220; top right is shot point 230. No processing has been applied to the data except for a display gain proportional to time. Bottom shows shot points 305 and 315. (Kjartansson, Gulf)

Kjartansson's Model for Lateral Variation in Amplitude

The Grand Isle data is incomprehensible in terms of the model based on layered media theory. Kjartansson proposed an alternative model. Figure 2 illustrates a geometry in which rays travel in straight lines from any source to a flat horizontal reflector, and thence to the receivers. The only complications are "pods" of some material that is presumed to disturb seismic rays in some anomalous way. Initially you may imagine that the pods absorb wave energy. (In the end it will be unclear whether the disturbance results from energy focusing or absorbing).

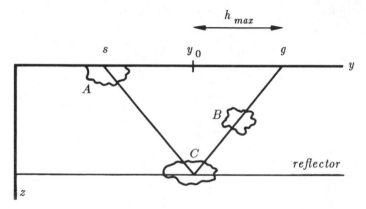

The model above produces the disturbed data space sketched below.

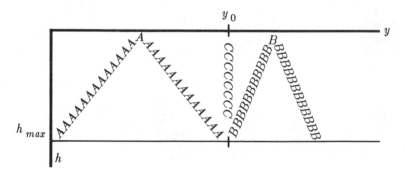

FIG. 3.1-2. Kjartansson's model. Anomalous material in pods A, B, and C may be detected by its effect on reflections from a deeper layer.

Pod A is near the surface. The seismic survey is affected by it twice — once when the pod is traversed by the shot and once when it is traversed by the geophone. Pod C is near the reflector and encompasses a small area of it. Pod C is seen at all offsets h but only at one midpoint, y_0. The raypath depicted on the top of figure 2 is one that is affected by all pods. It is at midpoint y_0 and at the widest offset h_{max}. Find the raypath on the lower diagram in figure 2.

Pod B is part way between A and C. The slope of affected points in the (y, h)-plane is part way between the slope of A and the slope of C.

Figure 3 shows a common-offset section across the gas field. The offset shown is the fifth trace from the near offset, 1070 feet from the shot point. Don't be tricked into thinking the water was deep. The first break at about .33 seconds is wide-angle propagation.

The power in each seismogram was computed in the interval from 1.5 to 3 seconds. The logarithm of the power is plotted in figure 4a as a function of midpoint and offset. Notice streaks of energy slicing across the (y, h)-plane at about a 45° angle. The strongest streak crosses at exactly 45° degrees through the near offset at shot point 170. This is a missing shot, as is clearly visible in figure 3. Next, think about the gas sand described as pod C in the model. Any gas-sand effect in the data should show up as a streak across all offsets at the midpoint of the gas sand — that is, horizontally across the page. I don't see such streaks in figure 4a. Careful study of the figure shows that the rest of the many clearly visible streaks cut the plane at an angle noticeably *less* than ±45°. The explanation for the angle of the streaks in the figure is that they are like pod B. They are part way between the surface and the reflector. The angle determines the depth. Being closer to 45° than to 0°, the pods are closer to the surface than to the reflector.

Figure 4b shows timing information in the same form that figure 4a shows amplitude. A CDP stack was computed, and each field seismogram was compared to it. A residual time shift for each trace was determined and plotted in figure 4b. The timing residuals on one of the common-midpoint gathers is shown in figure 5.

The results resemble the amplitudes, except that the results become noisy when the amplitude is low or where a "leg jump" has confounded the measurement. Figure 4b clearly shows that the disturbing influence on timing occurs at the same depth as that which disturbs amplitudes.

The process of *inverse slant stack*, to be described in Section 5.2 enables us to determine the depth distribution of the pods. This distribution is displayed in figures 4c and 4d.

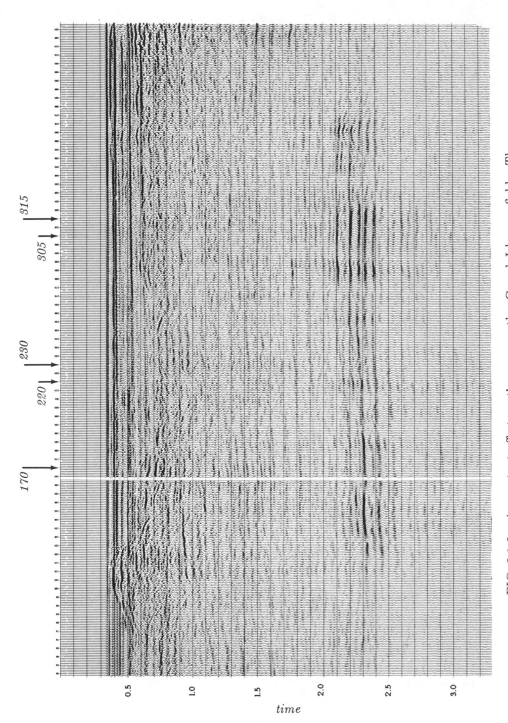

FIG. 3.1-3. A constant-offset section across the Grand Isle gas field. The offset shown is the fifth from the near trace. (Kjartansson, Gulf)

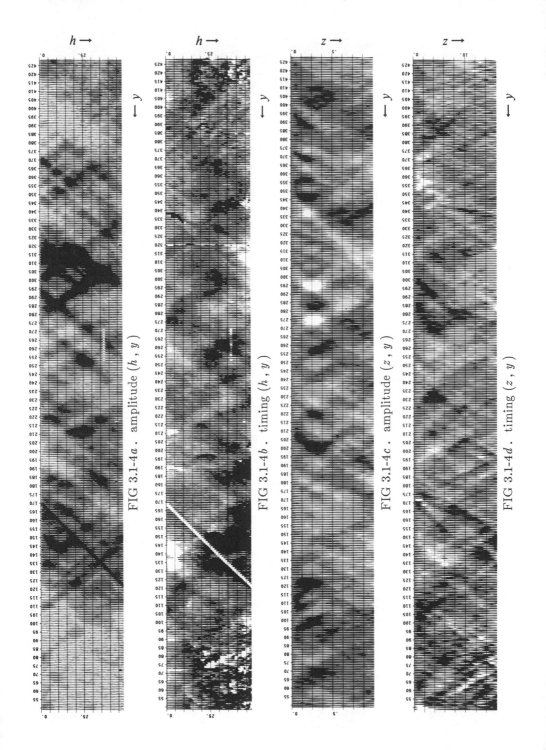

FIG 3.1-4a . amplitude (h, y)

FIG 3.1-4b . timing (h, y)

FIG 3.1-4c . amplitude (z, y)

FIG 3.1-4d . timing (z, y)

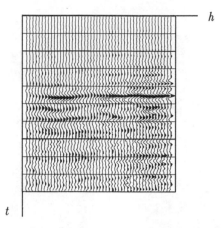

FIG. 3.1-5. Midpoint gather 220 (same as in figure 1b) after moveout. Shown is a one-second window centered at 2.3 seconds, time shifted according to an NMO for an event at 2.3 seconds, using a velocity of 7000 feet/sec. (Kjartansson)

Rotten Alligators

The sediments carried by the Mississippi River are dropped at the delta. There are sand bars, point bars, old river bows now silted in, a crow's foot of sandy distributary channels, and between channels, swampy flood plains are filled with decaying organic material. The landscape is clearly laterally variable, and eventually it will all sink of its own weight, aided by growth faults and the weight of later sedimentation. After it is buried and out of sight the lateral variations will remain as pods that will be observable by the seismologists of the future. These seismologists may see something like figure 6. Figure 6 shows a *three dimensional* seismic survey, that is, the ship sails many parallel lines about 70 meters apart. The top plane, a slice at constant time, shows buried river meanders. The data shown in figure 6 is described more fully by its donors, Dahm and Graebner [1982].

Focusing or Absorption?

Highly absorptive rocks usually have low velocity. Behind a low velocity pod, waves should be weakened by absorption. They should also be strengthened by focusing. Which effect dominates? How does the phenomenon depend on spatial wavelength? A full reconstruction of the physical model remains to be done. Maybe you can figure it out knowing that black on figure 4c denotes low amplitude or high absorption, and black on

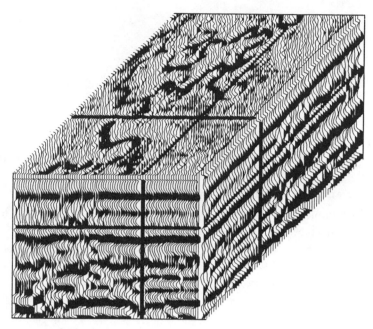

FIG. 3.1-6. Three-dimensional seismic data (Geophysical Services Inc.) from
the Gulf of Thailand. Data planes from within the cube are displayed on the
faces of the cube. The top plane shows ancient river meanders now sub-
merged.

figure 4d denotes low velocities.

EXERCISE

1. Consider waves converted from pressure P waves to shear S waves.
    Assume an S-wave speed of about half the P-wave speed. What would
    figure 2 look like for these waves?

3.2 Introduction to Dip

The study of seismic travel-time dependence upon source-receiver offset begins by calculating the travel times for rays in some ideal environments.

Sections and Gathers for Planar Reflectors

The simplest environment for reflection data is a single horizontal reflection interface, which is shown in figure 1. As expected, the zero-offset section mimics the earth model. The common-midpoint gather is a hyperbola whose asymptotes are straight lines with slopes of the inverse of the velocity v_1. The most basic data processing is called *common-depth-point stack* or CDP stack. In it, all the traces on the common-midpoint (CMP) gather are time shifted into alignment and then added together. The result mimics a zero-offset trace. The collection of all such traces is called the *CDP-stacked section*. In practice the CDP-stacked section is often interpreted and migrated as though it were a zero-offset section. In this chapter we will learn to avoid this popular, oversimplified assumption.

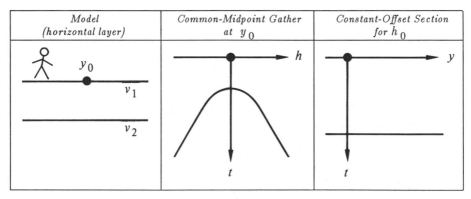

FIG. 3.2-1. Simplest earth model.

The next simplest environment is to have a planar reflector that is oriented vertically rather than horizontally. This is not typical, but is included here because the effect of earth dip is more comprehensible in an extreme case. Now the wave propagation is along the air-earth interface. To avoid confusion the reflector may be inclined at a slight angle from the vertical, as in figure 2.

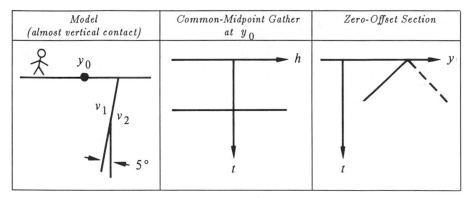

FIG. 3.2-2. Near-vertical reflector, a gather, and a section.

Figure 2 shows that the travel time does not change as the offset changes. It may seem paradoxical that the travel time does not increase as the shot and geophone get further apart. The key to the paradox is that midpoint is held constant, not shotpoint. As offset increases, the shot gets further from the reflector while the geophone gets closer. Time lost on one path is gained on the other.

A planar reflector may have any dip between horizontal and vertical. Then the common-midpoint gather lies between the common-midpoint gather of figure 1 and that of figure 2. The zero-offset section in figure 2 is a straight line, which turns out to be the asymptote of a family of hyperbolas. The slope of the asymptote is the inverse of velocity v_1.

The Dipping Bed

While the travel-time curves resulting from a dipping bed are simple, they are not simple to derive. Before the derivation, the result will be stated: for a bed dipping at angle α from the horizontal, the travel-time curve is

$$t^2 v^2 = 4(y - y_0)^2 \sin^2\alpha + 4h^2 \cos^2\alpha \tag{1}$$

For $\alpha = 45°$, equation (1) is the familiar Pythagoras cone — it is just like $t^2 = z^2 + x^2$. For other values of α, the equation is still a cone, but a less familiar one because of the stretched axes.

For a common-midpoint gather at $y = y_1$ in (h, t)-space, equation (1) looks like $t^2 = t_0^2 + 4h^2/v_{apparent}^2$. Thus the common-midpoint gather contains an *exact* hyperbola, regardless of the earth dip angle α. The effect of dip is to change the asymptote of the hyperbola, thus changing the apparent velocity. The result has great significance in applied work and is known as Levin's dip correction [1971]:

$$v_{apparent} = \frac{v_{earth}}{\cos(\alpha)} \tag{2}$$

(See also Slotnick [1959]). In summary, dip increases the stacking velocity.

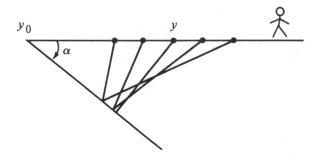

FIG. 3.2-3. Rays from a common-midpoint gather.

Figure 3 depicts some rays from a common-midpoint gather. Notice that each ray strikes the dipping bed at a different place. So a common-*midpoint* gather is not a common-*depth-point* gather. To realize why the reflection point moves on the reflector, recall the basic geometrical fact that an angle bisector in a triangle generally doesn't bisect the opposite side. The reflection point moves *up* dip with increasing offset.

Finally, equation (1) will be proved. Figure 4 shows the basic geometry along with an "image" source on another reflector of twice the dip. For convenience, the bed intercepts the surface at $y_0 = 0$. The length of the line $s'g$ in figure 4 is determined by the trigonometric Law of Cosines to be

$$t^2 v^2 = s^2 + g^2 - 2sg \cos 2\alpha$$

$$t^2 v^2 = (y - h)^2 + (y + h)^2 - 2(y - h)(y + h)\cos 2\alpha$$
$$t^2 v^2 = 2(y^2 + h^2) - 2(y^2 - h^2)(\cos^2\alpha - \sin^2\alpha)$$
$$t^2 v^2 = 4 y^2 \sin^2\alpha + 4 h^2 \cos^2\alpha$$

which is equation (1).

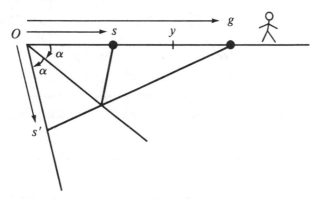

FIG. 3.2-4. Travel time from image source at s' to g may be expressed by the law of cosines.

Another facet of equation (1) is that it describes the constant-offset section. Surprisingly, the travel time of a dipping planar bed becomes curved at nonzero offset — it too becomes hyperbolic.

The Point Response

Another simple geometry is a reflecting point within the earth. A wave incident on the point from any direction reflects waves in all directions. This geometry is particularly important because any model is a superposition of such point scatterers. Figure 5 shows an example. The curves in figure 5 include flat spots for the same reasons that some of the curves in figures 1 and 2 were straight lines.

The point-scatterer geometry for a point located at (x, z) is shown in figure 6.

The equation for travel time t is the sum of the two travel paths

$$t v = \sqrt{z^2 + (s - x)^2} + \sqrt{z^2 + (g - x)^2} \tag{3}$$

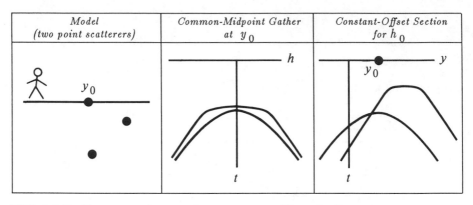

FIG. 3.2-5. Response of two point scatterers. Note the flat spots.

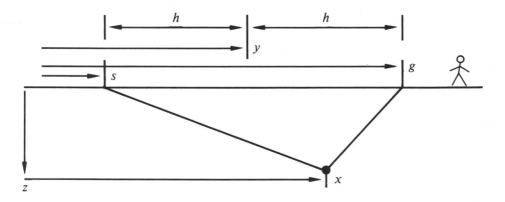

FIG. 3.2-6. Geometry of a point scatterer.

Cheops' Pyramid

Because of the importance of the point-scatterer model, we will go to considerable lengths to visualize the functional dependence among t, z, x, s, and g in equation (3). This picture is more difficult — by one dimension — than is the conic section of the exploding-reflector geometry.

To begin with, suppose that the first square root in (3) is constant because everything in it is held constant. This leaves the familiar hyperbola in (g, t)-space, except that a constant has been added to the time. Suppose instead that the other square root is constant. This likewise leaves a hyperbola in (s, t)-space. In (s, g)-space, travel time is a function of s plus a function of g. I think of this as one coat hanger, which is parallel to the s-

axis, being hung from another coat hanger, which is parallel to the g-axis.

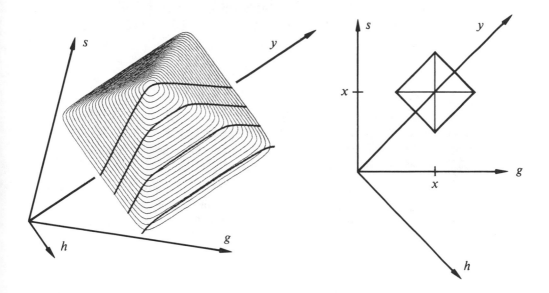

FIG. 3.2-7. Left is a picture of the travel-time mountain of equation (3) for fixed x and z. The darkened lines are constant-offset sections. Right is a cross section through the mountain for large t (or small z). (Ottolini)

A view of the travel-time mountain on the (s, g)-plane or the (y, h)-plane is shown in figure 7a. Notice that a cut through the mountain at large t is a square, the corners of which have been smoothed. A constant value of t is the square contoured in (s, g)-space, as in figure 7b. Algebraically, the squareness becomes evident for a point reflector near the surface, say, $z \to 0$. Then (3) becomes

$$v\,t \quad = \quad |\,s - x\,| \quad + \quad |\,g - x\,| \tag{4}$$

The center of the square is located at $(s, g) = (x, x)$. Taking travel time t to increase downward from the horizontal plane of (s, g)-space, the square contour is like a horizontal slice through the Egyptian pyramid of Cheops. To walk around the pyramid at a constant altitude is to walk around a square. Alternately, the altitude change of a traverse over g at constant s is simply a constant plus an absolute-value function, as is a traverse of s at constant g.

More interesting and less obvious are the curves on common-midpoint gathers and constant-offset sections. Recall the definition that the midpoint between the shot and geophone is y. Also recall that h is half the horizontal offset from the shot to the geophone.

$$y \quad = \quad \frac{g + s}{2} \tag{5a}$$

$$h \quad = \quad \frac{g - s}{2} \tag{5b}$$

A traverse of y at constant h is shown in figure 7. At the highest elevation on the traverse, you are walking along a flat horizontal step like the flat-topped hyperboloids of figure 5. Some erosion to smooth the top and edges of the pyramid gives a model for nonzero reflector depth.

For rays that are near the vertical, the travel-time curves are far from the hyperbola asymptotes. Then the square roots in (3) may be expanded in Taylor series, giving a parabola of revolution. This describes the eroded peak of the pyramid.

Random Point Scatterers

Figure 8 shows a synthetic constant-offset section (COS) taken from an earth model containing about fifty randomly placed point scatterers. Late arrival times appear hyperbolic. Earlier arrivals have flattened tops. The earliest possible arrival corresponds to a ray going horizontally from the shot to the geophone.

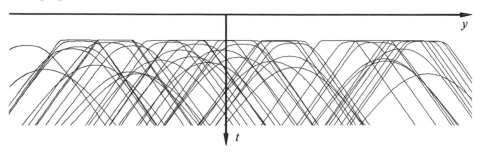

FIG. 3.2-8. Constant-offset section over random point scatterers.

Figure 9 shows a synthetic common-shot profile (CSP) from the same earth model of random point scatterers. Each scatterer produces a hyperbolic arrival. The hyperbolas are not symmetric around zero offset; their locations are random. They must, however, all lie under the lines $\mid g-s \mid = vt$. Hyperbolas with sharp tops can be found at late times as well as early times. However, the sharp tops, which are from shallow scatterers near the geophone, must lie near the lines $\mid g-s \mid = vt$.

FIG. 3.2-9. Common-shot profile over random point scatterers.

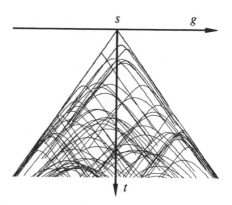

Figure 10a shows a synthetic common-midpoint gather (CMP) from an earth model containing about fifty randomly placed point scatterers. Because this is a common-*midpoint* gather, the curves are symmetric through zero offset. (The negative offsets of field data are hardly ever plotted). Some of the arrivals have flattened tops, which indicate scatterers that are not directly under the midpoint.

Normal-moveout (NMO) correction is a stretching of the data to try to flatten the hyperbolas. This correction assumes flat beds, but it also works for point scatterers that are directly under the midpoint. Figure 10b shows what happens when normal-moveout correction is applied on the random scatterer model. Some reflectors are flattened; others are "overcorrected."

Forward and Backward Scattering: Larner's Streaks

At some locations, near-surface waves overwhelm the deep reflections of geologic interest. Compounding our difficulty, the near-surface waves are usually irregular because the earth is comparatively more irregular at its surface than deeper down. On land, these interfering waves are called ground roll. At sea, they are called water waves (not to be confused with surface waves on

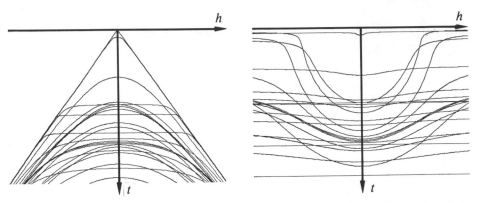

FIG. 3.2-10. Common-midpoint gather on earth of randomly located point scatterers (left). The same gathers after NMO correction (right).

water).

A model for such near-surface noises is suggested by the vertical reflecting wall in figure 2. In this model the waves remain close to the surface. Randomly placed vertical walls could result in a zero-offset section that resembles the field data of figure 11. Another less extreme model for the surface noises is the flat-topped curves in the random point-scatterer model.

In the random point-reflector model the velocity was a constant. In real life the earth velocity is generally slower for the near-surface waves and faster for the deep reflections. This sets the stage for some unexpected noise amplification.

CDP stacking enhances events with the stacking velocity and discriminates against events with other velocities. Thus you might expect that stacking at deeper, higher velocity would discriminate against low-velocity, near-surface events. Near-surface noises, however, are not reflections from horizontal layers; they are more like reflections from vertical walls or steeply dipping layers. But equation (2) shows that dip increases the apparent velocity. So it is not surprising that stacking at deep-sediment, high velocities can enhance surface noises. Occurrence of this problem in practice was nicely explained and illustrated by Larner et al. [1983]

Velocity of Sideswipe

Shallow-water noise can come from waves scattering from a sunken ship or from the side of an island or iceberg several kilometers to the side of the survey line. Think of boulders strewn all over a shallow sea floor, not only along the path of the ship, but also off to the sides. The travel-time curves

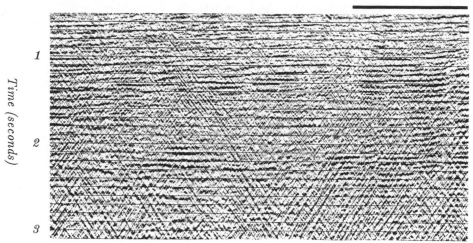

FIG. 3.2-11. CDP stack with water noise from the Shelikof Strait, Alaska. (by permission from *Geophysics,* Larner et al. [1983])

for reflections from the boulders nicely matches the random point-scatterer model. Because of the long wavelengths of seismic waves, our sending and receiving equipment does not enable us to distinguish waves going up and down from those going sideways.

Imagine one of these shallow scatterers several kilometers to the side of the ship. More precisely, let the scatterer be on the earth's surface, perpendicular to the midpoint of the line connecting the shot point to the geophone. A common-midpoint gather for this scatterer is a perfect hyperbola, as from the deep reflector contributions on figure 9. Since it is a water-velocity hyperbola, this scattered noise should be nicely suppressed by CDP stacking with the higher, sediment velocity. So the "streaking" scatterers mentioned earlier are not sidescatter. The "streaking" scatterers are those along the survey line, not those perpendicular to it.

The Migration Ellipse

Another insight into equation (3) is to regard the offset h and the total travel time t as fixed constants. Then the locus of possible reflectors turns out to describe an ellipse in the plane of $(y - y_0, z)$. The reason it is an ellipse follows from the geometric definition of an ellipse. To draw an ellipse, place a nail or tack into s on figure 6 and another into g. Connect the tacks by a string that is exactly long enough to go through (y_0, z). An ellipse going through (y_0, z) may be constructed by sliding a pencil along the string, keeping the string tight. The string keeps the total distance tv constant.

Recall (Section 1.3) that one method for migrating zero-offset sections is to take every data value in (y, t)-space and use it to superpose an appropriate semicircle in (y, z)-space. For nonzero offset the circle should be generalized to an ellipse (figure 1.3-1).

It is not easy to show that equation (3) can be cast in the standard mathematical form of an ellipse, namely, a stretched circle. But the result is a simple one, and an important one for later analysis, so here we go. Equation (3) in (y, h)-space is

$$ t \, v \;=\; \sqrt{z^2 + (y - y_0 - h)^2} + \sqrt{z^2 + (y - y_0 + h)^2} \qquad (6) $$

To help reduce algebraic verbosity, define a new y equal to the old one shifted by y_0. Also make the definitions

$$ t \, v_{rock} \;=\; 2 \, d \;=\; 2 \, t \, v_{half} \qquad (7a) $$

$$ a \;=\; z^2 + (y + h)^2 \qquad (7b) $$

$$ b \;=\; z^2 + (y - h)^2 \qquad (7c) $$

$$ a - b \;=\; 4 \, y \, h \qquad (7d) $$

With these definitions, (6) becomes

$$ 2 \, d \;=\; \sqrt{a} + \sqrt{b} \qquad (8) $$

Square to get a new equation with only one square root.

$$ 4 \, d^2 - (a + b) \;=\; 2 \sqrt{a} \sqrt{b} \qquad (9) $$

Square again to eliminate the square root.

$$ 16 \, d^4 - 8 \, d^2 (a + b) + (a + b)^2 \;=\; 4 \, a \, b \qquad (10a) $$

$$ 16 \, d^4 - 8 \, d^2 (a + b) + (a - b)^2 \;=\; 0 \qquad (10b) $$

Introduce definitions of a and b.

$$ 16 \, d^4 - 8 \, d^2 [2 \, z^2 + 2 \, y^2 + 2 \, h^2] + 16 \, y^2 \, h^2 \;=\; 0 \qquad (11) $$

Bring y and z to the right.

$$ d^4 - d^2 \, h^2 \;=\; d^2 (z^2 + y^2) - y^2 \, h^2 \qquad (12a) $$

$$ d^2 (d^2 - h^2) \;=\; d^2 \, z^2 + (d^2 - h^2) \, y^2 \qquad (12b) $$

$$ d^2 \;=\; \frac{z^2}{1 - \dfrac{h^2}{d^2}} + y^2 \qquad (12c) $$

Finally, recalling all earlier definitions,

$$t^2 \, v_{half}^2 \;\; = \;\; \frac{z^2}{1 - \dfrac{h^2}{t^2 \, v_{half}^2}} \; + \; (y - y_0)^2 \tag{13}$$

Fixing t, equation (13) is the equation for a circle with a stretched z-axis. Our algebra has confirmed that the "string and tack" definition of an ellipse matches the "stretched circle" definition. An ellipse in model space is the earth model given the observation of an impulse on a constant-offset section.

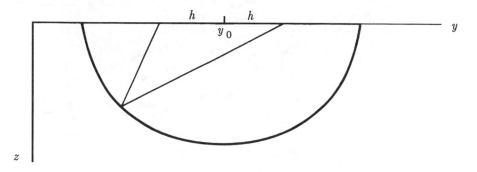

FIG. 3.2-12. Migration ellipse.

Seismometer Spacing

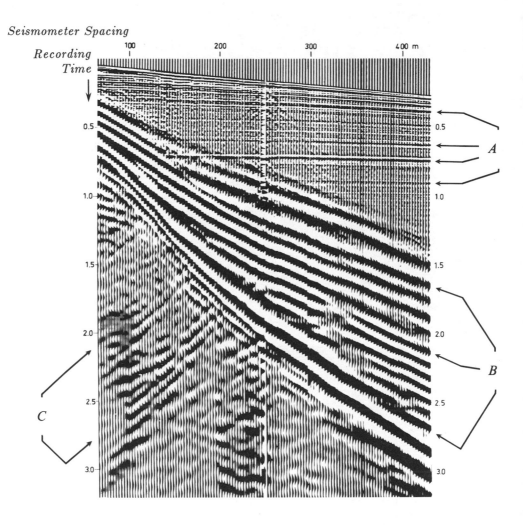

FIG. 3.2-13. Undocumented data from a recruitment brochure. This data may be assumed to be of textbook quality. The speed of sound in water is about 1500 m/sec. Identify the events at A, B, and C. Is this a common-shotpoint gather or a common-midpoint gather? (Shell Oil Company)

3.3 Survey Sinking with
the Double-Square-Root Equation

Exploding-reflector imaging will be replaced by a broader imaging concept, *survey sinking*. A new equation called the double-square-root (DSR) equation will be developed to implement survey-sinking imaging. The function of the DSR equation is to downward continue an entire seismic survey, not just the geophones but also the shots. After deriving the DSR equation, the remainder of this chapter will be devoted to explaining migration, stacking, migration before stack, velocity analysis, and corrections for lateral velocity variations in terms of the DSR equation.

Peek ahead at equation (9) and you will see an equation with two square roots. One represents the cosine of the wave *arrival* angle. The other represents the *takeoff* angle at the shot. One cosine is expressed in terms of k_g, the Fourier component along the geophone axis of the data volume in (s, g, t)-space. The other cosine, with k_s, is the Fourier component along the shot axis.

Our field seismograms lie in the (s, g)-plane. To move onto the (y, h)-plane inhabited by seismic interpreters requires only a simple rotation. The data could be Fourier transformed with respect to y and h, for example. Then downward continuation would proceed with equation (17) instead of equation (9).

The DSR equation depends upon the reciprocity principle which we will review first.

Seismic Reciprocity in Principle and in Practice

The principle of reciprocity says that the same seismogram should be recorded if the locations of the source and geophone are exchanged. A physical reason for the validity of reciprocity is that no matter how complicated a geometrical arrangement, the speed of sound along a ray is the same in either direction.

Mathematically, the reciprocity principle arises because the physical equations of elasticity are self adjoint. The meaning of the term *self adjoint* is illustrated in FGDP where it is shown that discretized acoustic equations yield a *symmetric* matrix even where density and compressibility are space variable. The inverse of any such symmetric matrix is another symmetric matrix called the impulse-response matrix. Elements across the matrix diagonal are equal to one another. Each element of any pair is a response to an impulsive source. The opposite element of the pair refers to the interchanged source and receiver.

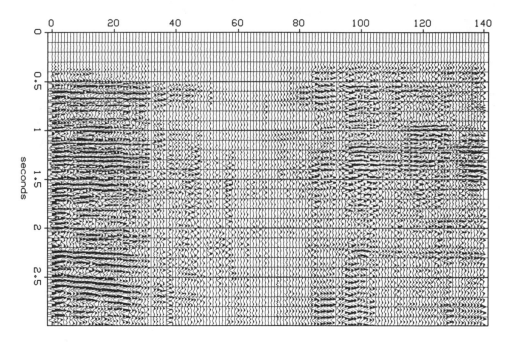

FIG. 3.3-1. Constant-offset section from the Central Valley of California. (Chevron)

A tricky thing about the reciprocity principle is the way antenna patterns must be handled. For example, a single vertical geophone has a natural antenna pattern. It cannot see horizontally propagating pressure waves nor vertically propagating shear waves. For reciprocity to be applicable, antenna patterns must not be interchanged when source and receiver are interchanged. The antenna pattern must be regarded as attached to the medium.

I searched our data library for split-spread land data that would illustrate reciprocity under field conditions. The constant-offset section in figure 1 was recorded by vertical vibrators into vertical geophones. The survey was not intended to be a test of reciprocity, so there likely was a slight lateral offset of the source line from the receiver line. Likewise the sender and receiver arrays (clusters) may have a slightly different geometry. The earth dips in figure 1 happen to be quite small although lateral velocity variation is known to be a problem in this area.

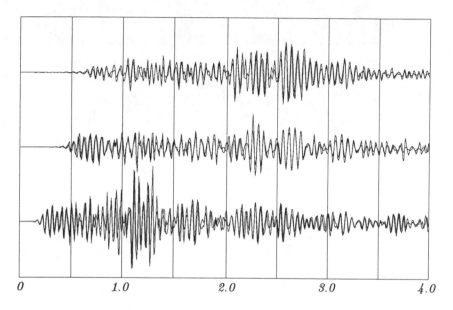

FIG. 3.3-2. Overlain reciprocal seismograms.

In figure 2, three seismograms were plotted on top of their reciprocals. Pairs were chosen at near offset, at mid range, and at far offset. You can see that reciprocal seismograms usually have the same polarity, and often have nearly equal amplitudes. (The figure shown is the best of three such figures I prepared).

Each constant time slice in figure 3 shows the reciprocity of many seismogram pairs. Midpoint runs horizontally over the same range as in figure 1. Offset is vertical. Data is not recorded near the vibrators leaving a gap in the middle. To minimize irrelevant variations, moveout correction was done before making the time slices. (There is a missing source that shows up on the left side of the figure). A movie of panels like figure 3 shows that the bilateral symmetry you see in the individual panels is characteristic of all times. Notice however that there is a significant departure from reciprocity on the one-second time slice around midpoint 120.

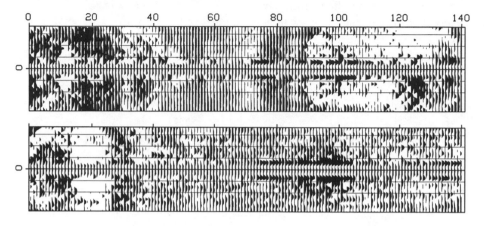

FIG. 3.3-3. Constant time slices at 1 second and 2.5 seconds.

In the laboratory, reciprocity can be established to within the accuracy of measurement. This can be excellent. (See White's example in FGDP). In the field, the validity of reciprocity will be dependent on the degree that the required conditions are fulfilled. A marine air gun should be reciprocal to a hydrophone. A land-surface weight-drop source should be reciprocal to a vertical geophone. But a buried explosive shot need not be reciprocal to a surface vertical geophone because the radiation patterns are different and the positions are slightly different. Fenati and Rocca [1984] studied reciprocity under varying field conditions. They reported that small positioning errors in the placement of sources and receivers can easily create discrepancies larger than the apparent reciprocity discrepancy. They also reported that theoretically reciprocal experiments may actually be less reciprocal than presumably nonreciprocal experiments.

Geometrical complexity within the earth does not diminish the applicability of the principle of linearity. Likewise, geometrical complexity does not reduce the applicability of reciprocity. Reciprocity does not apply to sound waves in the presence of wind. Sound goes slower upwind than downwind. But this effect of wind is much less than the mundane irregularities of field work. Just the weakening of echoes with time leaves noises that are not reciprocal. Henceforth we will presume that reciprocity is generally applicable to the analysis of reflection seismic data.

The Survey-Sinking Concept

The exploding-reflector concept has great utility because it enables us to associate the seismic waves observed at zero offset in many experiments (say 1000 shot points) with the wave of a single thought experiment, the exploding-reflector experiment. The exploding-reflector analogy has a few tolerable limitations connected with lateral velocity variations and multiple reflections, and one major limitation: it gives us no clue as to how to migrate data recorded at nonzero offset. A broader imaging concept is needed.

Start from field data where a survey line has been run along the x-axis. Assume there has been an infinite number of experiments, a single experiment consisting of placing a point source or shot at s on the x-axis and recording echoes with geophones at each possible location g on the x-axis. So the observed data is an upcoming wave that is a two-dimensional function of s and g, say $P(s,g,t)$. (Relevant practical questions about the actual spacing and extent of shots and geophones are deferred until Sections 3.6 and 4.3).

Previous chapters have shown how to downward continue the *upcoming* wave. Downward continuation of the upcoming wave is really the same thing as downward continuation of the geophones. It is irrelevant for the continuation procedures where the wave originates. It could begin from an exploding reflector, or it could begin at the surface, go down, and then be reflected back upward.

To apply the imaging concept of survey sinking, it is necessary to downward continue the sources as well as the geophones. We already know how to downward continue geophones. Since reciprocity permits interchanging geophones with shots, we really know how to downward continue shots too.

Shots and geophones may be downward continued to different levels, and they may be at different levels during the process, but for the final result they are only required to be at the same level. That is, taking z_s to be the depth of the shots and z_g to be the depth of the geophones, the downward-continued survey will be required at all levels $z = z_s = z_g$.

The image of a reflector at (x,z) is defined to be the strength and polarity of the echo seen by the closest possible source-geophone pair. Taking the mathematical limit, this closest pair is a source and geophone located together on the reflector. The travel time for the echo is zero. This survey-sinking concept of imaging is summarized by

$$Image\,(x\,,z\,) \;\;=\;\; Wave\,(s\!=\!x\,,\,g\!=\!x\,,\,z\,,\,t\!=\!0) \tag{1}$$

For good quality data, i.e. data that fits the assumptions of the downward-continuation method, energy should migrate to zero offset at zero travel time. Study of the energy that doesn't do so should enable improvement of the model. Model improvement usually amounts to improving the spatial distribution of velocity.

Review of the Paraxial Wave Equation

In Section 1.5 an equation was derived for paraxial waves. The assumption of a *single* plane wave means that the arrival time of the wave is given by a single-valued $t\,(x\,,z\,)$. On a plane of constant z, such as the earth's surface, Snell's parameter p is measurable. It is

$$\frac{\partial t}{\partial x} \;\;=\;\; \frac{\sin\theta}{v} \;\;=\;\; p \tag{2a}$$

In a borehole there is the constraint that measurements must be made at a constant x, where the relevant measurement from an *upcoming* wave would be

$$\frac{\partial t}{\partial z} \;\;=\;\; -\,\frac{\cos\theta}{v} \;\;=\;\; -\left[\frac{1}{v^{\,2}} - \left(\frac{\partial t}{\partial x}\right)^{2}\right]^{1/2} \tag{2b}$$

Recall the time-shifting partial-differential equation and its solution U as some arbitrary functional form f :

$$\frac{\partial U}{\partial z} \;\;=\;\; -\,\frac{\partial t}{\partial z}\,\frac{\partial U}{\partial t} \tag{3a}$$

$$U \;\;=\;\; f\!\left(t - \int_{0}^{z}\frac{\partial t}{\partial z}\,dz\right) \tag{3b}$$

The partial derivatives in equation (3a) are taken to be at constant x, just as is equation (2b). After inserting (2b) into (3a) we have

$$\frac{\partial U}{\partial z} \;\;=\;\; \left[\frac{1}{v^{\,2}} - \left(\frac{\partial t}{\partial x}\right)^{2}\right]^{1/2}\frac{\partial U}{\partial t} \tag{4a}$$

Fourier transforming the wavefield over $(x\,,t)$, we replace $\partial/\partial t$ by $-\,i\omega$. Likewise, for the traveling wave of the Fourier kernel $\exp(-\,i\omega t + ik_{x}\,x\,)$, constant phase means that $\partial t\,/\partial x = k_{x}\,/\omega$. With this, (4a) becomes

$$\frac{\partial U}{\partial z} \;\;=\;\; -\,i\omega\left[\frac{1}{v^{\,2}} - \frac{k_{x}^{\,2}}{\omega^{2}}\right]^{1/2} U \tag{4b}$$

The solutions to (4b) agree with those to the scalar wave equation unless v

is a function of z, in which case the scalar wave equation has both upcoming and downgoing solutions, whereas (4b) has only upcoming solutions. Chapter 2 taught us how to go into the lateral space domain by replacing ik_x by $\partial/\partial x$. The resulting equation is useful for superpositions of many local plane waves and for lateral velocity variations $v(x)$.

The DSR Equation in Shot-Geophone Space

Let the geophones descend a distance dz_g into the earth. The change of the travel time of the observed upcoming wave will be

$$\frac{\partial t}{\partial z_g} = -\left[\frac{1}{v^2} - \left(\frac{\partial t}{\partial g}\right)^2\right]^{1/2} \tag{5a}$$

Suppose the shots had been let off at depth dz_s instead of at $z=0$. Likewise then,

$$\frac{\partial t}{\partial z_s} = -\left[\frac{1}{v^2} - \left(\frac{\partial t}{\partial s}\right)^2\right]^{1/2} \tag{5b}$$

Both (5a) and (5b) require minus signs because the travel time decreases as either geophones or shots move down.

Simultaneously downward project both the shots and geophones by an identical vertical amount $dz = dz_g = dz_s$. The travel-time change is the sum of (5a) and (5b), namely,

$$dt = \frac{\partial t}{\partial z_g} dz_g + \frac{\partial t}{\partial z_s} dz_s = \left(\frac{\partial t}{\partial z_g} + \frac{\partial t}{\partial z_s}\right) dz \tag{6}$$

or

$$\frac{\partial t}{\partial z} = -\left\{\left[\frac{1}{v^2} - \left(\frac{\partial t}{\partial g}\right)^2\right]^{1/2} + \left[\frac{1}{v^2} - \left(\frac{\partial t}{\partial s}\right)^2\right]^{1/2}\right\} \tag{7}$$

This expression for $\partial t/\partial z$ may be substituted into equation (3a):

$$\frac{\partial U}{\partial z} = +\left\{\left[\frac{1}{v^2} - \left(\frac{\partial t}{\partial g}\right)^2\right]^{1/2} + \left[\frac{1}{v^2} - \left(\frac{\partial t}{\partial s}\right)^2\right]^{1/2}\right\} \frac{\partial U}{\partial t} \tag{8}$$

Three-dimensional Fourier transformation converts upcoming wave data $u(t,s,g)$ to $U(\omega, k_s, k_g)$. Expressing equation (8) in Fourier space gives

$$\frac{\partial U}{\partial z} = -i\omega\left\{\left[\frac{1}{v^2} - \left(\frac{k_g}{\omega}\right)^2\right]^{1/2} + \left[\frac{1}{v^2} - \left(\frac{k_s}{\omega}\right)^2\right]^{1/2}\right\} U \tag{9}$$

Recall the origin of the two square roots in equation (9). One is the cosine of the arrival angle at the geophones divided by the velocity at the geophones. The other is the cosine of the takeoff angle at the shots divided by the velocity at the shots. With the wisdom of previous chapters we know how to go into the lateral space domain by replacing ik_g by $\partial/\partial g$ and ik_s by $\partial/\partial s$. To incorporate lateral velocity variation $v(x)$, the velocity at the shot location must be distinguished from the velocity at the geophone location. Thus,

$$\frac{\partial U}{\partial z} = \left\{ \left[\left(\frac{-i\omega}{v(g)} \right)^2 - \frac{\partial^2}{\partial g^2} \right]^{1/2} + \right.$$

$$\left. + \left[\left(\frac{-i\omega}{v(s)} \right)^2 - \frac{\partial^2}{\partial s^2} \right]^{1/2} \right\} U \qquad (10)$$

Equation (10) is known as the double-square-root (DSR) equation in shot-geophone space. It might be more descriptive to call it the survey-sinking equation since it pushes geophones and shots downward together. Recalling the section on splitting and full separation (Section 2.4) we realize that the two square-root operators are commutative ($v(s)$ commutes with $\partial/\partial g$), so it is completely equivalent to downward continue shots and geophones separately or together. This equation will produce waves for the rays that are found on zero-offset sections but are absent (Section 1.1) from the exploding-reflector model.

The DSR Equation in Midpoint-Offset Space

By converting the DSR equation to midpoint-offset space we will be able to identify the familiar zero-offset migration part along with corrections for offset. The transformation between (g, s) recording parameters and (y, h) interpretation parameters is

$$y = \frac{g + s}{2} \qquad (11a)$$

$$h = \frac{g - s}{2} \qquad (11b)$$

Travel time t may be parameterized in (g, s)-space or (y, h)-space. Differential relations for this conversion are given by the chain rule for derivatives:

$$\frac{\partial t}{\partial g} = \frac{\partial t}{\partial y}\frac{\partial y}{\partial g} + \frac{\partial t}{\partial h}\frac{\partial h}{\partial g} = \frac{1}{2}\left(\frac{\partial t}{\partial y} + \frac{\partial t}{\partial h} \right) \qquad (12a)$$

$$\frac{\partial t}{\partial s} = \frac{\partial t}{\partial y}\frac{\partial y}{\partial s} + \frac{\partial t}{\partial h}\frac{\partial h}{\partial s} = \frac{1}{2}\left(\frac{\partial t}{\partial y} - \frac{\partial t}{\partial h} \right) \qquad (12b)$$

Having seen how stepouts transform from shot-geophone space to midpoint-offset space, let us next see that spatial frequencies transform in much the same way. Clearly, data could be transformed from (s, g)-space to (y, h)-space with (11) and then Fourier transformed to (k_y, k_h)-space. The question is then, what form would the double-square-root equation (9) take in terms of the spatial frequencies (k_y, k_h)? Define the seismic data field in either coordinate system as

$$U(s, g) \quad = \quad U'(y, h) \tag{13}$$

This introduces a new mathematical function U' with the same physical meaning as U but, like a computer subroutine or function call, with a different subscript look-up procedure for (y, h) than for (s, g). Applying the chain rule for partial differentiation to (13) gives

$$\frac{\partial U}{\partial s} \quad = \quad \frac{\partial y}{\partial s} \frac{\partial U'}{\partial y} + \frac{\partial h}{\partial s} \frac{\partial U'}{\partial h} \tag{14a}$$

$$\frac{\partial U}{\partial g} \quad = \quad \frac{\partial y}{\partial g} \frac{\partial U'}{\partial y} + \frac{\partial h}{\partial g} \frac{\partial U'}{\partial h} \tag{14b}$$

and utilizing (11) gives

$$\frac{\partial U}{\partial s} \quad = \quad \frac{1}{2} \left[\frac{\partial U'}{\partial y} - \frac{\partial U'}{\partial h} \right] \tag{15a}$$

$$\frac{\partial U}{\partial g} \quad = \quad \frac{1}{2} \left[\frac{\partial U'}{\partial y} + \frac{\partial U'}{\partial h} \right] \tag{15b}$$

In Fourier transform space where $\partial/\partial x$ transforms to ik_x, equation (15), when i and $U = U'$ are cancelled, becomes

$$k_s \quad = \quad \frac{1}{2} (k_y - k_h) \tag{16a}$$

$$k_g \quad = \quad \frac{1}{2} (k_y + k_h) \tag{16b}$$

Equation (16) is a Fourier representation of (15). Substituting (16) into (9) achieves the main purpose of this section, which is to get the double-square-root migration equation into midpoint-offset coordinates:

$$\frac{\partial}{\partial z} U = -i\frac{\omega}{v} \left\{ \left[1 - \left(\frac{vk_y + vk_h}{2\,\omega} \right)^2 \right]^{1/2} + \right.$$

$$\left. + \left[1 - \left(\frac{vk_y - vk_h}{2\,\omega} \right)^2 \right]^{1/2} \right\} U \tag{17}$$

Equation (17) is the takeoff point for many kinds of common-midpoint seismogram analyses. Some convenient definitions that simplify its appearance are

$$G \quad = \quad \frac{v \; k_g}{\omega} \tag{18a}$$

$$S \quad = \quad \frac{v \; k_s}{\omega} \tag{18b}$$

$$Y \quad = \quad \frac{v \; k_y}{2 \, \omega} \tag{18c}$$

$$H \quad = \quad \frac{v \; k_h}{2 \, \omega} \tag{18d}$$

Chapter 1 showed that the quantity $v \, k_x / \omega$ can be interpreted as the angle of a wave. Thus the new definitions S and G are the sines of the takeoff angle and of the arrival angle of a ray. When these sines are at their limits of ± 1 they refer to the steepest possible slopes in $(s \, , \, t \,)$- or $(g \, , \, t \,)$-space. Likewise, Y may be interpreted as the dip of the data as seen on a seismic section. The quantity H refers to stepout observed on a common-midpoint gather. With these definitions (17) becomes slightly less cluttered:

$$\frac{\partial}{\partial z} \, U \quad = \quad - \frac{i \, \omega}{v} \, \left[\, \sqrt{1 - (Y + H)^2} \, + \, \sqrt{1 - (Y - H)^2} \, \right] \, U \tag{19}$$

Most present-day before-stack migration procedures can be interpreted through equation (19). Further analysis of it will explain the limitations of conventional processing procedures as well as suggest improvements in the procedures.

EXERCISE

1.   Adapt equation (17) to allow for a difference in velocity between the shot and the geophone.

3.4 The Meaning of the DSR Equation

The double-square-root equation contains most nonstatistical aspects of seismic data processing for petroleum prospecting. This equation, which was derived in the previous section, is not easy to understand because it is an operator in a four-dimensional space, namely, (z, s, g, t). We will approach it through various applications, each of which is like a picture in a space of lower dimension. In this section lateral velocity variation will be neglected (things are bad enough already!). Begin with

$$\frac{dU}{dz} = \frac{-i\omega}{v} \left(\sqrt{1 - G^2} + \sqrt{1 - S^2} \right) U \tag{1a}$$

$$\frac{dU}{dz} = \frac{-i\omega}{v} \left(\sqrt{1 - (Y+H)^2} + \sqrt{1 - (Y-H)^2} \right) U \tag{1b}$$

Zero-Offset Migration (H = 0)

One way to reduce the dimensionality of (1b) is simply to set $H = 0$. Then the two square roots become the same, so that they can be combined to give the familiar paraxial equation:

$$\frac{dU}{dz} = -i\omega \frac{2}{v} \sqrt{1 - \frac{v^2 k_y^2}{4\omega^2}} \, U \tag{2}$$

In both places in equation (2) where the rock velocity occurs, the rock velocity is divided by 2. Recall that the rock velocity needed to be halved in order for field data to correspond to the exploding-reflector model. So whatever we did by setting $H = 0$, gave us the same migration equation we used in Chapter 1. Setting $H = 0$ had the effect of making the survey-sinking concept functionally equivalent to the exploding-reflector concept.

Zero-Dip Stacking (Y = 0)

When dealing with the offset h it is common to assume that the earth is horizontally layered so that experimental results will be independent of the midpoint y. With such an earth the Fourier transform of all data over y will vanish except for $k_y = 0$, or, in other words, for $Y = 0$. The two square roots in (1) again become identical, and the resulting equation is once more the paraxial equation:

$$\frac{dU}{dz} = -i\omega \frac{2}{v} \sqrt{1 - \frac{v^2 k_h^2}{4\omega^2}} \, U \tag{3}$$

Using this equation to downward continue hyperboloids from the earth's surface, we find the hyperboloids shrinking with depth, until the correct depth where best focus occurs is reached. This is shown in figure 1.

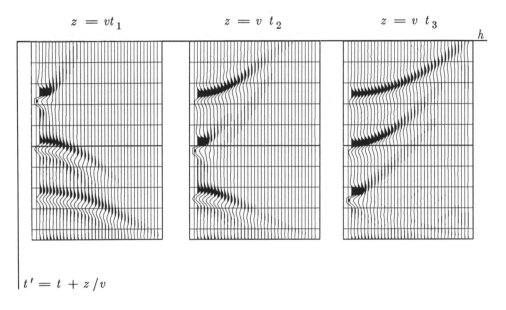

$$z = vt_1 \qquad z = v\,t_2 \qquad z = v\,t_3$$

h

$$t' = t + z/v$$

FIG. 3.4-1. With an earth model of three layers, the common-midpoint gathers are three hyperboloids. Successive frames show downward continuation to successive depths where best focus occurs.

The waves focus best at zero offset. The focus represents a downward-continued experiment, in which the downward continuation has gone just to a reflector. The reflection is strongest at zero travel time for a coincident source-receiver pair just above the reflector. Extracting the zero-offset value at $t = 0$ and abandoning the other offsets is a way of eliminating noise. (Actually it is a way of *defining* noise). Roughly it amounts to the same thing as the conventional procedure of summation along a hyperbolic trajectory on the original data. Naturally the summation can be expected to be best when the velocity used for downward continuation comes closest to the velocity of the earth. Later, offset space will be used to *determine* velocity.

Conventional Processing — the Separable Approximation

The *DSR* operator is now defined as the parenthesized operator in equation (1b):

$$DSR\,(Y,H) \;=\; \sqrt{1-(Y-H)^2} + \sqrt{1-(Y+H)^2} \qquad (4)$$

In Fourier space, downward continuation is done with the operator $\exp(i\,\omega v^{-1}\,DSR\;z)$.

There is a serious problem with this operator: it is *not separable* into a sum of an offset operator and a midpoint operator. *Nonseparable* means that a Taylor series for (4) contains terms like $Y^2 H^2$. Such terms cannot be expressed as a function of Y plus a function of H. Nonseparability is a data-processing disaster. It implies that migration and stacking must be done simultaneously, not sequentially. The only way to recover pure separability would be to return to the space of S and G. (That is a drastic alternative, far from conventional processing. We will return to it later).

Let us review the general issue of separability. The obvious way to get a separable approximation of the operator $\sqrt{1-X^2-Y^2}$ is to form a Taylor series expansion, and then drop all the cross terms. A more clever approximation is $\sqrt{1-X^2} + \sqrt{1-Y^2} - 1$, which fits all Y exactly when $X=0$ and all X exactly when $Y=0$. Applying this idea (though not the same equation) to the *DSR* operator gives

$$SEP\,(Y,H) \;=\; 2 + [DSR\,(Y,0)-2] + [DSR\,(0,H)-2] \qquad (5a)$$

$$SEP\,(Y,H) \;=\; 2\,[1 + (\sqrt{1-Y^2}-1) + (\sqrt{1-H^2}-1)] \qquad (5b)$$

Notice that at $H=0$ (5) becomes equal to the *DSR* operator. At $Y=0$ (5) also becomes equal to the *DSR* operator. Only when both H and Y are nonzero does *SEP* depart from *DSR*.

The splitting of (5) into a sum of three operators offers an advantage like the one offered by the 2-D Fourier kernel $\exp(ik_y\,y + ik_h\,h)$, which has a phase that is the sum of two parts. It means that Fourier integrals may have either y or h nested on the inside. So downward continuation with *SEP* could be done in (k_h, k_y)-space as implied by (1b), or we could choose to Fourier transform to (h, k_y), (k_h, y), or (y, h) by appropriate nesting operations.

It is convenient to give familiar names to the three terms in (5b). The first is associated with time-to-depth conversion, the second with migration, and the third with normal moveout.

$$SEP\,(Y,H) \;=\; TD + MIG\,(Y) + NMO\,(H) \qquad (5c)$$

The approximation (5) can be interpreted as "standard processing." The first stage in standard processing is NMO correction. In (5) the *NMO* operator downward continues all offsets at the earth's surface, to all offsets at depth. Selecting zero offset is no more than abandoning all other offsets. Like stacking over offset, selecting zero offset reduces the amount of data under consideration.

Ordinarily the abandoned offsets are not migrated. (Alternately, a clever procedure for changing stacking velocities after migration involves migrating several offsets near zero offset).

Since all terms in the *SEP* operator are interchangeable, it would seem wasteful to use it to migrate all offsets before stack. The result of doing so should be identical to after-stack migration.

Various Meanings of H = 0

Recall the various forms of the stepout operator:

Forms of stepout operator $2H/v$		
ray trace	*Fourier*	*PDE*
$\dfrac{dt}{dh}$	$\dfrac{k_h}{\omega}$	$\partial_h^{\,t} = \displaystyle\int_{-\infty}^{t} dt\, \dfrac{\partial}{\partial h}$

Reciprocity suggests that travel time is a symmetrical function of offset; thus dt/dh vanishes at $h = 0$. In that sense it seems appropriate to apply equation (2) to zero-offset sections. More precisely, the ray-trace expression dt/dh strictly applies only when a single plane wave is present. Spherical wavefronts are made from the superposition of plane waves. Then the Fourier interpretation of H is slightly different and more appropriate. To set $\omega = 0$ would be to select a zero frequency component, a simple integral of a seismic trace. To set $k_h = 0$ would be to select a zero spatial-frequency component, that is, an integration over offset. Conventional stacking may be defined as integration (or summation) over offset along a hyperbolic trajectory. Simply setting $k_h = 0$ is selecting a hyperbolic trajectory that is flat, namely, the hyperbola of infinite velocity. Such an integration will receive its major contribution from the top of the data hyperboloid, where the data events come tangent to the horizontal line of integration. (For some historical reason, such a data summation is often called *vertical stack).* Of the total contribution to the integral, most comes from a zone near the top, before the stepout equals a half-wavelength. The width of this zone, which is called a Fresnel zone, is the major factor contributing to the integral. The Fresnel zone concept was introduced in Section 1.2. The Fresnel zone has been extracted from a field profile in figure 2.

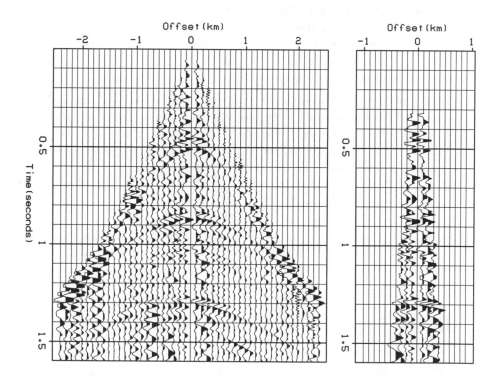

FIG. 3.4-2. (left) A land profile from Denmark (Western Geophysical) with the Fresnel zone extracted and redisplayed (right).

The definition of the Fresnel zone involves a frequency. For practical purposes we may just look at zero crossings. Examining figure 2 near one second we see a variety of frequencies. In the interval between $t=1.0$ and $t=1.1$ I see about two wavelengths of low frequencies and about 5 wavelengths of high frequencies. The highest frequencies are the main concern, because they define the limit of seismic resolution. The higher frequency has about 100 half wavelengths between time zero and a time of one second. As a rough generality, this observed value of 100 applies to all travel times. That is, at any travel time, the highest frequency that has meaningful spatial correlations is often observed to have a half period of about 1/100 of the total travel time. We may say that the quality factor Q of the earth's sedimentary crust is often about 100. So the angle that we are typically thinking about is $\cos 8° = .99$.

Theoretically, the main differences between a zero-offset section and a vertical stack are the amplitude and a small phase shift. In practical cases they are unlikely to migrate in a significantly different way. It would be nice if we could find an equation to downward continue data that is stacked at velocities other than infinite velocity.

The partial-differential-equation point of view of setting $H = 0$ is identical with the Fourier view when the velocity is a constant function of the horizontal coordinate; but otherwise the PDE viewpoint is a slightly more general one. To be specific, but not cluttered, equation (1) can be expressed in 15-degree, retarded, space-domain form. Thus,

$$\left[\frac{\partial}{\partial z} + \frac{v}{-i\,\omega\,8} \left(\frac{\partial^2}{\partial y^2} + \frac{\partial^2}{\partial h^2} \right) \right] U' = 0 \tag{6}$$

Integrate this equation over offset h. The integral commutes with the differential operators. Recall that the integral of a derivative is the difference between the function evaluated at the upper limit and the function evaluated at the lower limit. Thus,

$$\left(\frac{\partial}{\partial z} + \frac{v}{-i\,\omega\,8} \frac{\partial^2}{\partial y^2} \right) \left(\int U\,dh \right) + \frac{v}{-i\,\omega\,8} \frac{\partial U}{\partial h} \Big|_{h=-\infty}^{h=+\infty} = 0 \tag{7a}$$

The wave should vanish at infinite offset and so should its horizontal offset derivative. Thus the last term in (7a) should vanish. So, setting $H = 0$ has the meaning

$$(Paraxial\ operator)\ (vertical\ stack) = 0 \tag{7b}$$

A problem in the development of (7b) was that, twice, it was assumed that velocity is independent of offset: first, when the thin-lens term was omitted from (6), and second, when the offset integration operator was interchanged with multiplication by velocity. If the velocity depends on the horizontal x-axis, then it certainly depends on both midpoint and offset. In conclusion: If velocity changes slowly across a Fresnel zone, then setting $H = 0$ provides a valid equation for downward continuation of vertically stacked data.

Clayton's Cosine Corrections

A tendency exists to associate the sine of the earth dip angle with Y and the sine of the shot-geophone offset angle with H. While this is roughly valid, there is an important correction. Consider the dipping bed shown in figure 3.

FIG. 3.4-3. Geometry of a dipping bed. Note that the line
bisecting the angle 2β does not
pass through the midpoint
between g and s. (Clayton)

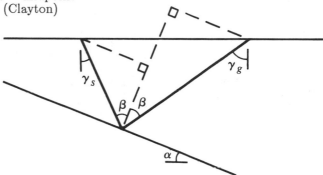

The dip angle of the reflector is α, and the offset is expressed as the offset angle β. Clayton showed, and it will be verified, that

$$Y = \sin \alpha \, \cos \beta \tag{8a}$$

$$H = \sin \beta \, \cos \alpha \tag{8b}$$

For small positive or negative angles the cosines can be ignored, and it is then correct to associate the sine of the earth dip angle with Y and the sine of the offset angle with H. At moderate angles the cosine correction is required. At angles exceeding 45° the sensitivities reverse, and conventional wisdom is exactly opposite to the truth. The reader should be wary of informal discussions that simply associate Y with dip and H with velocity. "Larner's streaks" in Section 3.2 were an example of mixing the effects of dip and offset. Indeed, at steep dips the usual procedure of using H to determine velocity should be changed somehow to use Y.

Next, (8) will be proven. The source takeoff angle is γ_s, and the incident receiver angle is γ_g. First, relate γ_s and γ_g to α and β. Adding up the angles of the smaller constructed triangle gives

$$(\frac{\pi}{2} - \gamma_s - \alpha) + \beta + \frac{\pi}{2} = \pi$$

$$\gamma_s = \beta - \alpha \tag{9a}$$

Adding up the angles around the larger triangle gives

$$\gamma_g \;=\; \beta + \alpha \qquad\qquad (9b)$$

To associate the angles at depth, α and β, with the stepouts dt/ds and dt/dg at the earth's surface requires taking care with the signs, noting that travel time increases as the geophone moves right and decreases as the shot moves right. Recall from Section 3.3 equations (16) and (18), the definitions of apparent angles Y and H,

$$Y - H \;=\; S \;=\; \frac{v\,k_s}{\omega} \;=\; v\,\frac{dt}{ds} \;=\; -\sin\gamma_s \;=\; \sin(\alpha - \beta)$$

$$Y + H \;=\; G \;=\; \frac{v\,k_g}{\omega} \;=\; v\,\frac{dt}{dg} \;=\; +\sin\gamma_g \;=\; \sin(\alpha + \beta)$$

Adding and subtracting this pair of equations and using the angle sum formula from trigonometry gives Clayton's cosine corrections (8):

$$Y \;=\; \frac{1}{2}\sin(\alpha + \beta) + \frac{1}{2}\sin(\alpha - \beta) \;=\; \sin\alpha \;\cos\beta$$

$$H \;=\; \frac{1}{2}\sin(\alpha + \beta) - \frac{1}{2}\sin(\alpha - \beta) \;=\; \sin\beta \;\cos\alpha$$

Snell-Wave Stacks and CMP Slant Stacks

Setting the takeoff angle S to zero also reduces the double-square-root equation to a single-square-root equation. The meaning of $S = 0$ is that $k_s = 0$ or equivalently that the data should undergo a summation (without time shifting) over shot s. Such a summation simulates a downgoing plane wave. The imaging principle behind the summation would be to look at the upcoming wave at the arrival time of the downgoing wave (Section 5.7). As explained further in Sections 5.2 and 5.3, S could also be set equal a constant, to simulate a downgoing Snell wave.

A Snell wave is a generalization of a downgoing plane wave at nonvertical incidence. The shots are not fired simultaneously, but sequentially at an inverse rate of $dt/ds = S/v$. This could be simulated with field data by summing across the (t, s)-plane along a line of slope dt/ds. Setting S to be some constant, say $S = v\, dt/ds$, also reduces the double-square-root equation to a paraxial wave equation, just the equation needed to downward continue the downgoing Snell wave experiment. Snell waves could be constructed for various $p = dt/ds$ values. Each could be migrated and imaged, and the images stacked over p. These ideas have been around longer than the DSR equation, yet they have gained no popularity. What could be the reason?

A problem with Snell wave simulation is that the wavefield is usually sampled at coarse intervals along a geophone cable, which itself never seems to extend as far as the waves propagate. Crafty techniques to interpolate and extrapolate the data are frustrated by the fact that on a common-geophone gather, the top of the hyperbola need not be at zero offset. For dipping beds the earliest arrival is often off the end of the cable. So the data processing depends strongly on the missing data.

These difficulties provide an ecological niche for the common-midpoint slant stack, namely, $H = p\ v$. (A fuller explanation of slant stack is in Section 5.2). At common *midpoint* the hyperbolas go through zero offset with zero slope. The data are thus more amenable to the interpolation and extrapolation required for integration over a slanted line. Setting $H = p\ v$ yields

$$k_z = -\frac{\omega}{v}\left[\sqrt{1 - (Y + pv)^2} + \sqrt{1 - (Y - pv)^2}\right] \tag{10}$$

This has not reduced the DSR equation to a paraxial wave equation, but it has reduced the problem to a form manageable with the available techniques, such as the Stolt or phase-shift methods. Details of this approach can be found in the dissertation of Richard Ottolini [1982].

Why Not Downward Continue in (S,G)-Space?

If the velocity were known and the only task were to migrate, then there would be no fundamental reason why the downward continuation could not be done in (S, G)-space. But the velocity really isn't well known. The sensitivity of migration to velocity error increases rapidly with angle, and angle accuracy is the presumed advantage of (S, G)-space. Furthermore, the finite extent of the recording cable and the tendency to spatial aliasing create the same problems with (S, G)-space migration as are experienced with Snell stacks. I see no fundamental reason why (S, G)-space migration should be any better than CMP slant stacks, and the aliasing and truncation situations seem likely to be worse. Less ambitious and more practical approaches to the wide-angle migration problem are found later in this chapter.

On the other hand, lateral velocity variation (if known) could demand that migration be done in (s, g)-space.

Still another reason to enter shot-geophone space would be that the shots were far from one another. Then the data would be aliased in both midpoint space and offset space. See Section 5.7.

3.5 Stacking and Velocity Analysis

Hyperbolic stacking over offset may be the most important computer process in the prospecting industry. It is more important than migration because
it reduces the data base from a volume in (s, g, t)-space to a plane in
(y, t)-space. At the present time few people who interpret seismic data have
computerized seismic data movies, so most interpreters must have their data
stacked before they can even look at it. Migration merely converts one plane
to another plane. Furthermore, migration has the disadvantage that it sometimes compounds the mess made by near-surface lateral velocity variation and
multiple reflections. Stacking can compound the mess too, but in bad areas
nothing can be seen until the data is stacked. In addition to its other drawing
points, stacking gives as a byproduct estimates of rock velocity.

Historically, stacking has been done using ray methods, and it is still
being done almost exclusively in this way. Migration, on the other hand, is
more often done using wave-equation methods, that is to say, by Fourier or
finite-difference methods.   Both migration and stacking are hyperbola-
recognition processes. The advantages of wave-equation methods in migration
have been many. Shouldn't these advantages apply equally to stacking? It
would seem so, but current industrial practice does not bear this out. The
reasons are not yet clear. So the latter part of this section really belongs to a
research monograph with the facetious title "Theory That Should Work Out
Soon." More advanced ideas of velocity estimation are in Sections 5.0-5.4.
Wave-equation stacking and velocity-determination methods are ingenious.
Perhaps they have not yet been satisfactorily tested, or perhaps they are just
imperfectly assembled. The reader can guess, and time will tell.

One possible reason why much of this theory is not in routine industrial
use is that the issue of stacking to remove redundancy may be more appropriately a statistical problem than a physical one. To allow for this contingency
I have included a bit on "wave-equation moveout," a way of deferring statistical analysis until after downward continuation. Another possibility is that
the problems of missing data off the ends of the recording cable and spatial
aliasing within the cable may be more flexibly attacked by ray methods than
by wave-equation methods. For this contingency I have included a brief subsection on data restoration. Whatever the case, the data-manipulation procedures in this chapter should be helpful.

Normal Moveout (NMO)

Normal moveout correction (NMO) is a stretching of the time axis to make all seismograms look like zero-offset seismograms. NMO was first discussed in Section 3.0. In its simplest form, NMO is based on the Pythagorean relation $t_{NMO}^2 = t^2 - x^2/v^2$. In a constant velocity earth, the NMO correction would take the asymptote of the hyperbola family and move it up to $t = 0$. This abandons anything on the time axis before the first arrival, and stretches the remainder of the seismogram. The stretching is most severe near the first arrival, and diminishes at later times. In the NMO example in figure 1 you will notice the low frequencies caused by the stretch.

NMO correction may be done to common-shot field profiles or to CMP gathers. NMO applied to a field profile makes it resemble a small portion of a zero-offset section. Then geologic structure is prominently exhibited. NMO on a CMP gather is the principal means of determining the earth's velocity-depth function. This is because CMP gathers are insensitive to earth dip.

Mathematically, the NMO transformation is a *linear* operation. It may seem paradoxical that a non-uniform axis-stretching operation is a linear operation, but axis stretching does satisfy the mathematical conditions of linearity. Do not confuse the widespread linearity condition with the less common condition of time invariance. Linearity requires only that for any decomposition of the original data P into parts (say P_1 and P_2) the sum of the NMOed parts is equal the NMO of the sum. Examples of decompositions include: (1) separation into early times and late times, (2) separation into even and odd time points, (3) separation into high frequencies and low frequencies, and (4) separation into big signal values and small ones.

To envision NMO as a linear operator, think of a seismogram as a vector. The NMO operator resembles a diagonal matrix, but the matrix contains *interpolation filters* along its diagonal, and the interpolation filters are shifted off from the diagonal to create the desired time delay.

Conventional Velocity Analysis

A conventional velocity analysis uses a collection of trial velocities. Each trial velocity is taken to be a constant function of depth and is used to moveout correct the data. Figure 2 (left) exhibits the CMP gather of figure 1 (left) after moveout correction by a constant velocity. Notice that the events in the middle of the gather are nearly flattened, whereas the early events are undercorrected and later events are overcorrected. This is typical because the amount of moveout correction varies inversely with velocity (by Pythagoras), and the earth's velocity normally increases with depth. A measure of the goodness of fit of the NMO velocity to the earth velocity is found by summing

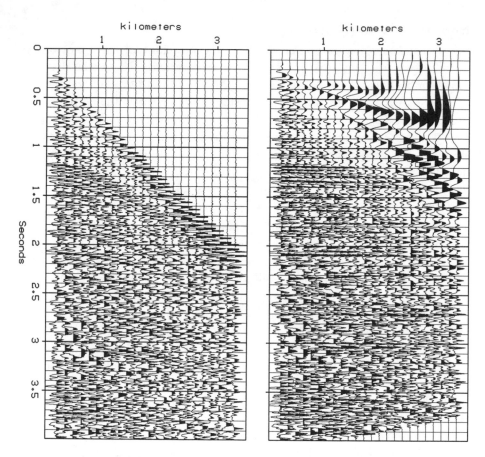

FIG. 3.5-1. CMP gather (Western Geophysical) from the Gulf coast shown at the left was NMO corrected and displayed at the right.

the CDP gather over offset. Presumably, the better the velocities match, the better (bigger) will be the sum. The process is repeated for many velocities. The amplitude of the sum, contoured as a function of time and velocity, is shown in figure 2 (right).

In practice additional steps may be taken before summing. The traces may be balanced (scaled to be equal) in their powers and in their spectra (see *deconvolution* in Section 5.5). Likewise the amplitude of the sum may be normalized and smoothed. (See Taner and Koehler [1969]). Also the data may be *edited* and *weighted* as explained in the next subsection.

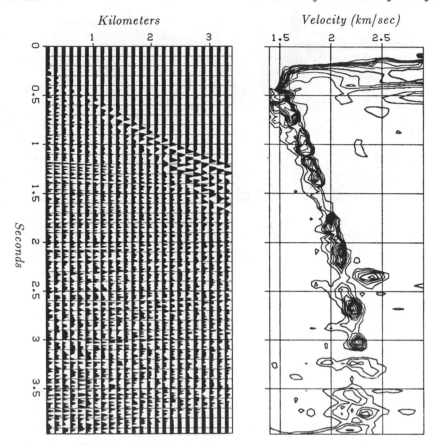

FIG. 3.5-2. NMO at constant velocity with velocity analysis. (Hale)

The velocity giving the best stack is an average of the earth's velocity above the reflector. The precise definition of this average is deferred till Section 5.4.

Mutes and Weights

An important part of conventional processing is the definition of a *mute*. A mute is a weighting function used to suppress some undesirable portions of the data. Figure 3 shows an example of a muted field profile. Weights and mutes have a substantial effect on the quality of a stack. So it is not surprising that in practice, they are the subject of much theorizing and experimentation.

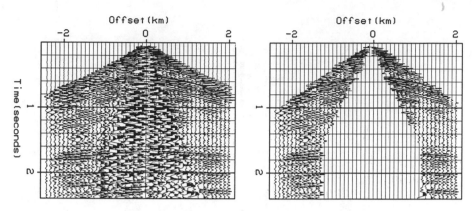

FIG. 3.5-3. Left is a land profile from Alberta (Western Geophysical). On the right it is muted to remove ground roll (at center) and head waves (the first arrivals).

Often the mute is a one-dimensional function of $r = h/t$. Reasons can be given to mute data at both large and small values of r.

At small values of r, energy is found that remains near the shot, such as falling dirt or water or slow ground roll.

At large values of r, there are problems with the first arrival. Here the NMO stretch is largest and most sensitive to the presumed velocity. The first arrival is often called a *head wave* or *refraction*. Experimentally, a head wave is a wave whose travel time appears to be a linear function of distance. Theoretically, a head wave is readily defined for layered media. The head wave has a ray that propagates horizontally along a layer boundary. In practice, a head wave may be weaker or stronger than the reflections. A strong head wave may be explained by the fact that reflected waves spread in three dimensions, while head waves spread in only two dimensions.

Muting may be regarded as weighting by zero. More general weights may be chosen to produce the most favorable CDP stack. A sophisticated analysis would certainly include noise and truncation. Let us do a simplified analysis. It leads to the most basic weighting function.

Ordinarily we integrate over offset along a hyperbola. Instead, think of the *three*-dimensional problem. You really wish to integrate over a hyperbola of revolution. Assume that the hyperboloid is radially symmetric. Weighting the integrand by h allows the usual line integral to simulate integration over the hyperboloid of revolution. A second justification for scaling data by offset h before stacking is that there is less velocity information near zero offset, where there is little moveout, and more velocity information at wider offset where $\Delta t / \Delta h$ is larger.

NMO Equations

The earth's velocity typically ranges over a factor of two or more within the depth range of a given data set. Thus the Pythagorean analysis needs reexamination. In practice, depth variable velocity is often handled by inserting a time variable velocity into the Pythagorean relation. (The classic reference, Taner and Koehler [1969], includes many helpful details). This approximation is much used, although it is not difficult to compute the correct nonhyperbolic moveout. Let us see how the velocity function $v(z)$ is mathematically related to the NMO. Ignoring dip, NMO converts common-midpoint gathers, one of which, say, is denoted by $P(h, t)$, to an earth model, say,

$$Q(h, z) \quad = \quad earth(z) \times const(h) \tag{1}$$

Actually, $Q(h, z)$ doesn't turn out to be a constant function of h, but that is the goal.

The NMO procedure can be regarded as a simple copying. Conceptually, it is easy to think of copying every point of the (h, t)-plane to its appropriate place in the (h, z)-plane. Such a copying process could be denoted as

$$Q[h, z(h, t)] \quad = \quad P(h, t) \tag{2}$$

Care must be taken to avoid leaving holes in the (h, z)-plane. It is better to scan every point in the output (h, z)-plane and find its source in the (h, t)-plane. With a table $t(h, z)$, data can be moveout corrected by the copying operation

$$Q(h, z) \quad = \quad P[h, t(h, z)] \tag{3}$$

Using the terminology of this book, the input $P(h, t)$ to the moveout correction is called a CMP gather, and the output Q is called a CDP gather.

In practice, the first step in generating the travel-time tables is to change the depth-variable z to a vertical travel-time-variable τ. So the required table is $t(h, \tau)$. To get the output data for location (h, τ) you take the input data at location (h, t). The most straightforward and reliable way to produce this table seems to be to march down in steps of z, really τ, and trace rays. That is, for various fixed values of Snell's parameter p, you compute $t(p, \tau)$ and $h(p, \tau)$ from $v(\tau)$ by integrating the following equations over τ:

$$\frac{dt}{d\tau} = \frac{dz}{d\tau}\frac{dt}{dz} = v\,\frac{1}{v\,\cos\theta} = \frac{1}{\sqrt{1 - p^2\,v(\tau)^2}} \tag{4}$$

$$\frac{dh}{d\tau} = \frac{dz}{d\tau}\frac{dh}{dz} = v\,\tan\theta = \frac{p\,v(\tau)^2}{\sqrt{1 - p^2\,v(\tau)^2}} \tag{5}\cdot$$

(In equations (4) and (5) dt/dz and dh/dz are based on rays, not wavefronts). Given $t(p,\tau)$ and $h(p,\tau)$, iteration and interpolation are required to eliminate p and find $t(h,\tau)$. It sounds awkward — and it is — because at wide angles there usually are head waves arriving in the middle of the reflections. But once the job is done you can save the table and reuse it many times. The multibranching of the travel time curves at wide offset motivates a wave-equation based velocity analysis. The greatest velocity sensitivity occurs just where the classic hyperbolic assumption and the single-arrival assumption break down.

Linearity Allows Postponing Statistical Estimation

The linearity of wave-equation data processing allows us to decompose a dataset into parts, process the parts separately, then recombine them. The result is the same as if they were never separated.

For example, suppose a CMP gather is divided into two parts, say, inner traces A and outer traces B. Let $(A,0)$ denote a CMP gather where the *outer* traces have been replaced by zeroes. Likewise, $(0,B)$ could be another copy of the gather where the *inner* traces have been replaced by zeroes. We could downward continue $(A,0)$ and separately downward continue $(0,B)$. After downward continuation, $(A,0)$ and $(0,B)$ could be added. Alternately, we could pause, do some thinking about statistics, and then choose to combine them with some weighting function. Figure 4 shows a dataset of three traces decomposed into three datasets, one for each trace. Semicircles depict the separate downward continuation of each trace. Each semicircle goes through zero offset, giving the appropriately stretched, NMOed trace.

The idea of using a weighting function is a drastic departure from our previous style of analysis. It represents a disturbing recognition that we have been neglecting something important in all scientific analysis, namely, statistics! What are the ingredients that go into the choice of a weighting function? They are many. Signal and noise variances play a role. Some channels may be noisy or absent. When final display is contemplated, it is necessary to consider human perception and the need to compress the dynamic range, so

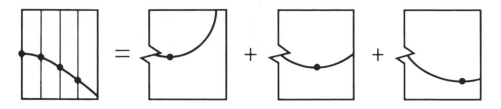

FIG. 3.5-4. A three-trace CMP gather decomposed by traces. At the left, impulses on the data are interpolated, depicting a hyperbola. At the right, data points are expanded into migration semicircles, each of which goes through zero offset at the apex of the hyperbola.

that small values can be perceived. Dynamic-range compression must be considered not only in the obvious (h, t)-space, but also in frequency space, dip space, or any other space in which the wavefields may get too far out of balance.

There are many ways to decompose a dataset. The choice depends on your statistical model and your willingness to repeat the processing many times. Perhaps the parts of the data gather should be decomposed not by their h values but by their values of $r = h/t$. Clearly, there is a lot to think about.

Lateral Interpolation and Extrapolation of a CMP Gather

Practical problems dealing with common-midpoint gathers arise because of an insufficient number of traces. *Truncation* problems are those that arise because the geophone cable has a fixed length that is not as long as the distance over which seismic energy propagates. Figure 5 shows why cable truncations are a problem for conventional, ray-trace, stacking methods as well as for wave-equation methods. *Aliasing* problems are those that arise because shots and geophones are not close enough together. Spatial aliasing of data on the offset axis seems to be a more serious problem for wave-equation methods than it is for ray-trace methods. The reason is that normal-moveout correction reduces the spatial frequencies. *Gaps* in the data, resulting from practical problems with the geophones, cable, and access to the terrain, are also frequently a snag.

Here these problems will all be attacked together with a systematic approach to estimating missing traces. The technique to be described is the simplest member of a more general family of *missing data* estimation procedures currently being developed at the Stanford Exploration Project.

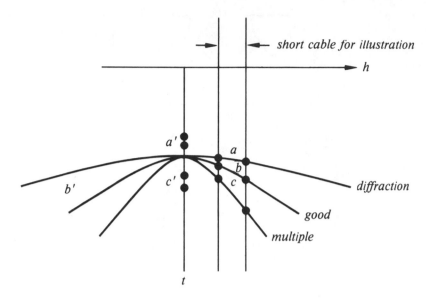

FIG. 3.5-5. Normal moveout at the earth velocity brings the cable truncations on good events to a good place, causing no problems. The cable truncations of diffractions and multiples, however, move to *a′* and *c′*, where they
could be objectionable. Such corruption could make folly of sophisticated
time-series analysis of the waveform found on a CDP stack.

First do normal-moveout correction, that is, stretch the time axis to
flatten hyperbolas. The initial question is what velocity to use for the
normal-moveout correction. For trace interpolation the appropriate moveout
velocity turns out to be that of the *dominating energy on the gather*. On a
given dataset this velocity could be primary velocity at some times and multiple velocity at other times. The reason for such a nonphysical velocity is this:
the strong events must be handled well, in order to save the weak ones.
Truncations of weak events can be ignored as a "second-order" problem. The
practical problem is usually to suppress strong water-velocity events in the
presence of weak sedimentary reflections, particularly at high frequencies. In
principle, we might be seeking weak P–SV waves in the presence of strong
P–P waves.

After NMO, the residual energy should have little dip, except of course
where missing data, now replaced by zeroes, forces the existing data to be
broad-banded in spatial frequency. In order to improve our view of this badly
behaved energy, we pass the data through a "badpass" filter, such as the
high-pass recursive dip filter discussed in Section 2.5.

$$\frac{\dfrac{k^2}{-i\omega}}{\alpha + \dfrac{k^2}{-i\omega}} \tag{6}$$

Notice that this filter greatly weakens the energy with small k, that is, the energy that was properly moveout corrected. On the other hand, near the missing traces, notice that the spectrum should be broad-band with k and that such energy passes through the filter with almost unit gain.

The output from the "badpass" filter is now ready to be subtracted from the data. The subtraction is done selectively. Where recorded data exists, nothing is subtracted. This completes the first iteration. Next the steps are repeated, and iterated. Convergence is finally achieved when nothing comes out of the badpass filter at the locations where data was not recorded. An example of this process can be found in figure 6.

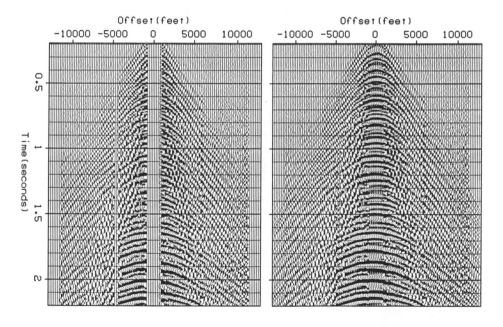

FIG. 3.5-6. Field profile from Alaska with missing channels on the left (Western Geophysical), restored by iterative spatial filtering on the right. (Harlan)

The above procedure has ignored the possibility of dip in the midpoint direction. The effect of dip on moveout is taken up in Section 3.6.

This procedure is also limited because it ignores the possibility that several velocities may be simultaneously present on a dataset. To really do a good job of extending such a dataset may require a parsimonious model and a velocity spectral concept such as the ones developed next and in Section 5.4.

In and Out of Velocity Space

Summing a common-midpoint gather on a hyperbolic trajectory over offset yields a stack called a *constant-velocity* or a *CV* stack. A velocity space may be defined as a family of CV stacks, one stack for each of many velocities. CV stacking is a transformation from *offset* space to *velocity* space. CV stacking creates a (t, v)-space velocity panel from a (t, h)-space common-midpoint gather. Conventional industrial velocity estimation amounts to CV stacking supplemented by squaring and normalizing. Linear transformations such as CV stack are generally *invertible,* but the transformation to velocity space is of very high dimension. Forty-eight channels and 1000 time points make the transformation 48,000-dimensional. With present computer technology, matrices this large cannot be inverted by algebraic means. However, there are some excellent approximate means of inversion.

For unitary matrices, the transpose matrix equals the inverse matrix. In wave-propagation theory, a *transpose* operator is often a good approximation to an *inverse* operator. Thorson [1984] pointed out that the transpose operation to CV stacking is just about the same thing as CV stacking itself. To do the operation transposed to CV stacking, begin with a velocity panel, that is, a panel in (t, v)-space. To create some given offset h, each trace in the (t, v)-panel must be first compressed to undo the original NMO stretch. That is, events must be pushed from the zero-offset time that they have in the (t, v)-panel to the time appropriate for the given h. Then stack the (t, v)-panel over v to produce the seismogram for the given h. Repeat the process for all desired values of h. The program for transpose CV stack is like the program for CV stack itself, except that the stretch formula is changed to a compensating compression.

The inversion of a CV stack is analogous to inversion of slant stack or Radon transformation (Section 5.2). That is, the CV stack is almost its own inverse, but you need to change a sign, and at the end, a filtering operation, like rho filtering, is also needed to touch up the spectrum, thereby finishing the job. It is the rho filtering that distinguishes *inverse* CV stack from the *transpose* of CV stack.

   The word *transpose* refers to *matrix transpose*. It is difficult to visualize why the word transpose is appropriate in this case because we are discussing data spaces that are two-dimensional and operators that are four-dimensional. But if you will map these two- and four-dimensional objects to familiar one- and two-dimensional objects by a transformation, such as equations (25) and (26) in Section 2.2, then you will see that the word *transpose* is entirely appropriate. The rho-type filtering required for CV stacks is slightly more complicated than ordinary rho filtering — refer to Thorson's thesis.

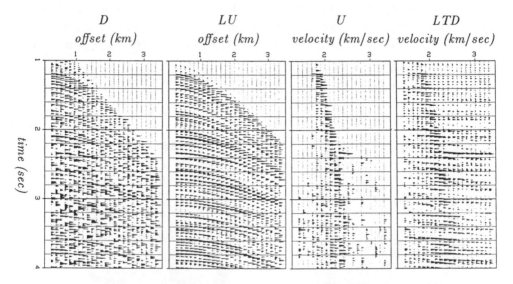

FIG. 3.5-7. Panel D at the left is a CMP gather from the Gulf of Mexico (Western Geophysical). The second panel (LU) is reconstructed data obtained from the third panel (U) by inverse NMO and stack. The last panel (LTD) is a CV stack of the first panel. (Thorson)

   Figure 7 shows an example of Thorson's velocity space inversion. Panel D is the original common-midpoint gather. Next is panel LU, the approximate reconstruction of D from velocity space. The hyperbolic events are reconstructed much better than the random noise. The random noise was not reconstructed so well because the range of velocities in the CV stack was limited between water velocity and 3.5 km/sec. The next two panels (U) and (LTD) are theoretically related by the "rho" filtering. LTD is the CV stack of D. LU is the transpose CV stack of U.

It is worth noting that there is a substantial amount of work in computing a velocity panel. A stack must be computed for each velocity. Velocity discrimination by wave-equation methods will be described next and in Section 5.4. The wave-equation methods are generally cheaper, though not fully comparable in effect.

The (z,t)-plane Method

In the 15° continuation equation $U_{zt} = -1/2\, v\ U_{hh}$, scaling the depth z is indistinguishable from scaling the velocity. Thus, downward continuation with the wrong velocity is like downward continuation to the wrong depth. Stephen M. Doherty [1975] used this idea in a velocity-estimation scheme — see figure 8.

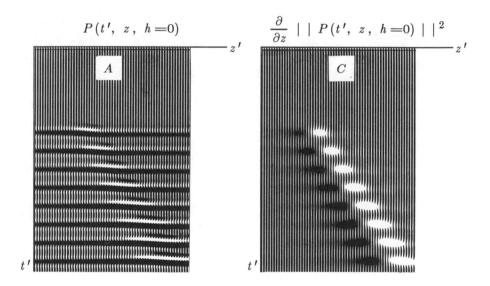

FIG. 3.5-8. Two displays of the (z, t)-plane at zero offset. The earth model is eight uniformly spaced reflectors under a water layer (a family of hyperboloids in (h, t) at $z = 0$). The left display is the zero-offset trace. The amplitude maximum at the focus is not visually striking, but the phase shift is apparent. The right display is the z-derivative of the envelope of the zero-offset trace. A linear alignment along $z' = vt'$ is more apparent. (Doherty)

The idea is to downward continue with a preliminary velocity model and to display the zero-offset trace, a function of t', at all travel-time depths τ. If the maximum amplitude occurs at $t' = \tau$, then your preliminary model is good. If the maximum is shifted, then you have some analysis to do before you can say what velocity should be used on the next iteration.

Splitting a Gather into High- and Low-Velocity Components

A process will be defined that can partition a CMP gather, *both reflections and head waves,* into one part with RMS velocity greater than that of some given model $\overline{v}(z)$ and another part with velocity less than $\overline{v}(z)$.

After such a partitioning, the low-velocity noise could be abandoned. Or the earth velocity could be found through iteration, by making the usual assumption that the velocity spectrum has a peak at earth velocity. As will be seen later, various data interpolation, lateral extrapolation, and other statistical procedures are also made possible by the linearity and invertibility of the partitioning of the data by velocity.

The procedure is simple. Begin with a common-midpoint gather, zero the negative offsets, and then downward continue according to the velocity model $\overline{v}(z)$. The components of the data with velocity less than $\overline{v}(z)$ will overmigrate through zero offset to negative offsets. The components of the data with velocity greater than $\overline{v}(z)$ will undermigrate. They will move toward zero offset but they will not go through. So the low-velocity part is at negative offset and the high-velocity part is at positive offset. If you wish, the process can then be reversed to bring the two parts back to the space of the original data.

Obviously, the process of multiplying data by a step function may create some undesirable diffractions, but then, you wouldn't expect to find an infinitely sharp velocity cutoff filter. Clearly, the false diffractions could be reduced by using a ramp instead of a step. An alternative to zeroing negative h would be to go into (k_h, ω)-space and zero the two quadrants of sign disagreement between k_h and ω.

This partitioning method unfortunately does not, by itself, provide a velocity spectrum. Energy away from $h = 0$ is unfocused and not obviously related to velocity. The need for a velocity spectrum motivates the development of other processes.

Reflected Head Waves on Sections

It is common for an interpreter looking at a stacked section to identify a reflected head wave. Experimentally it is just a hyperbolic asymptote seen in

(y, t)-space. Theoretically, it is a ray that moves away from a source along a horizontal interface until it encounters an irregularity, a fault perhaps, from which it reflects and returns toward the source. Reflected head waves are sometimes called reflected refractions. This event provides an easy velocity estimate, namely, $v = 2\, dy/dt$. From a processing point of view, such a velocity measurement is unexpected, because automatic processing extracts all velocity information in offset space, a space which many interpreters prefer to leave inside the computer. Of course, for a reflected head wave to be identified, a special geological circumstance must be present — a scatterer strong enough to have its hyperbolic asymptote visible. The point scatterer must also be strong enough to get through the typical suppression effects of shot and geophone patterns and CDP stacking. The most highly suppressed events, water velocity and ground roll, are just those whose velocities are most often apparent on stacked sections. (Recall Larner's streaks). Some strong reflected refraction energy was present on the common-shot profile shown in Section 3.2.

Velocity estimates made from reflections are averages of all the layers above the reflection point. To get depth resolution, it is necessary to subtract velocity estimates of different depth levels (Section 5.4). Because of the subtraction, accuracy is lost. So with reflected waves, there is naturally a trade-off between accuracy and depth resolution. On the other hand, velocity estimates from head waves naturally have a high resolution in depth.

Processing seems to ignore or discriminate against the backscattered head wave, yet it is often seen and used. There must be an explanation. Perhaps there is also a latent opportunity. From a theoretical point of view, Clayton's cosines showed that at wide angles the velocity and dip sensitivity of midpoint and offset are exchanged. At late times another factor becomes significant: the aperture of a cable length can be much less than the width of a migration hyperbola. So, although it is easy to find an asymptote in midpoint space, there is little time shift at the end of the cable in offset space.

What processing could take advantage of lateral reflectivity and could enhance, instead of suppress, our ability to determine velocity in this way? Start by stacking at a high velocity. Then use the idea that at any depth z, the power spectrum of the data $U(\omega, k_y)$ should have a cutoff at the evanescent stepout $p(z) = k_y/\omega = 1/v(z)$. This would show up in a plot of the power spectrum $U^* U$, or better yet of the dip spectrum, as a function of depth. Perhaps it would be still better to visually inspect the seismic section itself after filtering in dip space about the expected velocity.

The wave-extrapolation equation is an all-pass filter, so why does the power spectrum change with depth? It changes because at any depth z it is

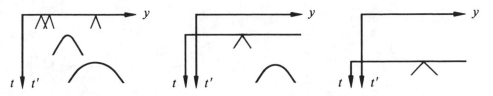

FIG. 3.5-9. The dip-spectrum method of velocity determination. To find the velocity at any depth, seek the steepest dip on the section at that depth. On the left, at the earth's surface, you see the surface ground roll. In frames B and C the slowest events are the asymptotes of successively faster hyperbolas.

necessary to exclude all the seismic data before $t=0$. This data should be zeroed before computing the dip spectrum. The procedure is depicted in figure 9.

To my knowledge this method has never been tried. I believe it is worth some serious testing. Even in the most layered of geological regions there are always faults and irregularities to illuminate the full available spectrum. Difficulty is unlikely to come from weak signals. More probably, the potential for difficulty lies in near-surface velocity irregularity.

EXERCISES

1. Assume that the data $P(y, h, t)$ is constant with midpoint y. Given a common-midpoint gather $P(h, t, z=0)$, define a Stolt-type integral transformation from $P(h, t, z=0)$ to $P(h=0, t, z)$ based on the double-square-root equation:

$$\frac{\partial}{\partial z} P = -i \frac{\omega}{v} \left[\sqrt{1 - (Y+H)^2} + \sqrt{1 - (Y-H)^2} \right] P$$

   As with Stolt migration, your answer should be expressed as a 2-D inverse Fourier transform.

2. Start with a CDP gather $u(h, t)$ defined (by reciprocity) at both positive and negative values of h. Describe the effect of the following operations: Fourier transform to $U(k_h, \omega)$; multiply by $1 + sgn(\omega) sgn(k_h)$; transform back to (h, t)-space.

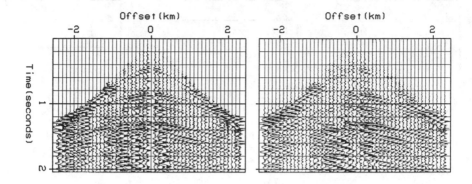

FIG. 3.5-E2. What is this?

3.6 Migration with Velocity Estimation

We often face the three complications dip, offset, and unknown velocity at the same time. The double-square-root equation provides an attractive avenue when the velocity is known, but when it isn't, we are left with velocity-estimation procedures, such as that in the previous section, *which assume no dip*. In this section a means will be developed of estimating velocity in the presence of dip.

Dip Moveout — Sherwood's Devilish

Recall (from Section 3.2) Levin's expression for the travel time of the reflection from a bed dipping at angle α from the horizontal:

$$t^2 v^2 = 4 (y - y_0)^2 \sin^2\alpha + 4 h^2 \cos^2\alpha \tag{1}$$

In (h, t)-space this curve is a hyperbola. Scaling the velocity by $\cos \alpha$ makes the travel-time curve identical to the travel-time curve of the dip-free case. This is the conventional approach to stacking and velocity analysis. It is often satisfactory. Sometimes it is unsatisfactory because the dip angle is not a single-valued function of space. For example, near a fault plane there will be diffractions. They are a superposition of all dips, each usually being weaker than the reflections. Many dips are present in the same place. They blur the velocity estimate and the stack.

In principle, migration before stack — some kind of implementation of the full DSR equation — solves this general problem. But where do we get the velocity to use in the migration equations? Although migration is some-what insensitive to velocity when only small angles are involved, migration becomes sensitive to velocity when wide angles are involved.

The migration process should be thought of as being interwoven with the velocity estimation process. J.W.C. Sherwood [1976] showed how the two processes, migration and velocity estimation, should be interwoven. The moveout correction should be considered in two parts, one depending on offset, the NMO, and the other depending on dip. This latter process was conceptually new. Sherwood described the process as a kind of filtering, but he did not provide implementation details. He called his process *Devilish*, an acronym for "dipping-event velocity inequalities licked." The process was later described more functionally by Yilmaz as *prestack partial migration*, but the process has finally come to be called simply *dip moveout* (DMO). We will first see Sherwood's results, then Rocca's conceptual model of the DMO pro-cess, and finally two conceptually distinct, quantitative specifications of the process.

Figure 1 contains a panel from a stacked section. The panel is shown several times; each time the stacking velocity is different. It should be noted that at the low velocities, the horizontal events dominate, whereas at the high velocities, the steeply dipping events dominate. After the *Devilish* correction was applied, the data was restacked as before. Figure 2 shows that the stack-ing velocity no longer depends on the dip. This means that after *Devilish*, the velocity may be determined without regard to dip. In other words, events with all dips contribute to the same consistent velocity rather than each dip-ping event predicting a different velocity. So the *Devilish* process should pro-vide better velocities for data with conflicting dips. And we can expect a better final stack as well.

Rocca's Smear Operator

Fabio Rocca developed a clear conceptual model for Sherwood's dip corrections. Figure 3 illustrates Rocca's concept of a prestack partial-migration operator. Imagine a constant-offset section $P(t, y, h = h_0)$ con-taining an impulse function at some particular (t_0, y_0). The earth model implied by this data is a reflector shaped like an ellipse, with the shot point at one focus and the receiver at the other. Starting from this earth model, a zero-offset section is made by forward modeling — that is, each point on the ellipse is expanded into a hyperbola. Combining the two operations —

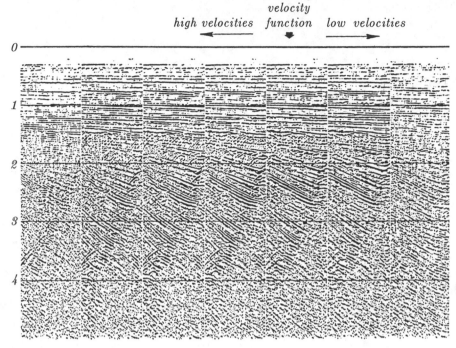

FIG. 3.6-1. Conventional stacks with varying velocity. (distributed by Digicon, Inc.)

constant-offset migration and zero-offset diffraction — gives the Rocca operator.

The Rocca operator is the curve of osculation in figure 3, i.e., the smile-shaped curve where the hyperbolas reinforce one another. If the hyperbolas in figure 3 had been placed everywhere on the ellipse instead of at isolated points, then the osculation curve would be the only thing visible (and you wouldn't be able to see where it came from).

The analytic expression for the travel time on the Rocca smile is the end of a narrow ellipse, shown in figure 4. We will omit the derivation of the equation for this curve which turns out to be

$$1 \;=\; \frac{(y - y_0)^2}{h^2} \;+\; \frac{t^2}{t_0^2} \tag{2}$$

The Rocca operator appears to be velocity independent, but it is not completely so because the curve cuts off at $dt/dy = 2/v$.

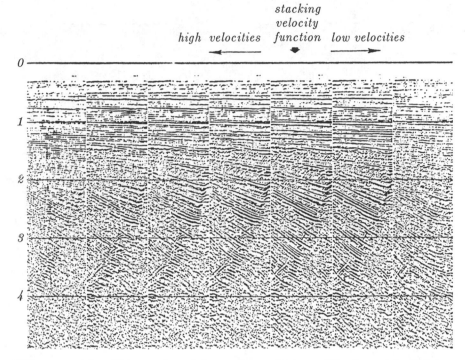

stacking
velocity
high velocities   function   low velocities

FIG. 3.6-2. *Devilish* stacks with varying velocity. (distributed by Digicon, Inc.)

The Rocca operator transforms a constant-offset section into a zero-offset section. This transformation achieves two objectives: first, it does normal-moveout correction; second, it does Sherwood's dip corrections. The operator of figure 3 is convolved across the midpoint axis of the constant-offset section, giving as output a zero-offset section at just one time, say, t_0. For each t_0 a different Rocca operator must be designed. The outputs for all t_0 values must be superposed. Figure 5 shows a superposition of several Rocca smiles for several values of t_0.

This operator is particularly attractive from a practical point of view. Instead of using a big, wide ellipse and doing the big job of migrating each constant-offset section, only the narrow, little Rocca operator is needed. From figure 5 we see that the energy in the dip moveout operator concentrates narrowly near the bottom. In the limiting case that h/vt_0 is small, the energy all goes to the bottom. When all the energy is concentrated near the bottom point, the Rocca operator is effectively a delta function. After compensating

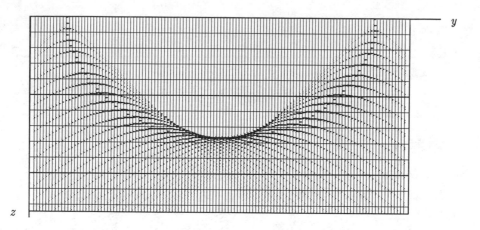

FIG. 3.6-3. Rocca's prestack partial-migration operator is a superposition of hyperbolas, each with its top on an ellipse. Convolving (over midpoint) Rocca's operator onto a constant-offset section converts the constant-offset section into a zero-offset section. (Gonzalez)

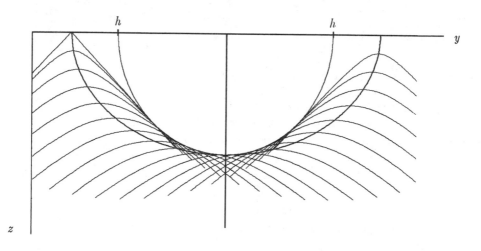

FIG. 3.6-4. Rocca's smile. (Ronen)

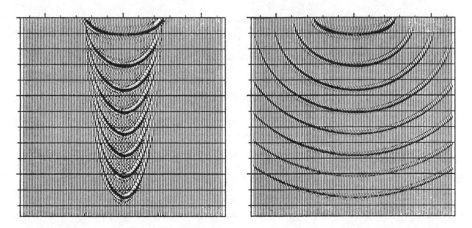

FIG. 3.6-5. Point response of dip moveout (left) compared to constant-offset migration (right). (Hale)

each offset to zero offset, velocity is determined by the normal-moveout residual; then data is stacked and migrated.

The *narrowness* of the Rocca ellipsoid is an advantage in two senses. Practically, it means that not many midpoints need to be brought into the computer main memory before velocity estimation and stacking are done. More fundamentally, since the operator is so compact, it does not do a lot to the data. This is important because the operation is done at an early stage, before the velocity is well known. So it may be satisfactory to choose the velocity for the Rocca operator as a constant, regional value, say, 2.5 km/sec.

An expression for the travel-time curve of the dip moveout operator might be helpful for Kirchhoff-style implementations. More to the point is a Fourier representation for the operator itself, which we will find next.

Hale's Constant-Offset Dip Moveout

Hale [1983] found a Fourier representation of dip moveout. Refer to the defining equations in table 1.

To use the dip-dependent equations in table 1, it is necessary to know the earth dip α. The dip can be measured from a zero-offset section. On the zero-offset section in Fourier space, the sine of the dip is $vk_y/2\omega$. To stress that this measurement applies only on the *zero*-offset section, we shall always write ω_0.

$$\sin \alpha \;=\; \frac{v\,k_y}{2\,\omega_0} \tag{3}$$

NMO	$t \rightarrow t_n$	$t = \sqrt{t_n^2 + 4h^2 v^{-2}}$
Levin's NMO	$t \rightarrow t_0$	$t = \sqrt{t_0^2 + 4h^2 v^{-2} \cos^2 \alpha}$
DMO	$t_n \rightarrow t_0$	$t_n = \sqrt{t_0^2 - 4h^2 v^{-2} \sin^2 \alpha}$

TABLE 3.6-1. Equations for normal moveout and dip moveout. Substituting the DMO equation into the NMO equation yields Levin's dip-corrected NMO.

In the absence of dip, NMO should convert any trace into a replica of the zero-offset trace. Likewise, in the presence of dip, the combination of NMO and DMO should convert any constant-offset section to a zero-offset section. Pseudo-zero-offset sections manufactured in this way from constant-offset sections will be denoted by $p_0(t_0, h, y)$. First take the midpoint coordinate y over to its Fourier dual k_y. Then take the Fourier transform over time letting ω_0 be Fourier dual to t_0.

$$P_0(\omega_0, h, k_y) = \int dt_0 \, e^{i \omega_0 t_0} P_0(t_0, h, k_y) \tag{4}$$

Change the variable of integration from t_0 to t_n.

$$P_0(\omega_0, h, k_y) = \int dt_n \, \frac{dt_0}{dt_n} \, e^{i \omega_0 t_0(t_n)} P_0(t_0(t_n), h, k_y) \tag{5}$$

Express the integrand in terms of NMOed data P_n. This is done by means of $P_n(t_n, h, k_y) = P_0(t_0(t_n), h, k_y)$.

$$P_0(\omega_0, h, k_y) = \int dt_n \, \frac{dt_0}{dt_n} \, e^{i \omega_0 t_0(t_n)} P_n(t_n, h, k_y) \tag{6}$$

As with Stolt migration, the Jacobian of the transformation, dt_0/dt_n scales things but doesn't do time shifts. The DMO is really done by the exponential term.

Omitting the Jacobian (which does little), the over-all process may be envisioned with the program outline:

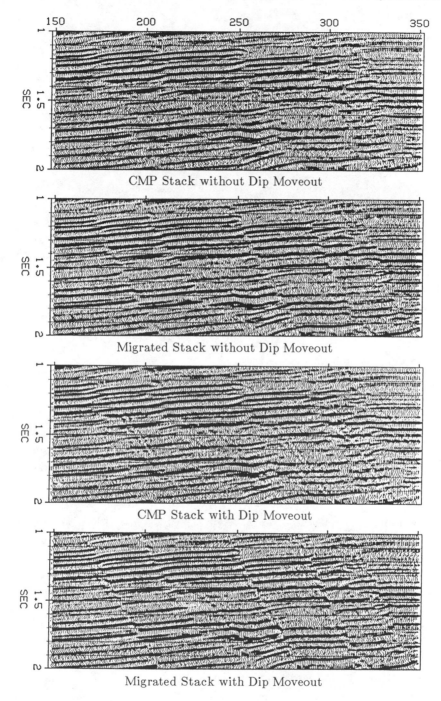

CMP Stack without Dip Moveout

Migrated Stack without Dip Moveout

CMP Stack with Dip Moveout

Migrated Stack with Dip Moveout

FIG. 3.6-6. Processing with dip moveout. (Hale)

$$P(k_y) = FT[P(y)]$$
$$P_n(t_n) = NMO[P(t)]$$

```
for all  k_y  {        # three nested loops, interchangeable
for all  h    {        # three nested loops, interchangeable
for all  ω_0  {        # three nested loops, interchangeable
        sum = 0
        for all  t_n  {
```

$$\text{sum} = \text{sum} + \exp\left[i\,\omega_0 \left(t_n{}^2 + \frac{h^2 k_y{}^2}{\omega_0^2} \right)^{1/2} \right] P_n(t_n, h, k_y)$$

```
        }
        P_0(ω_0, h, k_y) = sum
} } }
```

$$p_0(t_0, h, y) = FT\,2D\,[P_0(\omega_0, h, k_y)]$$

Notice that the exponential in the inner loop in the program does not depend on velocity. The velocity in the DMO equation in table 1 disappears on substitution of $\sin\alpha$ from equation (3). So dip moveout does not depend on velocity.

The procedure outlined above requires NMO before DMO. To reverse the order would be an approximation. This is unfortunate because we would prefer to do the costly, velocity-independent DMO step once, before the iterative, velocity-estimating NMO step.

Ottolini's Radial Traces

Ordinarily we regard a common-midpoint gather as a collection of seismic traces, that is, a collection of time functions, each one for some particular offset h. But this (h, t) data space could be represented in a different coordinate system. A system with some nice attributes is the radial-trace system introduced by Turhan Taner. In this system the traces are not taken at constant h, but at constant angle. The idea is illustrated in figure 7.

Besides having some theoretical advantages, which will become apparent, this system also has some practical advantages, notably: (1) the traces neatly fill the space where data is nonzero; (2) the traces are close together at early times where wavelengths are short, and wider apart where wavelengths are long; and (3) the energy on a given trace tends to represent wave propagation at a fixed angle. The last characteristic is especially important with multiple reflections (Section 5.6). But for our purposes the best attribute of radial traces is still another one.

Richard Ottolini noticed that a point scatterer in the earth appears on a radial-trace section as an *exact* hyperbola, not a flat-topped hyperboloid. The travel-time curve for a point scatterer, Cheops' pyramid, can be written as a "string length" equation, or a stretched-circle equation (Section 3.2).

FIG. 3.6-7. Inside the data volume of a reflection seismic traverse are planes called *radial-trace sections*. A point scatterer inside the earth puts a hyperbola on a radial-trace section.

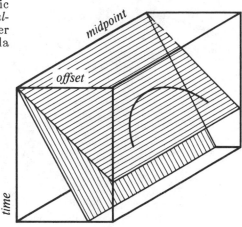

Making the definition

$$\sin \psi \;=\; \frac{2\,h}{v\,t} \tag{7}$$

and substituting into equation (13) of Section 3.2, yields

$$v\,t \;=\; 2 \left[\frac{z^{2}}{\cos^{2}\psi} + (y - y_0)^2 \right]^{1/2} \tag{8}$$

Scaling the z-axis by $\cos\psi$ gives the circle and hyperbola case all over again! Figure 8 shows a three-dimensional sketch of the hidden hyperbola.

We will see that the radial hyperbola of figure 8 is easier to handle than the flat-topped hyperboloid that is seen at constant h. Refer to the equations in table 2.

The second equation in table 2 is the usual exploding reflector equation for zero-offset migration. It may also be obtained from the DSR by setting $H = 0$. As written it contains the earth velocity, not the half velocity. Equation (8) says that hyperbolas of differing ψ values are related to one another by scaling the z-axis by $\cos\psi$. According to Fourier transform theory, scaling z by a $\cos\psi$ divisor will scale k_z by a $\cos\psi$ multiplier. This means the first equation in table 2 can be used for migrating and diffracting hyperbolas on radial-trace sections. Eliminating k_z from the first and second equations yields the middle equation $\omega{\rightarrow}\omega_0$ in table 2. This middle equation combines the operation of migrating all offsets (really any radial angle) and then diffracting out to zero offset. Thus the total effect is that of *offset continuation*, i.e. NMO and DMO. The last two equations in table 2 are a decomposition of the middle equation $\omega{\rightarrow}\omega_0$ into two sequential

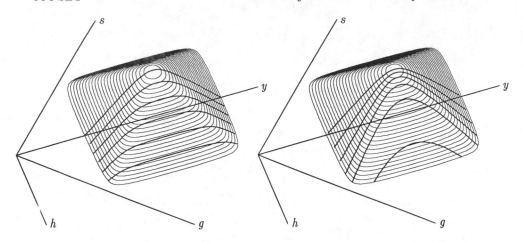

FIG. 3.6-8. An unexpected hyperbola in Cheops' pyramid is the diffraction hyperbola on a radial-trace section. (Harlan)

migration	$\omega \rightarrow k_z$	$k_y{}^2 + k_z{}^2\cos^2\psi \;=\; 4\omega^2/v^2$
zero-offset diff.	$k_z \rightarrow \omega_0$	$k_y{}^2 + k_z{}^2 \;=\; 4\omega_0^2/v^2$
DMO+NMO	$\omega \rightarrow \omega_0$	$.25\, v^2 k_y{}^2\sin^2\psi + \omega_0^2\cos^2\psi = \omega^2$
radial DMO	$\omega \rightarrow \omega_s$	$.25\, v^2 k_y{}^2\sin^2\psi + \omega_s{}^2 \;=\; \omega^2$
radial NMO	$\omega_s \rightarrow \omega_0$	$\omega_0 \cos\psi \;=\; \omega_s$

TABLE 3.6-2. Equations defining dip moveout and ordinary moveout in radial trace coordinates.

processes, $\omega \rightarrow \omega_s$ and $\omega_s \rightarrow \omega_0$. These two processes are like DMO and NMO, but the operations occur in radial space. Radial NMO is a simple time-invariant stretch; hence the notation ω_s.

Unlike the constant-offset case, dip moveout is now done *before* the stretching, velocity-estimating step. Let us confirm that the dip moveout is truly velocity-independent. Substitute (7) into the radial DMO transformation in table 2 to get the equation for transformation from time to stretched time:

$$\frac{h^2}{t^2} k_y^2 + \omega_s^2 = \omega^2 \tag{9}$$

We observe that the velocity v has dropped out of (9). Thus dip moveout in radial coordinates doesn't depend on velocity. Dip-moveout processing $\omega \to \omega_s$ does not require velocity knowledge. Radial coordinates offer the advantage that this comparatively costly process is done *before* the velocity is estimated $\omega_s \to \omega_0$.

The dip-moveout process, $\omega \to \omega_s$, can be conveniently implemented with a Stolt-type algorithm using (9).

The foregoing analysis has assumed a constant velocity. A useful practical approximation might be to revert to a $v(z)$ analysis after the dip moveout, just before conventional velocity analysis, stack, and zero-offset migration.

Both the radial-trace method and Hale's constant-offset method handle all angles exactly in a constant-velocity medium, but neither method treats velocity stratification exactly nor is it clear that this can be done — since neither method is rooted in the DSR. Yilmaz [1979] rooted his DMO work in the DSR, so his method can be expected to be exact for velocity stratification, but Yilmaz could not avoid angle-dependent approximations. So there remains theoretical work to be done.

Anti-Alias Characteristic of Dip Moveout

You might think that if (y,h,t)-space is sampled along the y-axis at a sample interval Δy, then any final migrated section $P(y,z)$ would have a spatial resolution no better than Δy. This is not the case.

The basic principle at work here has been known since the time of Shannon. If a time function and its derivative are sampled at a time interval $2\Delta t$, they can both be fully reconstructed provided that the original bandwidth of the signal is lower than $1/(2\Delta t)$. More generally, if a signal is filtered with m independent filters, and these m signals are sampled at an interval $m \Delta t$, then the signal can be recovered.

Here is how this concept applies to seismic data. The basic signal is the earth model. The various filtered versions of it are the constant-offset sections. Recall that the CDP reflection point moves up dip as the offset is increased. Further details can be found in a paper by Bolondi, Loinger, and Rocca [1982], who first pointed out the anti-alias properties of dip moveout. At a time of increasing interest in 3-D seismic data, special attention should be paid to the anti-alias character of dip moveout.

EXERCISE

1.   Describe the effect of the Jacobian in Hale's dip moveout process.

3.7 Lateral Velocity Variation in Bigger Doses

To the interpreting *geologist,* lateral velocity variation produces a strange distortion in the seismic section. And the distortion is worse than it looks. The *geophysicist* is faced with the challenge of trying to deal with lateral velocity variation in a quantitative manner. First, how can reliable estimates of the amount of lateral velocity variation be arrived at? Then, do we dare use these estimates for reprocessing data?

Our studies of *dip* and *offset* have resulted in straightforward procedures to handle them, even when they are simultaneously present. Unfortunately, increasing lateral velocity variation leads to increasing confusion — confusion we must try to overcome. Strong lateral velocity variation overlies the largest oil field in North America, Prudhoe Bay. Luckily, however, we have many idealized examples that are easy to understand. Any "ultimate" theory would have to explain these examples as limiting cases.

Let us review. The double-square-root equation presumably works if the square roots are expanded and if we accept the usual limitation of accuracy with angle. Our problem with the DSR is that it merely tells us how to migrate and stack *once the velocity is known.* Kjartansson's method of determining the distribution of (some function of) $v(x, z)$ assumes straight rays, no dip, and a single, planar reflector. On the other hand, stacking along with prestack partial migration allows any scattering geometry but enables determination of $v(z)$ only under the presumption that there is no lateral variation of velocity. Clearly, there are many gaps. We begin with comprehensible, special cases but ultimately sink into a sea of confusion.

Replacement Velocity: Freezing the Water

Sometimes you are lucky and you know the velocity. Maybe you know it because you are dealing with synthetic data. Maybe you know it because you

have already drilled 300 shallow holes. Or maybe you can make a good estimate because you have a profile of water depth and you are willing to guess
at the sediment velocity. Often the velocity problem is really a near-surface
problem. Perhaps you have been dragging your seismic streamer over the
occasional limestone reefs in the Red Sea.

Assuming that you know the velocity and that the lateral variations are
near the surface, then you should think about the idea of a *replacement velocity*. For example, suppose you could freeze the water in the Red Sea, just
until it is hard enough that the ice velocity and the velocity of the limestone
reefs are equal. That would remove the unnecessary complexity of the
reflections from deep targets. Of course you can't freeze the Red Sea, but you
can reprocess the data to try to mimic what would be recorded if you could.

First, downward continue the data to some datum beneath the lateral
variations. Then upward continue it back to the surface through the homogeneous replacement medium.

While in principle the DSR could be used for this job, in practice it
would be expensive and impractical. The best approach is to study the two
operations — going down, then going up — in combination. Since the two
operations are largely in opposition to each other, whatever is done to the
data should be just a function of the difference. For example, the equation

$$\frac{\partial P}{\partial z} \;=\; i\,\omega \left[\frac{1}{v\,(s)} \;+\; \frac{1}{v\,(g)} \;-\; \frac{2}{v_{avg}} \right] P \tag{1}$$

combines the downward continuation with the upward continuation and
makes little change to the wavefield P when the velocities are nearly the
same. Equation (1) is basically a time-shifting equation. There is an industry
process known as *static* corrections. The word *static* implies time-invariant
— the amount of time shift does not depend on time. When the appropriate
corrections are merely static shifts, then the earth model has lateral velocity
variations in the near surface only. This is often the case. Equation (1) also
has the ability to do time-variable time shifts because $v\,(s)$ and $v\,(g)$ can
be any function of depth z. Because of the wide-offset angle normally used,
it is desirable to extend (1) to a wider angle. Such extensions are found in
Lynn [1979]. Lynn also shows how partial differential equations can be written to describe the influence of lateral velocity variation on stacking velocity.
Berryhill [1979] illustrated the use of the Kirchhoff method for an irregular
datum.

In practice, the problem of estimating lateral velocity variations is usually more troublesome than the application of these velocities during migration. Static time shifts are estimated from a variety of measurements

including the elevation survey, travel times from the bottoms of shot holes to the surface, and crosscorrelation of reflection seismograms. Wiggins et al. [1976] provide an analysis to determine the static shifts from correlation measurements.

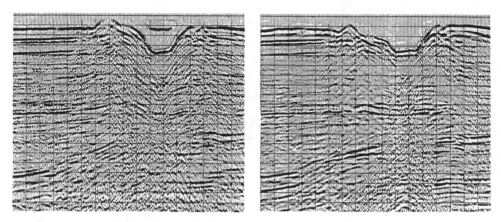

FIG. 3.7-1. Data (left) from Philippines with dynamic corrections (right). (by permission from *Geophysics*, Dent [1983])

Where the lateral variation runs deeper the time shifts become time-dependent. This is called the *dynamic* time-shift problem. To compute dynamic time shifts, dip is assumed to be zero. Rays are traced through a presumed model with laterally variable velocity. Rays are also traced through a reference model with laterally constant velocity. The difference of travel times of the two models defines the dynamic time shifts. See figure 1. Where the lateral variation runs deeper still, the problem looks more like a migration problem. Figure 2 illustrates a process called *REVEAL* by Digicon, Inc., who have not *revealed* whether a time-shift method or a wave-equation method was used.

Lateral Shift of the Hyperbola Top

Figure 3 shows a point scatterer below a dipping interface. As usual there is a higher velocity below. This is a simple prototype for many lateral-velocity-variation problems. Surface arrival times will be roughly hyperbolic with distortion because of the velocity jump at the interface. The minimum travel time (hyperboloid top) has been displaced from its usual location directly above the point scatterer. Observe that

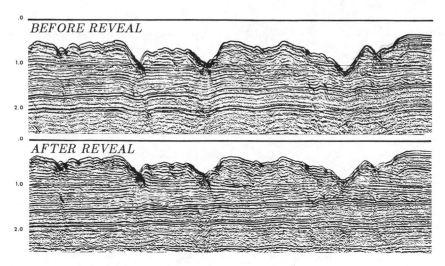

FIG. 3.7-2. Example of processing with a replacement velocity. Observe that deeper bedding is now flatter and more continuous. (distributed by Digicon, Inc.)

1. At minimum time, the ray emerges going straight up.

2. Minimum time is on the high-velocity side of the point scatterer.

3. Minimum time is displaced further from the scatterer as offset increases.

The travel-time curve is roughly hyperbolic, but the asymptote on the right side gives the velocity of the medium on the right side, and the asymptote on the left approximately gives the velocity on the left.

Let $T(x)$ denote the travel time from the point scatterer to the surface point x. The travel time for a constant-offset section is then $t(y) = T(y+h) + T(y-h)$. To find the earliest arrival, set $dt/dy = 0$. This proves that the slope at a on figure 3 is the negative of the slope at b. This shows why the displacement of the top of the hyperboloid from the scatterer increases with offset.

Lateral velocity variation causes hyperbolas to lose their symmetry. Computationally, it is the lens term that tilts hyperbolas, causing their tops to move laterally.

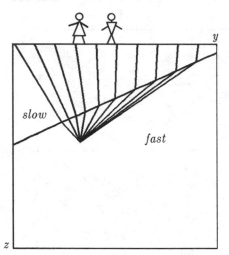 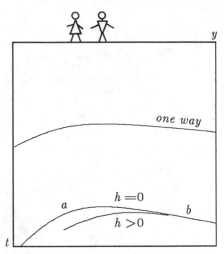

FIG. 3.7-3. Rays emerging from a point scatterer beneath a velocity wedge (left). Travel-time curve (right). The slope at a is the negative of that at b . The midpoint between a and b is at the top of the $h > 0$ curve.

Phantom Diffractor

A second example of lateral velocity variation is figure 4, also taken from Kjartansson's dissertation. The physical model shown on the inset in figure 4 is three constant velocity wedges separated by broken line segments representing reflectors. The bottom edge of the model also represents a reflector. The wavefield in figure 4 was made using the exploding-reflector calculation, which Kjartansson regarded as a reasonable approximation to a zero-offset section. Notice that under the tip of the 4 km/sec wedge is a small diffraction on the bottom horizontal reflector. Because such a diffraction has nothing to do with the flat reflector on which it is seen, it is termed a "phantom" diffraction. Phantom diffractions are not easy to recognize, but they do occur. In reality, the "bright spots" in Section 3.1 were probably phantom diffractions. It has been reported that phantom diffractions provide a means of prospecting for small, high-velocity, carbonate reefs.

Wavefront Healing

Figure 5 (also in FGDP) shows another example of ray bending. The first frame on the left shows a plane wave just after it has been distorted into a wavy shape by the thin-lens term. After this the thin-lens term vanishes. Later frames show the effect of increasing amounts of diffraction.

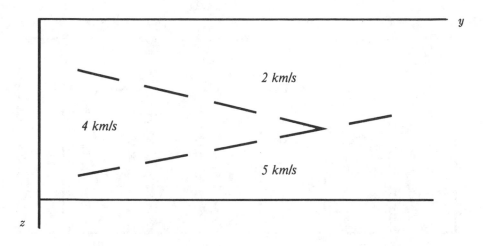

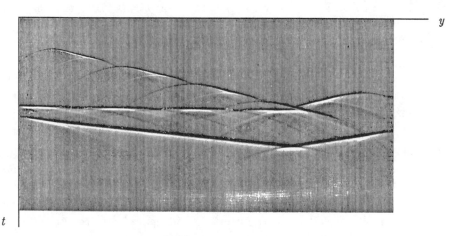

FIG. 3.7-4. The model in the upper panel was taken from Western Geophysical's *Depth Migration* brochure. The model is not physical because of the segmenting of the interface; however, the segments make it a good case for the study of lateral shifts. The synthetic data is in the lower panel (from Kjartansson). The phantom diffraction is on the latest arrival just below the tip of the wedge.

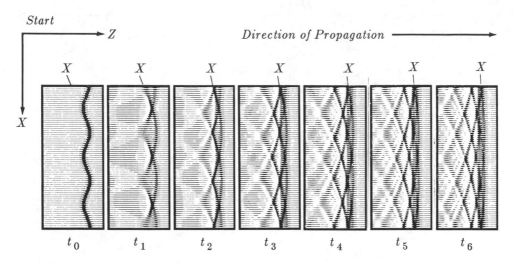

FIG. 3.7-5. The first frame on the left shows a plane wave just after it has been distorted into a wavy shape by the thin-lens term. After this the thin-lens term vanishes. Subsequent frames show the effect of increasing amounts of diffraction. Notice the lengthening of the wave packet and the *healing* of the first arrival. (FGDP, p. 213, figure 10-22)

Fault-Plane Reflection

Across a single vertical fault in the earth the velocity will be a simple step function of the horizontal coordinate. Rays traveling across such a fault suffer in amplitude because of reflection and transmission coefficients, depending on the angle. Since near-vertical rays are common, only small velocity contrasts are required to generate strong internal reflections. By this reasoning, steep faults should be more distorted, and hence more recognizable, on small-offset sections than on wide-offset sections or stacks.

This phenomenon is somewhat more confusing when seen in (x, t)-space. Figure 6 was computed by Kjartansson and used in a quiz. Study this figure and answer the questions in the caption. Here is a hint: A reflected ray beyond critical angle undergoes a phase shift. This will turn a pulse into a doublet that might easily be mistaken for two rays.

Figure 6 exhibits a geometry in which the exploding-reflector model fails to produce all the rays seen on a zero-offset section. The exploding-reflector model produces two types of rays: the ray that goes directly to the surface, and the ray that reflects from the fault plane before going to the surface. A zero-offset section has three rays: the two rays just mentioned, but moving at

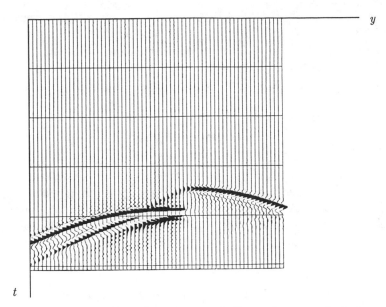

FIG. 3.7-6. Synthetic data from an exploding-reflector calculation for an earth model containing a point scatterer and a velocity jump $v_1/v_2 = 1.2$ across a vertical contact. (Kjartansson)

(a) Is the point scatterer in the slow or the fast medium?
(b) Identify four arrivals and diagram their raypaths.
(c) Identify and explain two kinds of computational artifacts.
(d) Find an evanescent wave.
(e) Find phase shift on a beyond-critical-angle reflection.
(f) A zero-offset section has ray not shown above. Where?

double travel time, once up, once down; and in addition the ray not present in figure 6, which hits the fault plane going one way but not the other way.

There is a simple way to make constant-offset sections in laterally variable media when the reflector is just a point. The exploding-reflector seismogram recorded at $x = s$ is simply time convolved with the one recorded at $x = g$. Convolution causes the travel times to add. Even the non-exploding-reflector rays are generated. Too bad this technique doesn't work for reflector models that are more complicated than a simple reflecting point.

Misuse of $v(x)$ for Depth Migration

The program that generated figure 6 could be run in reverse to do a migration. All the energy from all the interesting rays would march back to the impulsive source. Would this be an effective migration program in a field environment? It is unlikely that it would. The process is far too sensitive to quantitative knowledge of the lateral velocity jump. It is the quantitative value that determines the reflection coefficient and ultimately the correct recombination of all the wavefronts back to a pulse. To see how an incorrect value can result in further error, imagine using the hyperbola-summation migration method. Applied to this geometry this method implies weighted summation over all the raypaths in the figure. The incorrect value would put erroneous amplitudes on various branches. An erroneous location for the fault would likewise mislocate several branches.

The lesson to be learned from this example is clear. Unnecessary bumps in the velocity function can create imaginary fault-plane reflections. Consistent with known information, a presumed migration velocity should be as smooth as possible in the lateral direction. Unskilled and uninformed staff at a processing center remote from the decision making should not have the freedom to introduce rapid lateral changes in the velocity model.

First-Order Effects, the Lens Term

Now let us be specific about what is meant by the lens term in the present context of before-stack migration in the presence of lateral velocity variation. Specializing the DSR equation to 15° angles gives

$$\frac{\partial U}{\partial z} = -\left\{ \left[\frac{i\,\omega}{v\,(s\,)^2} + \frac{v\,(s\,)^2}{2\,i\,\omega} \frac{\partial^2}{\partial s^2} \right] + \left[\frac{i\,\omega}{v\,(g\,)^2} + \frac{v\,(g\,)^2}{2\,i\,\omega} \frac{\partial^2}{\partial g^2} \right] \right\} U \tag{2}$$

Rearranging the terms to group by behavior gives

$$\frac{\partial U}{\partial z} = -\left[\frac{i\,\omega}{v\,(s\,)^2} + \frac{i\,\omega}{v\,(g\,)^2} \right] U - \left[\frac{v\,(s\,)^2}{2\,i\,\omega} \frac{\partial^2}{\partial s^2} + \frac{v\,(g\,)^2}{2\,i\,\omega} \frac{\partial^2}{\partial g^2} \right] U \tag{3a}$$

$$\frac{\partial U}{\partial z} = \textit{lens term} + \textit{diffraction term} \tag{3b}$$

So you see the familiar type of lens term, but it has two parts, one for shifting at the shot, and one for shifting at the geophone.

The Migrated Time Section: An Industry Kludge

As geology becomes increasingly dramatic, reflection data gets more anomalous. The first thing noticeable is that the stacking velocity becomes unreasonable. In practice the available computer processes — based on inappropriate assumptions — will be tried anyway.

A stacking velocity will be chosen and a stack formed. How should the migration be done? Most basic migration programs omit the lens term. Although it is easy to include the lens term, the term is sensitive to lateral variation in velocity. Since estimates of lateral variation in velocity always have questionable reliability, use of a migration program with a lens term is usually limited to knowledgeable interpreters. The lens term is usually omitted from the basic migration utility program. Let us see what this means.

The migration equation is valid in some "local plane wave" sense, i.e.

$$k_z(y,z) = \frac{\omega}{v(y,z)} \left[1 - \left(\frac{v(y,z)\,k_y(y,z)}{\omega} \right)^2 \right]^{1/2} \tag{4}$$

A *migrated time section* is defined by transforming the depth variable z in (4) to a travel-time depth τ.

$$k_\tau(y,\tau) = \omega \left[1 - \left(\frac{v(y,\tau)\,k_y(y,\tau)}{\omega} \right)^2 \right]^{1/2} \tag{5}$$

The implementation of equation (5) requires no lens terms, so no large sensitivity to lateral velocity variation is expected. Unfortunately, there is a pitfall. The (y,z) coordinate system is an orthogonal coordinate system, but the (y,τ) system is not orthogonal [unless $v(y)=const$]. So equation (4), which says that $\cos\theta = \sqrt{1 - \sin^2\theta}$, is not correctly interpreted by (5). A hyperbola would migrate to its top when it should be migrating toward the low-velocity side.

In summary: In a production environment a great deal of data gets processed before anyone has a clear idea of how much lateral velocity variation is present. So the lens term is omitted. The results are OK if the lens term happens to commute with the diffraction term. The terms do commute when the lateral velocity variation is slow enough. Otherwise, you should reprocess with the lens term. The reprocessing will be sensitive to errors in velocity. Be careful!

4

The Craft of Wavefield Extrapolation

This chapter attends to those details that enable us to do a *high-quality* job of downward continuing wavefields. There will be few new seismic imaging concepts here. There will, however, be interesting examples of pitfalls. And in order to improve the quality of seismic images of the earth, several new and interesting mathematical concepts will be introduced. Toward the end of the chapter a program is prepared to simulate and compare various migration methods.

The Magic of Color

The first thing we will consider in this chapter is signal strength. Echoes get weaker with time. This affects the images, and requires compensation.

Next, seismic data is colored by filtering. This filtering can be done in space as well as time. Time-series analysis involves the concept of enhancing the signal-to-noise ratio by filtering to suppress some spectral regions and enhance others. Spectral weighting can also be used on wavefields in the space of ω and k. In the absence of noise, wave-equation theory tells us what filters to use. Loosely, the wave equation is a filter with a flat amplitude response in (ω,k)-space and a phase response that corresponds to the time delays of propagation. The different regions of (ω,k)-space have different amounts of noise. But the different regions need not all be displayed at the strength proposed by the wave equation, any more than data must be displayed with $\Delta x = \Delta z$.

An example of the mixture of filter theory and migration theory is provided by the behavior of the spatial Nyquist frequency. Because seismic data is often spatially aliased, this example is not without practical significance. Think of an impulse function with its Nyquist frequency removed. The removal has little relative effect on the impulse, but a massive *relative* effect on the zeroes surrounding the impulse. When migrating an impulse by frequency domain methods, spatial frequencies just below the spatial Nyquist are treated much differently from frequencies just above it. One is treated as left dip, the other as right dip. This discontinuity in the spatial frequency domain causes a spurious, spread-out response in the space domain shown in figure 1.

The spurious Nyquist noise is readily suppressed, not by excluding the Nyquist frequency from the display, but by a narrow band filter such as used in the display, namely $(1 + \cos k_x \Delta x)/(1 + .85 \cos k_x \Delta x)$ which goes smoothly to zero at the spatial Nyquist frequency. This filter has a simple tri-diagonal representation in the x-domain.

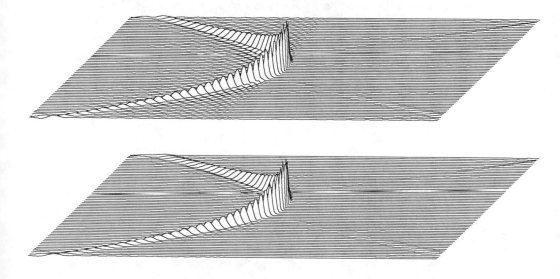

FIG. 4.0-1. Hyperbola amplified to exhibit surrounding Nyquist noise (top) removed by filtering (bottom).

Survey of Migration Technique Enhancements

In our quest for quality, we will also recall various approximations as we go. Now is the time to see how the use of approximations degrades results, and to discover how to improve those results. Five specific problems will be considered:

1   The frequency dispersion that results from the approximation of differential operators by difference operators

2   The anisotropy distortion of phase and group velocity that results from square-root approximations

3   The effect of truncation at the end of the survey line

4   Dips greater than ninety degrees

5   Wraparound problems of Fourier transformation

6   The effect of $v(z)$ upon the Stolt method and how to improve the result by stretching

Following study of these approximations, Section 4.6 is a penetrating study of causality, which covers much ground including how Fourier domain migration can simulate the causality intrinsic to time domain migrations. Section 4.7 is the grand summary of techniques. A single program is presented that can simulate diffraction hyperbolas from many different migration methods. This facilitates comparison of techniques and optimization of parameters. Figure 1 and many of the other figures in this chapter were produced with this program, so you should be able to reproduce them.

A Production Pitfall: Weak Instability from v(x)

Some quality problems cannot be understood in the Fourier domain. Unless carefully handled, lateral velocity variation can create instability.

The existence of lateral velocity jumps causes reflections from steep faults. A more serious problem is that the extrapolation equations themselves have not yet been carefully stated. The most accurate derivation of extrapolation equations included in this book so far was done from dispersion relations, which themselves imply velocity constant in x. The question of how a dispersion relation containing a $v\,k_x^2$ term should be represented was never answered. It might be represented by $v(x,z)\partial_{xx}$, $\partial_x v(x,z)\partial_x$, $\partial_{xx} v(x,z)$ or any combination of these. Each of these expressions, however, implies a different numerical value for the internal reflection coefficient. Worse still, by the time all the axes are discretized, it turns out that one of the most sensible representations leads to reflection coefficients greater than unity and to numerical instability.

A weak instability is worse than a strong one. A strong instability will be noticed immediately, but a weak instability might escape notice and later lead to incorrect geophysical conclusions. Fortunately, a stability analysis leads to a *bulletproof* method in Section 4.8.

4.1 Physical and Cosmetic Aspects of Wave Extrapolation

Frequency filtering, dip filtering, and gain control are three processes whose purposes seem to be largely cosmetic: they are used to improve the appearance of data. The criteria used to choose the quantitative parameters of these and similar processes are often vague and relate to human experience or visual perception. In principle, it should be possible to choose the parameters by invoking information theory and using objective criteria such as signal and noise dip spectra. But in routine practice this is not yet being done.

The importance of cosmetic processes is not to be underestimated. On many occasions, for example, a comparison of processing techniques (in order to choose a contractor perhaps?) has been frustrated by an accidental change in cosmetic parameters. These cosmetic processes arise naturally within wave-propagation theory. It seems best to first understand how they arise, and then to build them into the processing, rather than try to append them in some artificial way after the processing. The individual parts of the wave-extrapolation equations will now be examined to show their cosmetic effects.

t Squared

Echos get weaker with time. To be able to see the data at late times, we generally increase data amplification with time. I have rarely been disappointed by my choice of the function t^2 for the scaling factor. The t^2 scaling function cannot always be expected to work, because it is based on a very simple model. But I find t^2 to be more satisfactory than a popular alternative, the growing exponential. The function t^2 has no parameters whereas the exponential function requires two parameters, one for the time constant, and one for the time at which you must stop the exponential because it gets too large.

The first of the two powers of t arises because we are transforming three dimensions to one. The seismic waves are spreading out in three dimensions, and the surface area on the expanding spherical wave increases in proportion to the radius squared. Thus the area on which the energy is distributed is increasing in proportion to time squared. But seismic amplitudes are proportional to the square root of the energy. So the basic geometry of energy spreading predicts only a single power of time for the spherical divergence correction.

An additional power of t arises from a simple absorption calculation. Absorption requires a model. The model I'll propose is too simple to explain everything about seismic absorption, but it nicely predicts the extra power of t that experience shows we need. For the model we assume:

1 One dimensional propagation
2 Constant velocity
3 Constant absorption Q^{-1}
4 Reflection coefficients random in depth
5 No multiple reflections
6 White source

These assumptions immediately tell us that a monochromatic wave would decrease exponentially with depth, say, as $\exp(-\alpha\,\omega\,t)$ where t is travel-time depth and α is a decay constant which is inversely proportional to the wave quality factor Q. Many people go astray when they model real seismic data by such a monochromatic wave. A better model is that the seismic source is broad band, for example an impulse function. Because of absorption, high frequencies decay rapidly, eventually leaving only low frequencies, hence a lower signal strength. At propagation time t the original white (constant) spectrum is replaced by the forementioned function $\exp(-\alpha\,\omega\,t)$ which is a damped exponential function of frequency. The seismic energy available for the creation of an impulsive time function is just proportional to the area under the damped exponential function of frequency. As for the phase, all frequencies will be in phase because the source is assumed impulsive and the velocity is assumed constant. (See Section 4.6 for a causality problem lurking here). Integrating the exponential from zero to infinite frequency provides us with an inverse power of t thus completing the justification of a t^2 divergence correction.

It is curious that the shape of the expected seismogram envelope t^{-2} does not depend on the dissipation constant α. But changing the spectrum of the seismic source does change the shape of the envelope. It is left for an exercise to show that a seismic source with spectrum $|\omega|^{\beta}$ implies a divergence correction $t^{2+\beta}$.

The seismic velocity increases with depth, so sometimes people who know the velocity may improve the divergence correction by making it a function of velocity (and hence offset) as well as time.

In reality it may be fortuitous that t^2 fits data so well. Actually, Q generally increases with depth whereas reflection coefficients generally decrease with depth.

Noise, Surface Waves and Clip

If seismic data contained nothing but reflections, then there would be little trouble plotting it. You would simply multiply by t^2 and then scale so that the largest data values stayed in the available plotting area. In reality there are two problems: 1) noisy traces and 2) noise propagation modes. We have noisy traces because the people in the world won't all be quiet while we listen for echoes. Noise propagation modes are waves trapped in surface layers. So their divergence is in a two-dimensional space rather than the three-dimensional space for reflections. Water noises are additionally strong because of the homogeneity and low absorption of water.

Noises are handled by "clipping" data values at some level lower than the maximum. Clipping means that values larger than the clip value are replaced by the clip value. Since the size of the noise is generally unpredictable, the most reliable method is to use quantiles. Imagine the data points sorted in numerical order by the size of their absolute values. The nth quantile is defined as the absolute value that is $n/100$ of the way between the smallest and largest absolute value. So if data is clipped at the 99th percentile, then up to one percent of the data can be infinitely strong noise. I find that most field profiles have less than 10% noisy points. So I often clip at twice the 90th percentile. To find the quantile, it is not necessary to fully sort the data. That would be slow. Hoare's algorithm is much faster (see FGDP or Claerbout and Muir [1973] for full reference and more geophysical context).

Different plots have different purposes. It is often important to preserve linearity during processing, but at the last stage — plotting — linearity can be sacrificed to enable us to see all events, large and small. After all, human perceptions are generally logarithmic. In our lab we generally use power laws. I find that replacing data points by their signed square roots generally compresses all signals into a visible range. When plotting field profiles with a very close trace spacing, it may be better to use the signed cube roots. More generally, we do non-linear gain with

$$Display \;=\; \text{sgn}(Data) \; |Data|^{\gamma} \tag{1}$$

Gamma is a term in photography to describe nonlinearity of photographic film. Most of the data plots in this book use $\gamma = 1$, t^2 gain, and clip at the 99th percentile.

The industry standard approach seems to be AGC (Automatic Gain Control). AGC means to average the data magnitude in some interval and then divide by the magnitude. Although AGC is nonlinear, it is more linear than using γ so it is presumably better if you plan later processing. But with AGC, you lose reversibility and the sense of absolute gain.

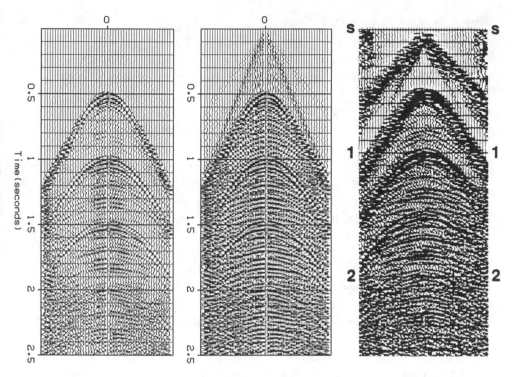

FIG. 4.1-1. Arctic profile from Western Geophysical. Left, with t^2. Center with t^2 and $\gamma=0.4$. Right, with Western's AGC.

Figure 1 is an interesting example. Since it is a split spread, you assume it to be land data. Ships can't push cables in front of them. But the left panel clearly shows marine multiples. The reverberation period is uniform, and there are no reflections before the water bottom. It must be data collected on ice over deep water (375m). From the non-linear gain in the center panel we clearly see a water wave, and before it a fast wave in the ice. There is also weak low-velocity, low-frequency "ground roll" on the ice. There are also some good reflections.

Complex Velocity in the 5° Equation

The 5° equation, namely,

$$\frac{\partial P}{\partial z} = \frac{1}{v} \frac{\partial P}{\partial t} \qquad (2a)$$

$$i \, k_z \;\; = \;\; \frac{-i\,\omega}{v} \qquad\qquad (2b)$$

states that a wavefront will take some time to get from one depth to another. With velocity v being a real constant, waves controlled by (2) propagate without change in form. In practice waveform changes are observed. So v should not be a real constant. An imaginary part of the velocity would cause attenuation. A frequency-dependent velocity would cause frequency dispersion.

Absorption

A basic model arises when $v(\omega)$ is defined by the equation

$$\frac{-i\,\omega}{v(\omega)} \;\; = \;\; \frac{\omega_0}{v_0} \left(\frac{-i\,\omega}{\omega_0} \right)^{1-\epsilon} \qquad\qquad (3)$$

For $\epsilon = 0$, equation (3) gives a constant velocity. Equation (3) models the so-called causal, constant Q attenuation where $Q^{-1} = \tan \pi\epsilon$ (see Section 4.6). Figure 2 shows an example of a synthetic seismogram generated by the exploding-reflector model using equations (2) and (3).

Equation (3) creates attenuation by introducing an imaginary part into the velocity. The main effect of this attenuation is to weaken the arrivals at late time. A secondary effect is to make the frequency content of late arrivals lower. A tertiary effect is this: It happens that the requirement of causality forces the real part of the velocity to be slightly frequency-dependent. In the figure, this slight frequency-dependence is evidenced by the "rise time" on each pulse being faster than the "fall time." This means that the high frequencies are traveling slightly faster than the low frequencies. In practice, this tertiary effect is rarely noticeable.

In making earth images, earth dissipation might be compensated by amplifying high-frequency energy during downward continuation. This might be done just like migration, except that $k_z = -\sqrt{\omega^2/v^2 - k_x{}^2}$ would be replaced by something like $ik_z = (-i\,\omega)^{1-\epsilon}$. In practice, however, no one would really do this, since it would amplify high frequency noise. This raises the issue of signal-to-noise ratio.

Noise isn't simply an ambient random fluctuation. It is mainly repeatable if the data is reshot. Noise is anything for which we have no satisfactory model. On a practical level, time-variable filters are often used to select pleasing time-variable passbands. Equations (2) and (3) could be used for this implementation of time-variable filters, but it would be an oversimplification

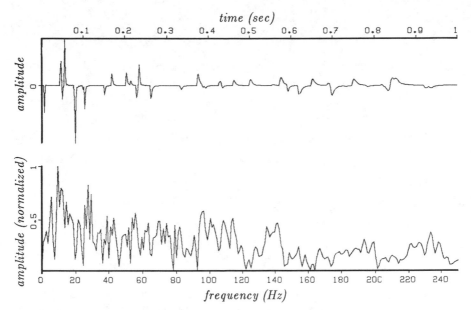

FIG. 4.1-2. Synthetic seismogram for an earth with $Q \approx 100$. (Hale)

to view their use as compensation for the earth Q.

Dispersion

The frequency-dependence of velocity in the case of surface waves is more dramatic. For example, a frequency-dependent velocity is given by the equation

$$\frac{-i\,\omega}{v\,(\omega)} \;=\; \frac{-i\,\omega}{v_0}\,\sqrt{1 + \omega_0^2/\omega^2} \tag{4}$$

Figure 3a contains some frequency-dispersive ground roll. In figure 3b the dispersion has been backed out by a migration-like process. One difference between this process and migration is that migration extrapolates down the z-axis whereas in figure 3b the extrapolation is along the x-axis. (The extrapolation direction is really just in the computer). Each trace in figure 3b is processed separately. In migration, data $p\,(t\,,z=0)$ is extrapolated to an image $p\,(t=0,z)$ using a dispersion relation $k_z = -\sqrt{\omega^2/v^2 - k_x^2}$. In this process, data $p\,(t\,,x=0)$ is extrapolated to an image $p\,(t=0,x)$ using a dispersion relation like $k_x = f\,(\omega/v)$. After this pseudomigration a pseudodiffraction is done with a *constant* velocity. The total effect is to undo the frequency dispersion. Finally, it is possible to see that the noise consists of two separate events. Techniques resembling this one were first used to locate faults in coal seams (Beresford-Smith and Mason [1980]).

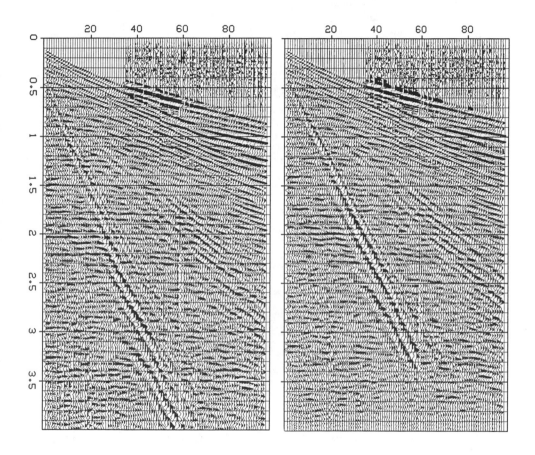

FIG. 4.1-3. Dispersive surface wave (left), with the frequency dispersion backed out (right). Bottom shows two arrivals, the direct, straight-line arrival, and a hyperbola flank. The hyperbola represents sidescatter that must come from some object on the earth's surface off to the side of the survey line. (Conoco, Sword)

False Semicircles in Migrated Data

Dip filtering can be used to suppress multiples. Section 5.5 will show that multiples are unlike primaries in one important respect: their strength may change rapidly in the horizontal direction. They need not be spread out into broad diffraction hyperbolas as primaries must. This difference arises because multiples often spend much time focusing themselves in the irregular, near-surface areas. Common evidence for this behavior is contained in the appearance of wide-angle migrated sections. Such sections often show semicircular arcs coming all the way up to the surface. These arcs warn that something is wrong. The arcs could result from multiples, statics, or unexplained impulsive noise. In any case, they could be partially suppressed without touching primaries.

Zapping Multiples in Dip Space

Think of the migration of a common-depth-point stack as downward continuation in (ω, k_x, z)-space. Ordinarily, velocity increases with depth. As the downward continuation proceeds, the velocity cutoff along the evanescent line bites out more and more area from the (ω, k_x)-space (Section 1.4). Energy beyond this cutoff does not fit the primary wave-propagation model, and it should be suppressed as soon as it is encountered. Such noise suppression can lead to a large drop in total power at late times.

Mixed Appearance of Dip-Filtered Data

An objection often raised against dip filtering is that it can give data a *mixed* appearance. *Mixed* means that adjacent channels appear to have been averaged and that they are no longer independent. This is indeed an effect of dip filtering, and it is inevitable at late times since the horizontal resolving power of reflection data decreases with time. There are two reasons for decreasing lateral resolution. First, dissipation causes high frequencies to disappear. Second, ray bending causes the angular aperture to decrease for deeper sources. (Section 1.2 and Section 1.5). It is unrealistic to ignore this fundamental limitation and imagine that adjacent channels should have an appearance of independence. If a mixed appearance is to be avoided for display purposes, then I advocate removing the low-velocity, coherent, signal-generated noise and replacing it by low-velocity, incoherent, Gaussian, random noise. Many plotters lose dynamic range at close trace spacing, and random noise can tend to restore it.

Accentuating Faults

It often happens that the location of oil is controlled by faulting. But the dominating effect of stratified reflectors may overwhelm the weak diffraction evidence of faulting. A cosmetic process could weaken the zero and small dips, accentuating dips in the range of 10° to 60°, and then suppress the wide angles and evanescent energy. As with frequency filtering, sharp cutoffs are not desirable because of the implied long (and in space, wide) impulse response.

Dip Filtering

Dip filtering is conveniently incorporated into the wave extrapolation equations. Instead of initializing the Muir expansion with $ik_z = -i\omega r_0$ we use $ik_z = \epsilon - i\omega r_0$. (Recall Section 2.1 that r_0 is the cosine of an exactly fitting angle). For the 15° equation we have

$$i\,k_z^{(15)}\,v \;=\; -\,i\,\omega + \frac{v^2\,k_x^2}{\epsilon - i\,\omega\,(r_0 + 1)} \tag{5a}$$

For the 45° equation we have

$$i\,k_z^{(45)}\,v \;=\; -\,i\,\omega + \cfrac{v^2\,k_x^2}{-\,i\,\omega\,2 + \cfrac{v^2\,k_x^2}{\epsilon - i\,\omega\,(r_0 + 1)}} \tag{5b}$$

Figures 4 and 5 show hyperbolas of diffraction for the 15° and 45° equations with and without the dip filtering parameter ϵ.

Gain Control Does Dip Filtering Too

Echoes arriving late are weaker than echoes arriving early. Thus data is ordinarily scaled for plotting using some time-variable scale. Should migration be done before or after this scaling? The results will differ in an interesting way. The top part of the hyperbola has flat dip, whereas the asymptotes, which come later, have steep dip. So, amplification of late information coincidentally amplifies the steep dips. I think the main effect of choosing to do migration before or after scaling is selection of the dip spectrum in the final display. A pedantically correct approach is to migrate first and scale second, but the result will be weaker in dip and fault information than the answer obtained by scaling first and migrating second. A side benefit of the latter method is that you can save computer memory by storing scaled values as

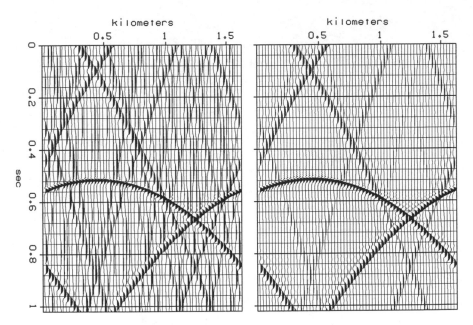

FIG. 4.1-4. Diffraction hyperbolas of the 15° equation without dip filtering (left), and with dip filtering (right).

short integers. I used 16 bit integer storage in my pioneering work. Computations and local storage used 32 bit floating point arithmetic. I see little justification for 32 bit storage generally used today. We can't interpolate between channels to 4 bits of accuracy.

Rejection by Incoherence or Rejection by Filtering?

It is a pitfall to judge a supposed noncosmetic process by a cosmetic effect. I once got caught. The process was migration before stack. The feature that was deemed desirable was the relative strength of the steepest clear event on the record, a fault-plane reflection. But even gain control can affect dip spectra! I hoped the process was working by correctly eliminating some of the rejection of steep dips by CDP stack. Perhaps it was, but how could I know whether this was happening or whether dips were being accidentally enhanced by spatial filtering?

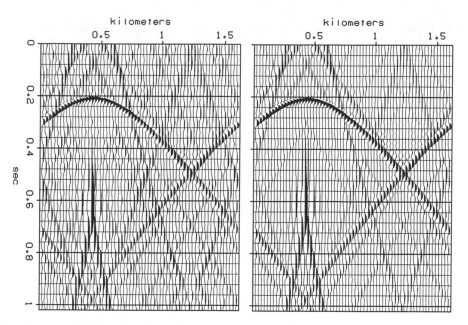

FIG. 4.1-5. Diffraction hyperbolas of the 45° equation without dip filtering (left), and with dip filtering (right).

Spatial Scaling before Migration

Scaling on the time axis before migration can be advantageous. What about scaling on the space axis? The traditional methods of scaling that are called *automatic gain control* (AGC) deduce a scaling divisor by smoothing the data envelope (or its square or its absolute value) over some window. Such scaling can vary rapidly from trace to trace, so concern is justified that diffractions might be caused by lateral jumps in the scaling function. On the other hand, there might be good reasons for the scale to jump rapidly from trace to trace. The shots and geophones used to collect land data normally have variable strength and coupling, and these problems affect the entire trace.

A model must be found that respects both physics and statistics. I suggest allowing for gain that is slowly time-variable and shots and geophones of arbitrarily variable strength, but I also prefer to regard an impulse as evidence that the earth really can focus. For example, data processing with this model can be implemented by smoothing the scaling *envelope* with the filter

$$1 \; - \; \frac{\omega^2}{\alpha + \omega^2} \; \frac{k_x^{\,2}}{\beta + k_x^{\,2}} \tag{6}$$

Filter cutoff parameters are α and β. When the scaling envelope has been smoothed with this filter, it no longer varies rapidly with both x and t, although it can vary with either one or the other. This filter (6) can be economically implemented using the tridiagonal algorithm.

Exponential Scaling

Exponential scaling functions have some ideal mathematical properties. (If you are not familiar with Z-transforms, you should read Section 4.6 or FGDP before proceeding.) Take the Z-transform of a time function a_t:

$$A(Z) \;=\; a_0 + a_1 Z + a_2 Z^2 + \cdots \tag{7}$$

The exponentially gained time function is defined by

$$\uparrow A(Z) \;=\; a_0 + a_1 e^\alpha Z + a_2 e^{2\alpha} Z^2 + \cdots \tag{8}$$

The symbol $\uparrow$ denotes exponential gain. Mathematically, $\uparrow$ means that Z is replaced by $e^\alpha Z$. Polynomial multiplication amounts to convolution of the coefficients:

$$C(Z) \;=\; A(Z) B(Z) \tag{9}$$

By direct substitution,

$$\uparrow C \;=\; (\uparrow A)(\uparrow B) \tag{10}$$

This means that exponential gain can be done either before or after convolution. You may recall from Fourier transform theory that multiplication of a time function by a decaying exponential $\exp(-\alpha t)$ is the equivalent of replacing $-i\omega$ by $-i\omega + \alpha$ in the transform domain.

Specialize the downward-continuation operator $\exp(ik_z z)$ to some fixed z and some fixed k_x. The operator has become a function of ω that may be expressed in the time domain as a filter a_t. Hyperbola flanks move *upward* on migration. So the filter is *anticausal*. This is denoted by

$$A(Z) \;=\; a_0 + a_{-1}\frac{1}{Z} + a_{-2}\frac{1}{Z^2} + \cdots \tag{11}$$

The large negative powers of Z are associated with the hyperbola flanks. Exponentially boosting the coefficients of positive powers of Z is associated with diminishing negative powers — so $\uparrow A$ is A with a weakened tail — and tends to attenuate flanks rather than move them. Thus $\uparrow A$ may be described as viscous.

From a purely physical point of view cosmetic functions like gain control and dip filtering should be done after processing, say, $\uparrow(AB)$. But $\uparrow(AB)$ is equivalent to $(\uparrow A)(\uparrow B)$, and the latter operation amounts to using a viscous operator on exponentially gained data. In practice, it is common to forget the viscosity and create $A(\uparrow B)$. Perhaps this means that dipping events carry more information than flat ones.

The Substitution Operator

The $\uparrow$ operator has been defined as the substitution $Z \rightarrow Z\,e^{\alpha}$. The main property of this operator is that if $C = A\,B$, then $\uparrow C = (\uparrow A)(\uparrow B)$. This property would be shared by any algebraic substitution for Z, not just the one for exponential gain. Another simple substitution can be used to achieve time-axis stretching or compression. For example, replacing Z by Z^2 stretches the time axis by two. Yet another substitution, which has a deeper meaning than either of the previous two, is the substitution of the constant Q dissipation operator $(-i\omega)^{\gamma}$. In summary:

Substitutions for Z-Transform Variable Z [*all preserve* $C(Z){=}A(Z)B(Z)$]	
Exponential growth	$Z \rightarrow Ze^{\alpha}$ $(i\omega \rightarrow i\omega + \alpha)$
Time expansion $(\alpha > 1)$	$Z \rightarrow Z^{\alpha}$
(Inverse) Constant Q dissipation	$-i\omega \rightarrow (-i\omega)^{\gamma}$

EXERCISES

1. Use a table of integrals to show that a seismic source with spectrum $|\omega|^{\beta}$ implies a divergence correction $t^{2+\beta}$.

2. Assuming that t^2 is a suitable divergence correction for field profiles, what divergence correction should be applied to CDP stacks?

3. How is the t^2 correction altered for water of travel time depth t_0? Assume the Q of water is infinite.

4. Consider a source spectrum $e^{-\beta|\omega|}$. How is the t^2 correction altered?

5. The spectrum in figure 2 shows high frequencies smoother than low frequencies. Explain.

6. State some criteria that can be used in the selection of the cutoff parameters α and β for the filter (6).

4.2 Anisotropy Dispersion and Wave-Migration Accuracy

Two distinct types of errors are made in wave migration. Of greater practical importance is *frequency dispersion*, which occurs when different frequencies propagate at different speeds. This may be reduced by improving the accuracy of finite-difference approximations to differentials. Its cure is refinement of the differencing mesh. See Section 4.3.

Of secondary importance, and the subject of this section, is *anisotropy dispersion.* Anisotropic wave propagation is waves going different directions with different speeds. In principle, anisotropic dispersion is remedied by the Muir square-root expansion. In practice, the expansion is generally truncated at either the 15° or 45° term, creating anisotropy error in data processing. The reasons often given for truncating the series and causing the error are (1) the cost of processing and (2) the larger size of other errors in the overall data collection and processing activity. Anisotropy error should be studied in order to (1) recognize the problem when it occurs and (2) understand the basic trade-off between cost and accuracy.

Anisotropy is often associated with the propagation of light in crystals. In reflection seismology, anisotropy is occasionally invoked to explain small discrepancies between borehole velocity measurements (vertical propagation) and velocity determined by normal moveout (horizontal propagation). These fundamental, physical anisotropies and the subject of this section, anisotropy in data processing, share a common mathematical and conceptual basis.

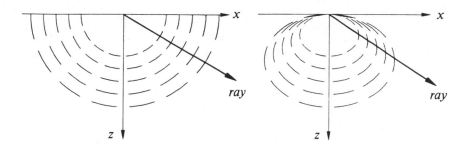

FIG. 4.2-1. Wavefronts in an isotropic medium (left) and an anisotropic medium (right). Note that on the right, the rays are not perpendicular to the wavefronts. (Rothman)

Rays not Perpendicular to Fronts

Anisotropy means that waves propagating in different directions propagate at different speeds. Anisotropy does *not* mean that velocity is a function of spatial location, and thus anisotropy does not cause rays to bend. The peculiar thing about anisotropy is that rays are not perpendicular to wavefronts. Figure 1 illustrates this idea. The diagram on the left shows spherical wavefronts emanating from a point source at the origin. This is the usual, isotropic case. The diagram on the right shows the nonspherical wavefronts of the 15° migration equation. Note that near the z-axis they are nearly spherical, but further away they do a poor job of matching a sphere with its center at the origin.

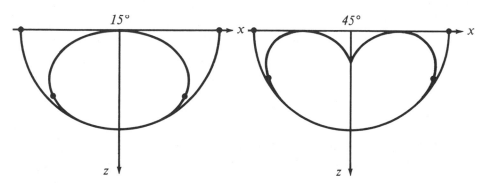

FIG. 4.2-2. Wavefronts of 15° (left) and 45° (right) extrapolation equations, inscribed within the exact semicircle. Waves with $\sin \theta = v k_x / \omega = \pm 1$ are marked with small dots. Evanescent energy lies beyond the dots. (Rothman)

The ideal wavefront from a Huygens secondary source is a semicircle. The secondary source that results from the 15° extrapolation equation is an ellipse. The secondary source that results from the 45° extrapolation equation is an interesting, heartlike shape. These are drawn in figure 2. In practice, the top parts of the ellipse and the heart are rarely observed because they are in the evanescent zone, and the x-axis is seldom refined enough for them to be below the aliasing frequency. The center of the heart is sometimes seen in the (x, t)-plane when the 45° program is used. It is shown by a line drawing in figure 3 and shown using a 45° diffraction program in figure 4.

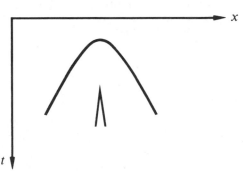

FIG. 4.2-3.   45°   heart   theory. The cusp arises in the evanescent region. (Rothman)

Wavefront Direction and Energy Velocity

In ordinary wave propagation, energy propagates perpendicular to the wavefront. When there is anisotropy dispersion, the angle won't be perpendicular.

The apparent horizontal velocity seen along the earth's surface is dx/dt. The apparent velocity along a vertical, e.g., as seen in a borehole, is dz/dt. By geometry, both of these apparent speeds exceed the wave speed. The vector perpendicular to the wavefront with a magnitude inverse to the velocity is called the slowness vector:

$$slowness\ vector\ =\ \left(\frac{dt}{dx}, \frac{dt}{dz} \right)$$

The phase velocity vector is defined to go in the direction of the slowness vector, but have the speed of the wavefront normal. More precisely, the phase velocity vector is the slowness vector divided by its squared magnitude:

$$phase\ velocity\ =\ \frac{\left(\frac{dt}{dx}, \frac{dt}{dz} \right)}{\left(\frac{dt}{dx} \right)^2 + \left(\frac{dt}{dz} \right)^2}$$

FIG. 4.2-4. Impulse response of the 45° wave-extrapolation equation. The arrival before t_0 is a wraparound.

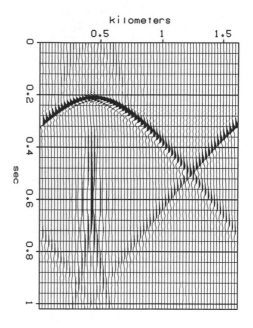

For a disturbance of sinusoidal form, namely, $\exp(i\,\phi) = \exp(-i\,\omega t + ik_x\,x + ik_z\,z)$, the phase ϕ may be set equal to a constant:

$$0 \;=\; d\,\phi \;=\; -\,\omega\,dt \,+\, k_x\,dx \,+\, k_z\,dz$$

Thus, in Fourier space the slowness vector is

$$slowness\ vector \;=\; \left(\,\frac{k_x}{\omega}\,,\;\frac{k_z}{\omega}\,\right) \tag{1a}$$

The direction of energy propagation is somewhat more difficult to derive, but it comes from the so-called group velocity vector:

$$group\ velocity \;=\; \left(\,\frac{\partial}{\partial k_x}\,,\;\frac{\partial}{\partial k_z}\,\right)\omega(k_x,\,k_z) \tag{1b}$$

For the scalar wave equation $\omega^2/v^2 = k_x{}^2 + k_z{}^2$, the group velocity vector and the phase velocity vector turn out to be the same, as can be verified by differentiation and substitution. The most familiar type of dispersion is frequency dispersion, i.e. different frequencies travel at different speeds. Later in this section it will be shown that the familiar (15°, 45°, etc.) extrapolation equations do not exhibit frequency dispersion. That is, as functions of ω and angle k_x/ω, the velocities in these equations do not depend on ω. In other words, the elliptical and heart shapes in figure 2 are not frequency-dependent.

An interesting aspect of anisotropy dispersion is that energy appears to be going in one direction when it is really going in another. An exaggerated instance of this occurs when the group velocity has a downward component and the phase velocity has an upward component. Figure 5, depicting the dispersion relation of the 45° extrapolation equation, shows an example. A slowness vector, which is in the direction of the wavefront normal, has been selected by drawing an arrow from the origin to the dispersion curve. The corresponding direction of group velocity may now be determined graphically by noting that group velocity is defined by the gradient operator in equation (1b). Think of ω as the height of a hill from which k_z points south and k_x points east. Then the dispersion relation is a contour of constant altitude. Different numerical values of frequency result from drawing figure 5 to different scales. The group velocity, in the direction of the gradient, is perpendicular to the contours of constant ω.

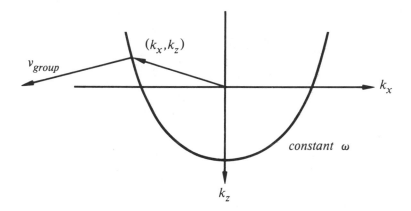

FIG. 4.2-5. Dispersion relation for downgoing extrapolation equation showing group velocity vector and slowness vector. (Rothman)

The anisotropy-dispersion phenomenon can be most clearly recognized in a movie, although it can be understood on a single frame, as in figure 6. The line drawing interprets energy flow from the top, through the prism, reflecting at the 45° angle, reflecting from the side of the frame, and finally entering an area of the figure that is sufficiently large and uncluttered for the phase fronts to be recognizable as energy apparently propagating upward but really propagating downward.

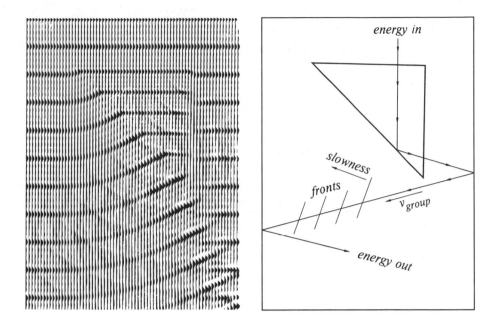

FIG. 4.2-6. Plane waves of four different frequencies propagating through a right, 45° prism. Left is the wavefield. Right is a ray interpretation that illustrates different directions of energy and wavefront normal. (Estevez)

That neither energy nor information can propagate upward in figure 6 should be obvious when you consider the program that calculates the wavefield. The program does not have the entire frame in memory; it produces one horizontal strip at a time from the strip just above. Thus the movie's phase fronts, which appear to be moving upward, seem curious. Theoretically, wave extrapolation using the 45° equation is not expected to handle angles to 90°. Yet the example in figure 6 shows that these extreme cases are indeed handled, although in a somewhat perverted way.

I once saw a similar circumstance on reflection seismic data from a geologically overthrusted area. The data could not be made available to me at the time, and by now is probably long lost in the owner's files, so I can only offer the line drawing in figure 7, which is from memory. The increasing velocity with depth causes the ray to bend upward and reflect from the underside of the overthrust. To see what is happening in the wave equation, it is helpful to draw the dispersion curve at two different velocities, as in figure 8.

Downward continuation of a bit of energy with some particular stepout $dt/dx = k_x/\omega$ begins at an ordinary angle on the near-surface, slow-velocity

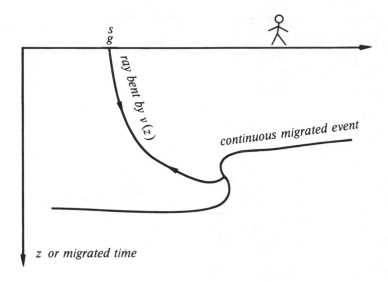

FIG. 4.2-7. Ray reflected from the underside of an overthrust.

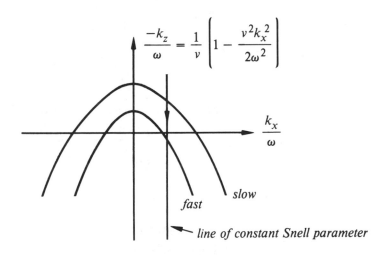

FIG. 4.2-8. Dispersion curve at two different velocities, v_{fast} and v_{slow}.

dispersion curve. But as deeper velocity material is encountered, that same stepout implies a negative phase velocity. Although the thrust angle is unlikely to be quantitatively correct, the general picture is appropriate. It is like figure 6. If you want a quantitatively correct migration, see Section 4.5, or for something completely different, see the method of Kosloff et al. [1983] and Baysal et al. [1983].

Analyzing Errors of Migration

A dipping reflector that is flat and regular can be analyzed in its entirety using the phase velocity concept. The group velocity concept is required only when more than one angle is simultaneously present. This simultaneity occurs with the point scatterer response. It also occurs when there is variable reflection amplitude along a dipping bed. The group velocity is needed because representation of either a curved event or an amplitude anomaly requires a range of plane-wave angles. Analogously, in time-series analysis the Fourier representation of an amplitude-modulated sinusoid requires a bandwidth of sinusoids.

Figure 9 depicts a smooth, flat, dipping bed that has been undermigrated because the $\hat{k}_z$ defined by some rational square-root approximation or some numerical approximation did not match the correct square-root value of k_z.

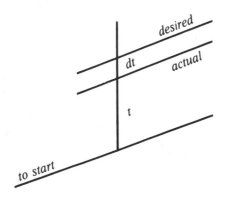

FIG. 4.2-9. Undermigrated dipping reflector.

The error in figure 9 is entirely a time-shift error. Since the reflection coefficient is constant along the reflector, no lateral shift error can be recognized. The time error may be theoretically determined by

$$\frac{dt}{t} \approx \frac{dz}{z} \approx \frac{\hat{k}_z - k_z}{k_z} \tag{2}$$

For the so-called 15° equation, it turns out that about a half-percent phase error is made at 25°.

Next, the error in the collapse of a hyperbola will be determined. Figure 10 depicts the downward continuation of a hyperbola. For clarity, the downward continuation was not taken all the way to the focus. Select a ray of some Snell's parameter $p = dt/dx$ by choosing some slope p. Imagine a tangent line segment of slope p to each of the hyperboloids. If there were a little amplitude anomaly where the slope is p, you would be able to identify it on each of the hyperboloids.

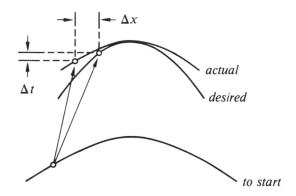

FIG. 4.2-10. Error of hyperbola collapse. Note that the actual curve is above the desired curve, but the actual point is below the desired point.

In figure 10 the amount of time moved is too little; likewise, the lateral distance moved is too small. In practice, errors of the 15° equation with $r_0 = 1$ are sometimes compensated by an increase in either z or v of about 6%. The amounts of the errors may be calculated from

$$\frac{\Delta t}{t} = \frac{\frac{\partial}{\partial \omega}(\hat{k}_z - k_z)}{\frac{\partial}{\partial \omega} k_z} \tag{3a}$$

$$\frac{\Delta x}{x} = \frac{\frac{\partial}{\partial k_x}(\hat{k}_z - k_z)}{\frac{\partial}{\partial k_x} k_z} \tag{3b}$$

where k_z is taken to be a function of ω and k_x. It turns out that for the 15° equation, about a half-percent group velocity error occurs at 20°. Thus the group velocity error is generally worse than the phase velocity error.

Derivation of Group Velocity Equation

An impulse function at the origin in (x, z)-space is a superposition of Fourier components:

$$\iint e^{+i k_x x + i k_z z} \, dk_x \, dk_z \tag{4}$$

Physics (and perhaps numerical analysis) leads to a dispersion relation that is a functional relation between ω, k_x, and k_z, say, $\omega(k_x, k_z)$. The most common example of such a dispersion relation is the scalar wave equation $\omega^2 = v^2 (k_x{}^2 + k_z{}^2)$. The solution to the equation is

$$e^{-i \omega t + i k_x x + i k_z z} \tag{5}$$

Integrating (5) over (k_x, k_z) produces a monochromatic time function that at $t=0$ is an impulse at $(x, z)=(0, 0)$. This expression at some very large time t is

$$\iint e^{-i t [\omega(k_x, k_z) - k_x x/t - k_z z/t]} \, dk_x \, dk_z \tag{6}$$

At t very large, the integrand is a very rapidly oscillating function of unit magnitude. Thus the integral will be nearly zero unless the quantity in square brackets is found to be nearly independent of k_x and k_z for some sizable area in (k_x, k_z)-space. Such a flat spot can be found in the same way that the maximum or minimum of any two dimensional function is found, by setting derivatives equal to zero. This analytical approach is known as the stationary phase method. It gives

$$0 = \frac{\partial}{\partial k_x}[\ ] = \frac{\partial \omega}{\partial k_x} - \frac{x}{t} \tag{7a}$$

$$0 = \frac{\partial}{\partial k_z}[\ ] = \frac{\partial \omega}{\partial k_z} - \frac{z}{t} \tag{7b}$$

So, in conclusion, at time t the disturbances will be located at

$$(x, z) \quad = \quad t \left(\frac{\partial \omega}{\partial k_x} , \; \frac{\partial \omega}{\partial k_z} \right) \tag{8}$$

which justifies the definition of group velocity.

Now let us see how figure 2a was calculated. The 15° dispersion relation was solved for ω and inserted into (8). The resulting (x, z) turned out to be a function of k_x / ω. Trying all possible values of k_x / ω gave the curve.

Derivation of Energy Migration Equation

Energy migration in (x, t)-space is analyzed in a fashion similar to the way the group velocity was derived. Take depth to be large in the integral

$$\iint e^{i \; z \; [k_z(\omega, \, k_x) - \omega \, t / z + k_x \, x / z]} \; d\omega \; dk_x \tag{9}$$

The result is that the energy goes to

$$(x, t) \quad = \quad z \left(- \frac{\partial k_z}{\partial k_x} , \; \frac{\partial k_z}{\partial \omega} \right) \tag{10}$$

This justifies our previous assertion that (3) can be used to analyze energy propagation errors. Equation (10) was also used to calculate the curve in figure 3. The validity of the stationary phase concept is confirmed by figure 4, which was produced using inverse Fourier transformation.

Extrapolation Equations are not Frequency-Dispersive.

To prove that the familiar 15°, 45°, etc. wave extrapolators are not frequency-dispersive, recall from Section 2.2 that the dispersion relations all have the form $k_z / \omega = f(k_x / \omega)$, where f is a semicircle approximation, say, 15° or 45°. No dispersion relation of this form can be frequency-dispersive. Performing the derivatives required by (10), you see that while the (x, t)-coordinates of a wavefront depend on the dip angle through the parameter $v k_x / \omega$, they do not depend explicitly on ω. So any frequency dispersion observed in practice does not arise from a 15° or 45° approximation.

4.3 Frequency Dispersion
and Wave-Migration Accuracy

Frequency dispersion results from different frequencies propagating at
different speeds. The physical phenomenon of frequency dispersion is rarely
heard in daily life, although many readers may have heard it while ice skating
on lakes and rivers. Elastic waves caused by cracking ice propagate disper-
sively, causing pops to change into percussive notes. Frequency dispersion is
generally observable on seismic waves that propagate along the earth's surface
but frequency dispersion is hardly ever perceptible on internally reflected
waves. In seismic data processing, frequency dispersion is a nuisance and an
embarrassment to process designers. It arises mainly with the finite
differencing method because differential operators and difference operators do
not coincide at high frequencies. Frequency dispersion can always be
suppressed by sampling more densely, and it is the job of the production
analyst to see that this is done. Figure 1 depicts some dispersed pulses.

FIG. 4.3-1. (a) A pulse. (b) A
pulse slightly dispersed as by the
physical dissipation of high frequencies. (c) A pulse with a substantial   amount   of   frequency
dispersion, such as could result
from careless data processing.

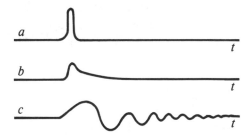

Frequency dispersion caused by data processing can be a useful warning
that the data is in danger of being aliased. Frequency-domain methods do
not depend on difference operators, so they have the advantage of not showing dispersion. The penalties that go along with this advantage are (1) limitation to constant material properties, (2) wraparound, and (3) the occurrence of
spatial aliasing without the warning of dispersion.

Figure 2 shows an example of frequency dispersion in migrated data. At
the top of the figure is a CDP stack. In the middle is the data after

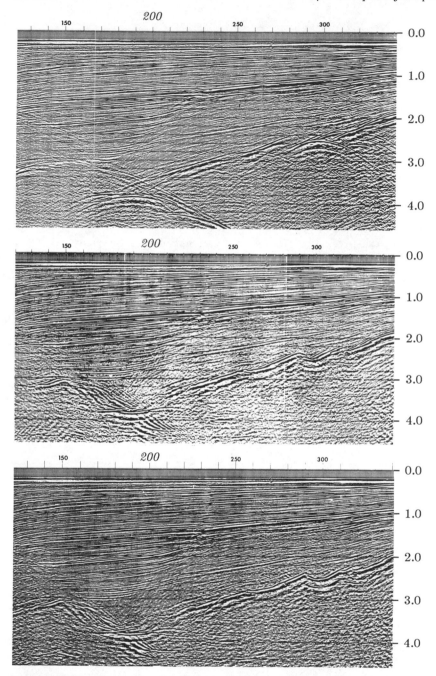

FIG. 4.3-2.   Conquering frequency dispersion. (Taner and Koehler, distributed by Seiscom Delta Inc.)

processing with no attempt to control frequency dispersion. The worst dispersion is near shot point 200 at 4 seconds. The bottom shows the data after reprocessing with greater attention to dispersion.

Spatial Aliasing

Aliasing can occur on the axes of time, depth, geophone, shot, midpoint, offset, or crossline. Aliasing is the worst on the horizontal space axes. Section 1.3, figure 3 provides an illustration. Looking at that figure, you get confused about whether the dip is to the left or right. Mathematical analysis has the same difficulty. The dispersion relation of the wave equation enables us to compute the vertical spatial frequency k_z from the temporal frequency ω, the velocity v, and the horizontal spatial frequency k_x using the semicircle relation $k_z(\omega, k_x) = \sqrt{\omega^2/v^2 - k_x^2}$. Sampling on the x-axis sets an upper limit on k_x equal to the Nyquist frequency $\pi/\Delta x$. Both frequency-domain methods and finite-difference methods treat higher frequencies as though they were folded back at the Nyquist frequency. Thus the semicircle dispersion relation is replicated above the Nyquist frequency, as shown in figure 3.

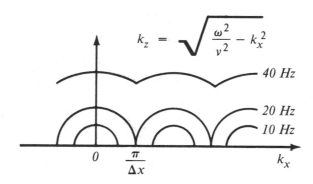

FIG. 4.3-3. The effective dispersion relation of the wave equation when the horizontal axis is sampled. Frequencies are given for typical zero-offset migration.

The problem of spatial aliasing begins when two circles touch each other, as shown at 20 Hz in figure 3. This occurs when a half-wavelength $v/2f$ equals the spatial sample rate Δx. The exploding-reflector model implies that the velocity to use is half the rock velocity. Thus the aliasing problem is avoided if $2f \, \Delta x < 1/2 \, v_{rock}$. For a rock velocity equal to 2 km/sec, the safe frequencies are listed in the table below:

	Δx	safe frequency
standard	25 m	< 20 Hz
reconnaissance	50 m	< 10 Hz
3-D cross line	100 m	< 5 Hz

Another view of the spatial aliasing problem is that steeply dipping waves are suppressed by the geophone group. (This disregards shot-space aliasing). From this standpoint the limit past which spatial aliasing begins should be thought of in terms of angles at which energy is missing from the data. Taking the ray angle to be 30° instead of 90° doubles horizontal wavelengths. Thus, for 30° and a rock velocity of 2 km/sec, to ensure safety from aliasing, frequencies should be in the ranges listed below:

	Δx	safe frequency
standard	25 m	< 40 Hz
reconnaissance	50 m	< 20 Hz
3-D cross line	100 m	< 10 Hz

Because data usually has good signal above 40 Hz, wide-angle processing is often frustrated by spatial aliasing.

The problem of spatial aliasing usually overshadows the difference between the 15° and the 90° equations. Aliased energy does not move between hyperbola flanks and the apex. Aliased energy tends to stay in place. This is illustrated on figure 4 which shows a 90° hyperbola and a 15° hyperboloid from a finite difference equation. Overall, there is little difference. Look at the amplitude of the hyperbolic arrival. It is dropping off faster than predicted by spherical spreading and the obliquity function. This is because the dispersion curve semicircles overlap one another. There can be no angles of propagation beyond that which aliases x. Since waves can't go so steeply, they don't. The pulse doesn't spread properly.

Second Space Derivatives

The defining equation for a second *difference* operator is

$$\frac{\delta^2}{\delta x^2} P = \frac{P(x + \Delta x) - 2P(x) + P(x - \Delta x)}{(\Delta x)^2} \tag{1}$$

The second *derivative* operator is defined by taking the limit

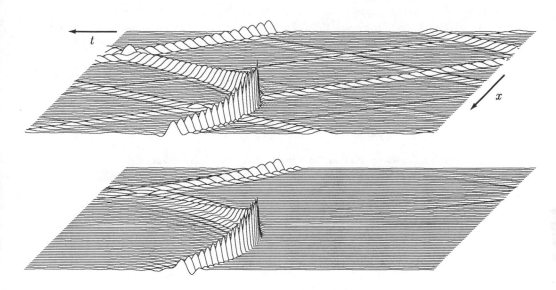

FIG. 4.3-4.   One   second   of   synthetic   hyperbolas   with $\Delta t = 4$ ms,
$\Delta x = 25$ meters, and velocity 2 km/sec. Fourier domain 90° hyperbola (top)
and 15° finite difference hyperboloid (bottom).

$$\frac{\partial^2}{\partial x^2} P \quad \underset{\lim \Delta x \to 0}{=} \quad \frac{\delta^2}{\delta x^2} P \tag{2}$$

Many different definitions can all go to the same limit as Δx goes to zero.
The problem is to find an expression that is accurate when Δx is larger
than zero and, on a practical level, is not too complicated. Our first objective
is to see how the accuracy of equation (1) can be evaluated quantitatively.
Second, we will look at an expression that is slightly more complicated than
(1) but much more accurate.

The basic method of analysis we will use is Fourier transformation. Take
the derivatives of the complex exponential $P = P_0 \exp(ikx)$ and look at
any errors as functions of the spatial frequency k. For the second derivative,

$$\frac{\partial^2}{\partial x^2} P \quad = \quad -k^2 P \tag{3}$$

Define $\hat{k}$ by an expression analogous to the difference operator:

$$\frac{\delta^2}{\delta x^2} P \quad = \quad -\hat{k}^2 P \tag{4}$$

Ideally $\hat{k}$ would equal k. Inserting the complex exponential $P = P_0 \exp(ikx)$ into (1), we see that the definition (4) gives an expression for $\hat{k}$ in terms of k:

$$-\hat{k}^2 P = \frac{P_0}{\Delta x^2}\left[e^{ik(x+\Delta x)} - 2 e^{ikx} + e^{ik(x-\Delta x)}\right] \tag{5a}$$

$$-\frac{\delta^2}{\delta x^2} = \hat{k}^2 = \frac{2}{\Delta x^2}\left[1 - \cos(k\,\Delta x)\right] \tag{5b}$$

It is a straightforward matter to make plots of $\hat{k}\,\Delta x$ versus $k\,\Delta x$ from (5b). The half-angle trig formula allows us to take an analytic square root of (5b), which is

$$\frac{\hat{k}\,\Delta x}{2} = \sin\frac{k\,\Delta x}{2} \tag{5c}$$

Series expansion shows that for low frequencies $\hat{k}$ is a good approximation to k. At the Nyquist frequency, defined by $k\,\Delta x = \pi$, the approximation $\hat{k}\,\Delta x = 2$ is a poor approximation to π.

The 1/6 Trick

Increased absolute accuracy may always be purchased by reducing Δx. Increased accuracy relative to the Nyquist frequency may be purchased at a cost of computer time and analytical clumsiness by adding higher-order terms, say,

$$\frac{\partial^2}{\partial x^2} \approx \frac{\delta^2}{\delta x^2} - \frac{\Delta x^2}{12}\frac{\delta^4}{\delta x^4} + etc. \tag{6}$$

As Δx tends to zero (6) tends to the basic definitions (1) and (2). Coefficients like the 1/12 in (6) may be determined by the Taylor-series method if great accuracy is desired at small k. Or somewhat different coefficients may be determined by curve-fitting techniques if accuracy is desired over some range of k. In practice (6) is hardly ever used, because there is a less obvious expression that offers much more accuracy at less cost! The idea is indicated by

$$\frac{\partial^2}{\partial x^2} \approx \frac{\dfrac{\delta^2}{\delta x^2}}{1 + b\,\Delta x^2\dfrac{\delta^2}{\delta x^2}} \tag{7a}$$

where b is an adjustable constant. The accuracy of (7a) may be numerically evaluated by substituting from (5b) to get

$$\left(\frac{\hat{k}\ \Delta x}{2} \right)^2 = \frac{\sin^2 \dfrac{k\ \Delta x}{2}}{1 - b\ 4\ \sin^2 \dfrac{k\ \Delta x}{2}} \tag{7b}$$

The square root of (7b) is plotted in figure 5 for a value of $b = 1/6$.

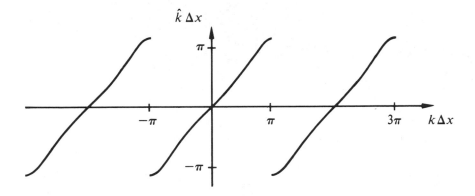

FIG. 4.3-5. Accuracy of the second-derivative representation (7) (for $b = 1/6$) as a function of spatial wavenumber. The sign of the square root of (7b) was chosen to agree with k in the range $-\pi$ to π and to be periodic outside the range. (Hale)

Taking b in (7) to be $1/12$, then (7) and (6) would agree to second order in Δx. The $1/12$ comes from series expansion, but the $1/6$ fits over a wider range and is a value in common use. Francis Muir has pointed out that the value $1/4 - 1/\pi^2 \approx 1/6.726$ gives an *exact* fit at the Nyquist frequency and an accurate fit over all lower frequencies! Few explorationists consider the remaining accuracy deficiency of (7) to be sufficient to warrant interpolation of field-recorded values. Figure 6 compares hyperbolas for various values of b. Observe in figure 6 that the longest wavelengths travel at the same speed regardless of b. The time axis in figure 6 is only 256 points long, whereas in practice it would be a thousand or more. So figure 6 exaggerates the frequency dispersion attributable in practice to finite differencing the x-axis.

Let us be sure it is clear how (7) is put into use. Take $b = 1/6$. The simplest prototype equation is the heat-flow equation:

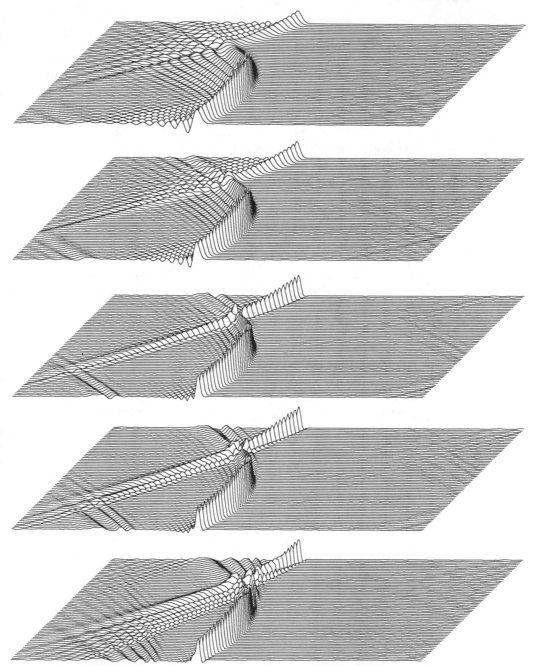

FIG. 4.3-6. Hyperbolas for $b = 0, 1/12, 1/6.726, 1/6, 1/5.$

$$\frac{\partial}{\partial t} q = \frac{\partial^2}{\partial x^2} q \approx \frac{\dfrac{\delta^2}{\delta x^2}}{1 + \dfrac{\Delta x^2}{6} \dfrac{\delta^2}{\delta x^2}} q \tag{8}$$

Multiply through the denominator:

$$\left[1 + \frac{\Delta x^2}{6} \frac{\delta^2}{\delta x^2} \right] \frac{\partial}{\partial t} q \approx \delta_{xx} q \tag{9}$$

Time and Depth Derivatives — the Bilinear Transform

You might be inclined to think a second derivative is a second derivative and that there is no mathematical reason to do time derivatives differently than space derivatives. This is not the case. A hint of disparity between t and x derivatives comes from boundary conditions. With time derivatives (and often with the depth z derivative) we must consider causality — which means the future is determined solely from the present and past. Appropriate boundary conditions on the time axis are initial conditions — the function (and perhaps some derivatives) is specified at *one* point, the initial point in time. For depth z that special point is the earth's surface at $z=0$. But lateral space derivatives are different: they require boundary conditions at two widely separated points, usually at the left and right sides of the volume.

The differential equation

$$\frac{dq}{dz} = i k_z(\omega, k_x) q \tag{10}$$

is associated with the very definition of k_z. The analogous difference equation will define $\hat{k}_z$:

$$\frac{q_{z+\Delta z} - q_z}{\Delta z} = i \hat{k}_z \frac{q_{z+\Delta z} + q_z}{2} \tag{11}$$

Inserting the solution of (10) $q = q_0 \exp(ik_z z)$ into (11) gives us the relation between the desired k_z and the actual $\hat{k}_z$.

$$i \hat{k}_z \Delta z = 2 \frac{e^{ik_z \Delta z} - 1}{e^{ik_z \Delta z} + 1} = 2 \frac{e^{ik_z \Delta z/2} - e^{-ik_z \Delta z/2}}{e^{ik_z \Delta z/2} + e^{-ik_z \Delta z/2}} \tag{12}$$

This equation is known as the bilinear transform (Section 4.6).

$$i \hat{k}_z \Delta z = 2 i \frac{\sin k_z \Delta z/2}{\cos k_z \Delta z/2} \tag{13}$$

$$\frac{\hat{k}_z \, \Delta z}{2} \;=\; \tan \frac{k_z \, \Delta z}{2} \tag{14}$$

Equation (14) gives the accuracy of first derivatives obtained using the Crank-Nicolson method. Recall the migration differencing schemes in Section 2.7. We did the time differencing in the same way that we did the depth differencing. So the same accuracy limitation must apply, namely,

$$\frac{\hat{\omega} \, \Delta t}{2} \;=\; \tan \frac{\omega \, \Delta t}{2} \tag{15}$$

Series expansion shows that $\hat{\omega}$ goes to ω as Δt goes to zero. Relative errors in ω at (4, 10, and 20) points per wavelength are (30%, 3%, and 1%). These errors are quite large, calling for either a choice of small Δt or a more accurate method than (14).

The bad news is that there does not seem to exist a representation of causal differentiation that is any more accurate than the Crank-Nicolson representation. There is nothing like the 1/6 trick. Thus the sample intervals of Δz and Δt must be reduced considerably from the Nyquist criterion. The practical picture may not be as bleak as the one I am painting. Many people are pleased with both the speed and accuracy of time-domain migrations at $\Delta t = 4$ milliseconds.

Stolt's classic paper [1978] besides introducing the fast Fourier transform migration method, points out that more accuracy can be achieved when the requirement of causality is dropped. Stolt shows how dropping causality at the known depth level while retaining it at the next level allows stable finite differencing. With the depth z-axis we are stuck with causal derivatives, although Fourier methods could be used for discrete layers. The depth axis is not so troublesome as the x- and t-axes, however, because it affects computer time only, not data storage.

Finite difference solutions don't just approximate the frequency — what they really do is to approximate $\exp ik_z \, \Delta z$. Solve (11) for the unknown.

$$q_{z+\Delta z} \;=\; \left[\frac{1 + i \, \hat{k}_z \, \Delta z / 2}{1 - i \, \hat{k}_z \, \Delta z / 2} \right] q_z \tag{16}$$

So for N_z layers in depth $z = N \, \Delta z$ we have the approximation

$$e^{ik_z N \Delta z} \;\approx\; \left[\frac{1 + i \, \hat{k}_z \, \Delta z / 2}{1 - i \, \hat{k}_z \, \Delta z / 2} \right]^N \tag{17}$$

which will be of later use for Fourier domain simulations of finite difference programs. Such simulations, coming in Section 4.7, enable us to compare the accuracy of various migration methods.

4.4 Absorbing Sides

Computer memory cells are often used to model points in a volume that contains propagating waves. Though we often wish to model an infinite volume, the number of computer cells is, regrettably, finite. Waves in the computer reflect back from the boundaries of the finite computer memory when we would prefer that the waves had gone away to infinity. To avoid the need for infinite computer capacity this section develops the theory of absorptive side boundary conditions.

There are two kinds of side boundary difficulties. First is where we are at the end of our observations. Second is where we somehow decided to limit the extent of our calculations. These boundaries might be the same. But to avoid confusion, let us presume that the data is of more limited extent than the computer memory. So alongside the data, which comprises the initial conditions for the calculation, is a region which will be called the *data padding*.

Data Padding

The crudest assumption is that additional zero-valued data may be presumed for the padded area. To avoid an edge diffraction artifact the data must merge smoothly with the padding. So zero padding is a good assumption only if the data is already small around the side boundary. When data is zero padded, it is debatable whether or not it should be tapered (gradually scaled to zero) to match up smoothly with the zero padding. I prefer to avoid tapering the data. That amounts to falsifying it. Instead I prefer to pad the data not with zeroes, but with something that looks more like the data. A simple way is to replicate the last trace, scaling it downward with distance from the boundary. This works best when the stepout of the data matches the stepout of the extension. Any theory for optimum data padding has two important ingredients: a noise model and a signal model. An ideal data extrapolation is rarely, if ever, available in practice. Section 3.5 contains suggestions for more elaborate models for extensions of gathers.

Truncation at Cable Ends and at Survey Ends

In exploration there are two kinds of horizontal truncation problems. The first, which is at the end of the geophone cable, affects mainly common-midpoint stacking. The second is at the geographical boundaries of the survey and affects mainly migration. In both migration and stacking, hyperboloids are collapsed to points. But the processes differ because of the data itself. With stacking, it is predictable that energy will dip downward toward the far-offset cable truncation. With migration, reflectors can dip either downward or upward at the ends of the section. The downward-dipping case is better behaved. There seismic events move smoothly from the boundary to the interior.

The troublesome case occurs with migration when the seismic events dip upward at the edges of the survey. Then downward continuation moves seismic energy toward the boundary. On arriving at the boundary, it bounces back with opposite dip, and interferes with energy still moving toward the boundary. The problem may be reduced by appending space to the sides of the dataset, thus providing the dipping energy with a place to go. (You have previously decided what initial data padding to put in this space).

Engquist Boundaries for the Scalar Wave Equation

The simplest "textbook" boundary condition is that a function should vanish on the boundary. A wave incident onto such a boundary reflects with a change in polarity (so that the incident wave plus the reflected wave will vanish on the boundary). The next-to-simplest boundary condition is the zero-slope condition. It is also a perfect reflector, but the reflection coefficient is $+1$ instead of -1. Two points at the edge of the differencing mesh are required to represent the zero-slope boundary. The most general boundary condition usually considered is a linear combination of function value and slope. This is also a two-point boundary condition. It so happens that our extrapolation equations (Section 2.2) contain only a single depth derivative, so that on the z-axis they are a two-point condition. Observing this, Björn Engquist recognized a new application for our extrapolation equations. Many researchers in other disciplines are interested in forward modeling, that is, evolving forward in time with an equation like the scalar wave equation, say, $P_{xx} + P_{zz} = P_{tt}/v^2$. These people suffer severely the consequences of limited memory. Engquist's idea was that they should use our extrapolation equations for their boundary conditions. (This idea led to his winning the SIAM prize). Suppose they desire an infinite absorbing volume surrounding a box in the (x, z)-plane. Then they need a boundary condition that goes all the way around the box. They could use the downgoing wave equation on the bottom of the box and the upcoming wave equation on the top edge. The sides could be handled analogously with an interchange of x and z. This idea was thoroughly tested and confirmed by Robert Clayton. For an example of one of his comparisons, see figure 1.

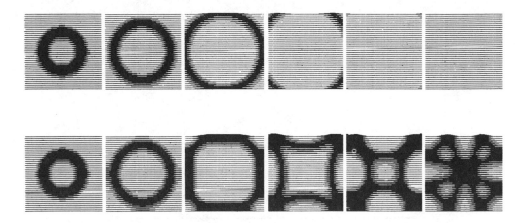

FIG. 4.4-1. Expanding circular wavefront in a box with absorbing sides (top) and zero-slope sides (bottom). (Clayton)

Engquist Side Conditions for the Extrapolation Equations

In data processing an extrapolation equation is used in the interior of the region under study. This is unlike forward modeling, in which the full scalar wave equation is used in the interior and an extrapolation equation can be used on the boundary. The scalar wave equation has a circular dispersion relation, whereas the extrapolation equation has, ideally, a semicircular one. Reasoning by analogy, Engquist speculated that a quarter-circular dispersion relation might be an ideal side boundary for wave-extrapolation problems. To make his idea more specific and immediately applicable, he proposed that the quarter-circle be approximated by a straight line. This is shown in figure 2.

The advantage of the straight-line dispersion relation is that in the space domain it is represented by a simple, first-order, differential equation. A first-order equation has first derivatives that can be expressed over just two data points, and thus it can be used as a conventional, two-point, side-boundary condition. The right-side equation in figure 2 defines the boundary dispersion relation D:

$$0 \;=\; \frac{vk_z}{\omega} - 1 + const \; \frac{k_x}{\omega} \;=\; D(\omega, k_x, k_z) \tag{1}$$

In (t, x, z)-space this equation is

$$0 \;=\; (v\,\frac{\partial}{\partial z} + \frac{\partial}{\partial t} + const\,\frac{\partial}{\partial x})\,P \tag{2}$$

In retarded time, $\partial/\partial z$ may be eliminated by substitution from the interior equation.

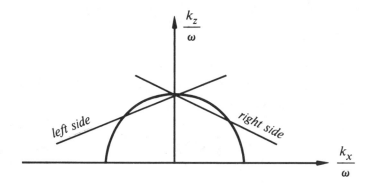

FIG. 4.4-2. Dispersion relation of simple absorbing side conditions.

For a mathematical, nonphysical point of view, imagine some peculiar physics which prescribe that the physical equation that applies in some region is just the equation which has the dispersion relation of the absorbing side condition. Beside this fictitious region imagine another in which the usual extrapolation equation applies. At the point of contact between the regions the solutions would match. It may come as no great surprise that the smallest boundary reflections would occur where the two dispersion relations were a good match to each other. So the slope of the straight line would be selected to form a good fit over the range of angles of interest. An example of side-boundary absorption during migration is shown in figure 3.

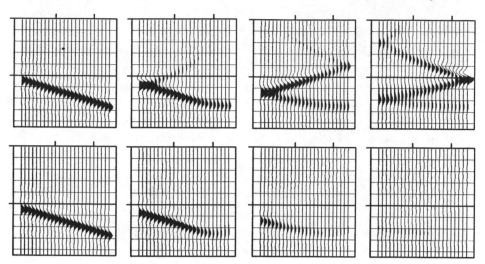

FIG. 4.4-3. Downward continuation with zero slope side boundaries (top), and absorbing side boundaries (bottom). (Toldi)

Size of the Reflection Coefficient

Let us look at some of the details of the reflection coefficient calculation. A unit amplitude, monochromatic plane wave incident on the side boundary generates a reflected wave of magnitude c. The mathematical representation is:

$$P(x, z) = e^{-i \omega t + i k_z z} \left(e^{+i k_x x} + c \, e^{-i k_x x} \right) \tag{3}$$

In equation (3) ω and k_x are arbitrary, and k_z is determined from ω and k_x using the dispersion relation of the *interior* region, i.e., a semicircle approximation. Assuming this interior solution is applicable at the side boundary, you insert equation (3) into the differential equation (2), which represents the side boundary. As a result, $\partial/\partial x$ is converted to $+i k_x$ on the incident wave, and $\partial/\partial x$ is converted to $-i k_x$ on the reflected wave. Also, $\partial/\partial z$ is converted to $i k_z$. Thus the first term in (3) produces the dispersion relation $D(\omega, k_x, k_z)$ times the amplitude P. The second term produces the reflection coefficient c times $D(\omega, -k_x, k_z)$ times P. So (2) with (3) inserted becomes:

$$c = \frac{-D(\omega, k_x, k_z)}{D(\omega, -k_x, k_z)} \tag{4}$$

271

The case of zero reflection arises when the numerical value of k_z selected by the interior equation at (ω, k_x) happens also to satisfy exactly the dispersion relation D of the side boundary condition. This explains why we try to match the quarter-circle as closely as possible. The straight-line dispersion relation does *not* correspond to the most general form of a side boundary condition, which is expressible on just two end points. A more general expression with adjustable parameters b_1, b_2, and b_3, which fits even better, is

$$D(\omega, k_x, k_z) = \left(1 - b_3\frac{vk_x}{\omega}\right)\frac{vk_z}{\omega} - \left(b_1 - b_2\frac{vk_x}{\omega}\right)$$

The absolute stability of straight-line absorbing side boundaries for the 15° equation can be established, including the discretization of the x-axis. Unfortunately, an airtight analysis of stability seems to be outside the framework of the Muir impedance rules. As a consequence, I don't believe that stability has been established for the 45° equation.

4.5 Tuning up Fourier Migrations

First we will see how to migrate dips greater than 90°. Then we will attack the main two disadvantages of Fourier migrations, namely, their periodicity and their poor tolerance of space-variable velocity.

Dips Greater than 90°

Migration of dips greater than 90° requires careful handling of evanescent energy. As this is being written, most migration-by-depth-extrapolation programs ignore or set to zero the energy that turns evanescent. The proper thing to do with energy becoming evanescent at depth z is to save it for a second pass *upward*. The upward pass begins from the bottom of the section with a zero downgoing wave. As the downgoing wave is extrapolated upward, the saved evanescent energy is reintroduced. As usual, the images are withdrawn from the wave at time $t = 0$.

To illustrate the concept, a program will be sketched that makes two images, first the usual image of the top side of the reflector, and second the image of the under side. The images may be viewed separately or summed.

The program makes the simplifying restriction on the velocity that $dv/dz \geq 0$. Because of this assumption, evanescent energy can be stored "in place" and ignored until the return pass. It is worth noting that the second pass is cheaper than the first pass because the region in which evanescence never occurred, $|k| < |\omega|/v(\tau_{max})$, need not be processed.

```
# first pass of conventional phase-shift migration.
P(ω, k_z) = FT[u(t, x)]
For τ = Δτ, 2Δτ, ···, τ_max {
     For all k_z {
          Uimage(k_z, τ) = 0.
          For all ω > |k| v(τ) {
               C = exp(- i ω Δτ √(1 - v(τ)² k_z²/ω²))
               P(ω, k_z) = P(ω, k_z) * C
               Uimage(k_z, τ) = Uimage(k_z, τ) + P(ω, k_z)
          }
     }
     uimage(x, τ) = FT[Uimage(k_z, τ)]
}


# Second pass for underside image.
For τ = τ_max, τ_max-Δτ, τ_max-2Δτ, ···, 0 {
     For all k_z {
          Dimage(k_z, τ) = 0.
          For ω = |k| v(τ) to ω = |k| v(τ_max) {
               # The wave changes direction but so does Δτ
               C = exp(- i ω Δτ √(1 - v(τ)² k_z²/ω²))
               P(ω, k_z) = P(ω, k_z) * C
               Dimage(k_z, τ) = Dimage(k_z, τ) + P(ω, k_z)
          }
     }
     dimage(x, τ) = FT[Dimage(k_z, τ)]
}
```

Stopping Phase-Shift Migration Wraparound (S. Levin [1984])

Figure 1 shows a family of hyperbolas. Notice that these hyperbolas do not extend to infinite time but they truncate at a cut-off time t_c. A Fourier method will be described to create such time-truncated data. The method leads to a phase-shift migration program without wraparound artifacts.

When the fast Fourier transformation algorithm first came into use people noticed that it could be used for filtering. Transient filtering could be done exactly in the periodic Fourier domain if signals and filters were

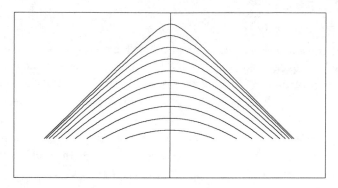

FIG. 4.5-1. Hyperbolas truncated at a particular time.

surrounded by enough zero padding. The same concept applies with migration. If field data and migration hyperbolas are surrounded by enough zeroes in the time- and space-domain then migration can be done in the Fourier domain with *no* wraparound. The trick is to see how the truncated hyperbolas in figure 1 can be constructed in the Fourier domain.

To have truncations at time t_c, special point sources must be used. The deeper the source, the narrower must be its angular aperture. Take a hyperbola with first arrival at time t_0 to be truncated at some time t_c. The propagation angle θ of energy at the cutoff is given by $\cos\theta = t_0/t_c$. So exploding reflectors have their k_x-spectrum truncated at $\sin\theta = v\,k_x/\omega$. A 90° aperture implies echoes with an infinite time delay. Here is a sketch of the program.

\# Modeling with time truncation at t_c

$Model(k_x, z) = FT[model(x, z)]$

For all ω and all k_x

  $U(\omega, k_x) = 0.$

For $z = z_{max}, z_{max}{-}\Delta z, z_{max}{-}2\Delta z, \cdots, 0$ {

    For all ω {

        For all $|k_x| < |\omega|/v$ {

           if ($z < v \ t_c$){

              $sine = \sqrt{1{-}z^2/v^2 t_c^2}$

              if($|v \ k_x| < |\omega| \ sine$)

                  $aperture = 1.$

              else

                  $aperture = 0.$

           }

        else

           $aperture = 0.$

        $U(\omega, k_x) = U(\omega, k_x) \ e^{-i \, \Delta z \, \omega \sqrt{v^{-2} - k_x^2/\omega^2}} + aperture \ * \ Model(k_x, z)$

        }

    }

}

$u(t, x) = FT2D[U(\omega, k_x)]$

The above modeling program may be converted to a migration program (as in Section 1.3) by running the depth z loop down instead of up and by multiplying the downward continued data by the aperture function. The modifications to the program not only improve the quality of the migration, but the calculation is faster.

Controlling Stolt Migration Wraparound

As with the phase-shift algorithm, the key to reducing the computational artifacts of the Stolt algorithm is to suppress the time-domain wraparound. We will see that this amounts to accurate frequency-domain interpolation.

First, consider an impulse function at time t_0. Its Fourier transform is $\exp(-i\omega t_0)$. If t_0 is large, then the Fourier transform is a rapidly oscillating function of ω. Rapidly oscillatory functions are always difficult to interpolate. It is better to shift backward the time function, thereby smoothing its frequency function, then to interpolate the frequency function, and finally to undo the shift. Given seismic data on the time interval $0 < t < T$, the frequency function will be smoother if the data is shifted to an interval $-T/2 < t < T/2$. So the first proposed improvement to the Stolt migration program is to multiply in the frequency domain by $\exp(i\omega T/2)$, then interpolate, and finally multiply by $\exp(-i\omega T/2)$.

Linear interpolation is almost the easiest form of interpolation. On the other hand, Fourier transform theory suggests interpolation with the *sinc* function (by definition sinc $u = (sin\ u)/u$). The *sinc* function of frequency, when brought back into the time domain is a rectangle function of time. Take this rectangle function to be nonzero on the interval $-T/2 < t < T/2$. Recall that the fast (inverse) Fourier transform algorithm sums at uniform intervals in the frequency domain. This implicitly assumes zero between sample points, which in turn assumes that the time-domain function is periodic outside the given time interval. Now take the rectangle function of time to be the multiplier in the time domain that converts the periodic time function to the observed transient one. This multiplication in the time domain is equivalent to a convolution in the frequency domain with the appropriate *sinc* function. Convolution of the continuous *sinc* function with the given discrete-interval frequency function is really interpolation. Unfortunately, the *sinc* function extends infinitely down the frequency axis. Worse yet, it decays slowly. So some approximation or truncation of the *sinc* is used. Bill Harlan showed that tapered *sinc* functions achieve satisfactory accuracy more cheaply than zero padding. It seems, however, that the best approach is both to zero pad and to use some *sinc*-like interpolation. A definitive study of interpolation is that of Rosenbaum and Boudreaux [1981].

Stolt Stretch

The great strength and the great weakness of the Stolt migration method is that it uses Fourier transformation over depth. This is a strength because it makes his method much faster than all other methods. And it is a weakness because it requires a velocity that is a constant function of depth. The earth velocity typically ranges over a factor of two within the seismic section, and the effect of velocity on migration tends to go as its square. To ameliorate this difficulty, Stolt suggested stretching the time axis to make the data look more like it had come from a constant-velocity earth. Stolt proposed the stretching function

$$\tau(t) = \left[\frac{2}{v_0^2} \int_0^t t\ v_{RMS}^2(t)\ dt \right]^{1/2} \tag{1a}$$

where

$$v_{RMS}^2(t) = \frac{1}{t} \int_0^t v^2(t)\ dt \tag{1b}$$

At late times, which are associated with high velocities, Stolt's stretch implies that τ grows faster than t. The τ-axis is uniformly sampled to allow the fast Fourier transform. Thus at late time the samples are increasingly dense on the t-axis. This is the opposite of what earth Q and the sampling theorem suggest, but most people consider this a fair price.

The most straightforward derivation of (1) is based on the idea of matching the curvature of ideal hyperbola tops to the curvature on the stretched data. The equation of an ideal hyperbola in (x, τ)-space is

$$v_0^2 \, \tau^2 \; = \; x^2 + z^2 \tag{2}$$

Simple differentiation shows that the curvature at the hyperbola top is

$$\left. \frac{d^2 \tau}{dx^2} \right|_{x=0} = \frac{1}{\tau \, v_0^2} \tag{3}$$

It can be shown that in a stratified medium, equation (3) applies, except that the velocity must be replaced by the RMS velocity:

$$\left. \frac{d^2 t}{dx^2} \right|_{x=0} = \frac{1}{t \, v_{RMS}^2} \tag{4}$$

We seek a stretched time $\tau(t)$. We would like to match the curves $t(x)$ and $\tau(x)$ for all x. But that would overdetermine the problem. Instead we could just match the derivatives at the hyperbola top, i.e., the second derivative of $\tau[t(x)]$ with respect to x at $x=0$. With the substitutions (3) and (4), this would give an expression for $\tau \, d\tau/dt$ which, after integrating and taking its square root, yields (1).

A different derivation of the stretch gives a more accurate result at steeper angles. Instead of matching hyperbola curvature at the top, we go some distance out on the flank and match the slope and value. It is the flanks of the hyperbola that actually migrate, not the tops, so this result is more accurate. Algebraically the derivation is also easier, because only *first* derivatives are needed. Differentiating equation (2) with respect to x for a reflector at any depth z_j gives

$$\frac{d\tau}{dx} = \frac{x}{\tau \, v_0^2} \tag{5}$$

There is an analogous expression in a stratified medium. To obtain it, solve $x = \int v \sin\theta \, dt = p \int v^2 \, dt$ for $p = dt/dx$:

$$\frac{dt}{dx} = \frac{x}{\displaystyle\int_0^t v^2(p, t) \, dt} \tag{6}$$

Expressions (5) and (6) play the same role as (3) and (4), but (5) and (6) are valid everywhere, not just at the hyperbola top. Differentiating $\tau(t)$ gives

$$\frac{d\tau}{dx} = \frac{d\tau}{dt}\frac{dt}{dx} \tag{7}$$

Inserting (5) and (6) into (7) gives

$$\frac{x}{\tau v_0^2} = \frac{d\tau}{dt}\frac{x}{\int_0^t v^2(p,t)\,dt} \tag{8}$$

$$\tau\,d\tau = \left[\frac{1}{v_0^2}\int_0^t v^2(p,t')\,dt'\right]dt \tag{9}$$

Integrating (9) gives $\tau^2/2$ on the left. Then, taking the square root gives (1a) but with a new definition for RMS velocity:

$$v_{RMS}^2(t) = \frac{1}{t}\int_0^t v^2(p,t)\,dt \tag{1c}$$

The thing that is new is the presence of the Snell parameter p. In a stratified medium characterized by some velocity, say, $v'(z)$, the velocity $v(p,t)$ is defined for the tip of the ray that left the surface at an angle with a stepout p. In practice, what value of p should be used? The best procedure is to look at the data and measure the $p = dt/dx$ of those events that you wish to migrate well. A default value is $p = 2(sin\,30°)/(2.5$ km/sec$) = .4$ millisec/meter. The factor of 2 is from the exploding-reflector model.

Gazdag's v(x) Method

The phase-shift method of migration is attractive because it allows for arbitrary depth variation in velocity and arbitrary angles of propagation up to $90°$. Unfortunately, lateral variation in velocity is not permitted because of the Fourier transformation over the x-axis. To alleviate this difficulty, Gazdag and Sguazzero [1984] proposed an interpolation method. Recall from Section 1.3 that the phase-shift method 2-D Fourier transforms the data $p(x,t)$ to $P(k_x,\omega)$. Then $P(k_x,\omega)$ is downward continued in steps of depth by multiplication with $\exp[ik_z(\omega,k_x)\,\Delta z]$. Gazdag proposed several reference velocities, say, v_1, v_2, v_3, and v_4. He downward continued one depth step with each of the velocities, obtaining several reference copies of the downward-continued data, say, P_1, P_2, P_3, and P_4. Then he inverse Fourier transformed each of the P_j over k_x to $p_j(x,\omega)$. At each x, he interpolated the reference waves of nearest velocity to get a final value, say, $p(x,\omega)$

which he retransformed to $P(k_z, \omega)$ ready for another step. This appears to be an inefficient method since it duplicates the usual migration computation for each velocity. Surprisingly, the method seems to be successful, perhaps because of the peculiar nature of computation using an array processor.

EXERCISE

1. To obtain a sharp cutoff in time t_c requires a broad bandwidth in the spectral domain. Given that figure 1 is expressed on a 1000×1000 mesh, deduce the uncertainty in the cutoff t_c.

2. The phase-shift method tends to produce a migration that is periodic with z because of the periodicity of the Fourier transform over t. Ordinarily, this is not troublesome because we do not look at large z. The upcoming wave at great depth should be zero *before* $t=0$. Kjartansson pointed out that periodicity in z could be avoided if the wave at $t=0$ is subtracted from the wavefield before the computation descends further. Thus, information could never get to negative time and "wrap around." Indicate how the program should be changed.

4.6 Impedance

Classical physics gives much attention to energy conservation and dissipation. Engineering filter theory gives much attention to causality — that there can be no response before the excitation. In geophysics we often need to ensure both causality and energy loss. We need to incorporate both, not only in theoretical derivations, but also in computations, and sometimes in computations that are discretized in time. There is a special class of mathematical functions called *impedance* functions that describe causal, linear disturbances in physical objects that dissipate energy.

Nature evolves forward in time. Naturally, impedance functions play a fundamental role in any modeling calculation where time evolves from past to future. Besides their use in physical modeling, impedances also find use in the depth extrapolation of waves. We geophysicists take data on the earth's surface and extrapolate downward to get information at depth. It is not the same as nature's extrapolation in time. In principle we don't require impedance functions to extrapolate in depth. But depth extrapolations made without impedance functions could exhibit growing oscillations, much like a physical system receiving energy from an external source. In fact, "straight-forward" implementations of physical equations often exhibit unstable extrapolations. By formulating our extrapolation problems with impedance functions, we ensure stability. Of all the virtues a computational algorithm can have — stability, accuracy, clarity, generality, speed, modularity, etc. — the most important seems to be stability.

In this section we examine the theory of impedance functions, their precise definition, their computation in the world of discretized time, and the rules for combining simple impedances to get more complicated ones. We will also examine other special functions, the *minimum-phase* filter and the *reflectance* filter in their relation to the impedance filter. Wide-angle wave extrapolation and migration in the time domain will be formulated with impedance functions. Rocks are unlike "pure" substances because they contain irregularities at all scales. A particularly simple impedance function will be found that mimics the dissipation of energy in rocks, unlike the classical equations of Newtonian viscoelasticity.

Beware of Infinity!

To prove that one equals zero take an infinite series, for example, 1, –1, +1, –1, +1, $\cdots$ and group the terms in two different ways, and add them in this way:

$$(1{-}1) + (1{-}1) + (1{-}1) + \cdots \quad = \quad 1 + (-1{+}1) + (-1{+}1) + \cdots$$

$$0 \;+\; 0 \;+\; 0 \;+ \cdots \quad = \quad 1 + \;\; 0 \;\; + \;\; 0 \;\; + \quad \cdots$$

$$0 = 1$$

Of course this does not prove that one equals zero: it proves that care must be taken with infinite series. Next, take another infinite series in which the terms may be regrouped into any order without fear of paradoxical results. Let a pie be divided into halves. Let one of the halves be divided in two, giving two quarters. Then let one of the two quarters be divided into two eighths. Continue likewise. The infinite series is 1/2, 1/4, 1/8, 1/16, $\cdots$. No matter how the pieces are rearranged, they should all fit back into the pie plate and exactly fill it.

The danger of infinite series is not that they have an infinite number of terms but that they may sum to infinity. Safety is assured if the sum of the absolute values of the terms is finite. Such a series is called *absolutely convergent*.

Z - transform

The Z-transform of an arbitrary, time-discretized function x_t is defined by

$$X(Z) = \cdots x_{-2} Z^{-2} + x_{-1} Z^{-1} + x_0 + x_1 Z + x_2 Z^2 + \cdots \qquad (1)$$

Give Z the physical interpretation of time delay by one time unit. Then Z^2 delays two time units. Expressions like $X(Z) U(Z)$ and $X(Z) U(1/Z)$ are useful because they imply convolution and cross-correlation of the time-domain coefficients. (See FGDP).

Going on to consider numerical values for the delay operator Z, we discover that it is useful to ask whether $X(Z)$ is finite or infinite. Numerical values of Z that are of particular interest are $Z=+1$, $Z=-1$, and all those complex values of Z which are unit magnitude, say, $|Z|=1$ or

$$Z = e^{i \omega \Delta t} \qquad (2)$$

where ω is the real Fourier transform variable. Taking ω to be real means that Z is on the unit circle. Then the Z-transform is a discrete Fourier transform. Our attention can be restricted to time functions with a finite amount of energy by demanding that $U(Z)$ be finite for all values of Z on the unit circle $|Z|=1$. Filter functions are always restricted to have finite energy.

The most straightforward way to say that a filter is causal is to say that its time domain coefficients vanish before zero lag, that is $u_t=0$ for $t<0$. Another way to say it is to say that $U(Z)$ is finite for $Z=0$. At $Z=0$ the Z-transform would be infinite if the coefficients u_{-1}, u_{-2}, etc. were not zero. For a causal function, each term in $|U(Z)|$ will be smaller if Z is taken inside the disk $|Z|<1$ rather than on it. Thus convergence at $Z=0$ and on the circle $|Z|=1$ implies convergence everywhere inside the unit disk. So boundedness combined with causality means convergence in the unit disk. Convergence at $Z=0$ but not on the circle $|Z|=1$ would refer to a causal function with infinite energy, a case of no practical interest. What kind of function converges on the circle, at $Z=\infty$, but not at $Z=0$? What function converges at all three places, $Z=0$, $Z=\infty$, and $|Z|=1$?

The filter $1/(1-2Z)$ can be expanded into powers of Z in (at least) two different ways. These are

$$\frac{1}{1-2Z} \;=\; 1 + 2Z + 4Z^2 + 8Z^3 + \cdots \tag{3}$$

$$=\; -\frac{1}{2Z}\,\frac{1}{1-\dfrac{1}{2Z}} \;=\; -\frac{1}{2Z}\left[\,1 + \frac{1}{2Z} + \frac{1}{4Z^2} + \cdots\,\right]$$

Which of these two infinite series converges depends on the numerical value of Z. For $|Z|=1$ the first series diverges, but the second converges. So the only acceptable filter is *anticausal*. Is a series expansion unique? It is if it converges. Complex-variable theory proves this.

Let b_t denote a filter. Then a_t is its inverse filter if the convolution of a_t with b_t is a delta function. In the Fourier domain, we would say that filters are inverse to one another if their Fourier transforms are inverse to one another. Z-transforms can be used to define the inverse filter, say, $A(Z)=1/B(Z)$. Whether the filter $A(Z)$ is causal depends on whether it is finite everywhere inside the unit disk, or really on whether $B(Z)$ vanishes *anywhere* inside the disk. For example, $B(Z)=1-2Z$ vanishes at $Z=1/2$. There $A(Z)=1/B(Z)$ must be infinite, that is to say, the series $A(Z)$ must be nonconvergent at $Z=1/2$. Thus — as we have just seen — a_t is noncausal. A most interesting case, called *minimum phase,* occurs when both a filter $B(Z)$ and its inverse are causal. In summary:

causal	$|B(Z)| < \infty$ for $|Z| \leq 1$
causal inverse	$|1/B(Z)| < \infty$ for $|Z| \leq 1$
minimum phase	*both above conditions*

Review of Impedance Filters

Use Z-transform notation to define a filter $R(Z)$, its input $X(Z)$, and its output $Y(Z)$. Then

$$Y(Z) \;=\; R(Z)\,X(Z) \tag{4}$$

The filter $R(Z)$ is said to be *causal* if the series representation of $R(Z)$ has no negative powers of Z. In other words, y_t is determined from present and past values of x_t. Additionally, the filter $R(Z)$ is *minimum phase* if $1/R(Z)$ has no negative powers of Z. This means that x_t can be determined from present and past values of y_t by straightforward polynomial division in

$$X(Z) = \frac{Y(Z)}{R(Z)} \tag{5}$$

Given that $R(Z)$ is already minimum phase, it can in addition be an impedance function if positive energy or work is represented by

$$0 \leq work = \sum_t force \times velocity = \sum_t voltage \times current \tag{6a}$$

$$= \frac{1}{2} \sum_t (\bar{x}_t\, y_t + \bar{y}_t\, x_t) \tag{6b}$$

$$= coef\ of\ Z^0\ of\ \left[\bar{X}\left(\frac{1}{Z}\right) Y(Z) + \bar{Y}\left(\frac{1}{Z}\right) X(Z) \right] \tag{6c}$$

$$= \frac{1}{2\pi} \int_0^{2\pi} Re\,(\bar{X}\ Y)\, d\omega \tag{6d}$$

$$= \int Re\,(\bar{X}\ R\ X)\, d\omega = \int \bar{X}\ X\ Re\,(R)\, d\omega \tag{6e}$$

Since $\bar{X}X$ could be an impulse function located at any ω, it follows that $Re\,[R(\omega)] \geq 0$ for all real ω. In summary:

Definition of an Impedance	
causality	$r_t = 0$ for $t < 0$ i.e. $\|R(Z)\| < \infty$ for $\|Z\| \leq 1$
causal inverse	$\|1/R(Z)\| < \infty$ for $\|Z\| \leq 1$
dissipates energy	$2\,Re\,R(\omega) = R(Z) + \bar{R}(1/Z) \geq 0$ *real* ω

Adding an impedance to its Fourier conjugate gives a real positive function (the imaginary part of which is zero) like a power spectrum, say,

$$\left(r_0 + r_1 Z + r_2 Z^2 + \cdots \right) + \tag{7a}$$

$$\left(\bar{r}_0 + \bar{r}_1 \frac{1}{Z} + \bar{r}_2 \frac{1}{Z^2} + \cdots \right) \geq 0 \quad for\ real\ \omega$$

$$R(Z) + \bar{R}\left(\frac{1}{Z}\right) \geq 0 \quad for\ real\ \omega \tag{7b}$$

which is the basis for the remarkable fact that every impedance time function
is one side of an autocorrelation function.

Impedances also arise in economic theory when X and Y are price
and sales volume. I suppose that there the positivity of the impedance means
that in the game of buying and selling you are bound to lose!

Causal Integration

Begin with a function in discretized time p_t. The Fourier transform
with the substitution $Z = \exp(i\omega\,\Delta t)$ is the Z-transform

$$P(Z) \quad = \quad \cdots p_{-2}\,Z^{-2} + p_{-1}\,Z^{-1} + p_0 + p_1\,Z + p_2\,Z^2 + \cdots \tag{8}$$

Define $-i\,\hat{\omega}$ (which will turn out to be an approximation to $-i\,\omega$) by

$$\frac{1}{-i\,\hat{\omega}\,\Delta t} = \frac{1}{2}\frac{1+Z}{1-Z} \tag{9}$$

Define another time function q_t with Z-transform $Q(Z)$ by applying the
operator to $P(Z)$:

$$Q(Z) \quad = \quad \frac{1}{2}\frac{1+Z}{1-Z}\,P(Z) \tag{10}$$

Multiply both sides by $(1-Z)$:

$$(1-Z)\,Q(Z) \quad = \quad \frac{1}{2}(1+Z)\,P(Z) \tag{11}$$

Equate the coefficient of Z^t on each side:

$$q_t - q_{t-1} \quad = \quad \frac{p_t + p_{t-1}}{2} \tag{12}$$

Taking p_t to be an impulse function, we see that q_t turns out to be a step
function, that is,

$$p \quad = \quad \cdots\, 0,0,0,0,0,1,0,0,0,0,0,0,\,\cdots \tag{13a}$$

$$q \quad = \quad \cdots\, 0,0,0,0,0,\frac{1}{2},1,1,1,1,1,1,\,\cdots \tag{13b}$$

So q_t is the discrete domain representation of the integral of p_t from
minus infinity to time t. It is the same as a Crank-Nicolson-style numerical
integration of the differential equation $dQ/dt = P$. The operator
$(1+Z)/(1-Z)$ is called the bilinear transform. The accuracy of the approximation to differentiation can be seen by multiplication on top and bottom by
$Z^{-.5}$ and substitution of $Z = e^{i\omega\Delta t}$.

$$-i\ \frac{\hat{\omega}\ \Delta t}{2}\ =\ \frac{1-Z}{1+Z}\ =\ \frac{Z^{-.5}-Z^{+.5}}{Z^{-.5}+Z^{+.5}}\ = \tag{14a}$$

$$-i\ \frac{\hat{\omega}\ \Delta t}{2}\ =\ -i\ \frac{\sin \omega\ \Delta t/2}{\cos \omega\ \Delta t/2}\ =\ -i\ \tan \frac{\omega\ \Delta t}{2} \tag{14b}$$

The integration operator has a pole at $Z=1$, which is exactly on the unit circle. This raises the possibility of the paradox of infinity. In other words, there are other noncausal expansions too. For example, taking $1/(-i\omega)$ to be an imaginary, antisymmetric function of ω implies a real, antisymmetric time function, namely, $sgn(t) = t/|t|$, which is not usually regarded as the integration operator. To avoid any ambiguity, we introduce here a small positive number ϵ and define $\rho = 1 - \epsilon$. The integration operator becomes

$$I\ =\ \frac{1}{2}\ \frac{1+\rho Z}{1-\rho Z} \tag{15a}$$

$$=\ \frac{1}{2}\ (1+\rho Z)\ \left[1+\rho Z + (\rho Z)^2 + (\rho Z)^3 + \ \cdots \ \right] \tag{15b}$$

$$=\ \frac{1}{2}\ +\ \rho Z + (\rho Z)^2 + (\rho Z)^3 + \ \cdots \tag{15c}$$

Because ρ is slightly less than one, this series converges for any value of Z on the unit circle. If ϵ had been slightly negative instead of positive the expansion would have come out in negative instead of positive powers of Z.

Now the big news is that the causal integration operator is an example of an impedance function. The operator is clearly causal with a causal inverse. Let us check in the frequency domain that the real part is positive. Rationalizing the denominator gives

$$I\ =\ \frac{1}{2}\ \frac{(1+\rho Z)}{(1-\rho Z)}\ \frac{(1-\rho/Z)}{(1-\rho/Z)}\ =\ \frac{(1-\rho^2)+\rho(Z-1/Z)}{positive} \tag{16a}$$

$$=\ \frac{(1-\rho^2)+2\,i\,\rho \sin \omega\Delta t}{positive} \tag{16b}$$

Again, it is the choice of a positive ϵ that has caused $1-\rho^2$, and hence the real part to be positive for all ω, as shown in figure 1.

As multiplication by $-i\omega$ in the frequency domain is associated with differentiation d/dt in the time domain, so is division by $-i\omega$ associated with integration. People usually associate the asymmetric operator $(1, -1)$ with differentiation. But notice that the inverse to the causal integration operator, namely,

$$I^{-1}\ =\ 2\ \frac{1-\rho Z}{1+\rho Z} \tag{17a}$$

$\rho = .92$

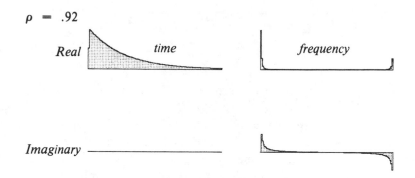

FIG. 4.6-1. The causal integration operator I. The frequency axis is represented by a discrete Fourier transform over 256 points. Zero time and zero frequency are on the left end of their respective axes.

$$I^{-1} = 2 - 4\rho Z + 4(\rho Z)^2 - 4(\rho Z)^3 + \cdots \qquad (17b)$$

also represents differentiation, although it is completely causal and not at all asymmetric. In linear systems analysis this representation of discrete differentiation is often the preferred one. The construction of higher-order, stable differential equations is subject to certain rules, to be covered, for combining impedances.

Occasionally it is necessary to have a *negative* real part for the differentiation operator. This can be achieved by taking ϵ to be negative, which means taking $\rho > 1$, and doing the infinite series expansion in powers of Z^{-1}, that is, anticausally instead of causally. In either the anticausal or the causal case the imaginary part will still be $-i\omega$, but the real part will have the opposite sign.

Muir's Rules for Combining Impedances

For every physical system that conserves or dissipates energy there is an impedance function. Impedance functions are *special* combinations of differential operators and positive-valued physical constants. We will see just what combinations are allowed.

To ensure stable computations, it is important to be able to ensure that a supposed impedance function really is an impedance function. A difficulty in applied geophysics is this: Although you might require results only over a limited range of frequencies, and you might make approximations that are reasonable within that range, if the calculated impedance becomes negative outside the applicable range (it often happens near the Nyquist frequency), then the impedance filter will yield a numerically divergent output. So even though the impedance is almost correct, it is not usable.

Francis Muir provided† three rules for combining simple impedances to get more complicated ones. These rules are especially useful because we can start from the discrete-time forms of the differentiation and integration operators. Let R' denote a new impedance function generated from known impedance functions R, R_1, and R_2. These three ways of combining impedance are

    1.   Multiplication by positive scalar a $R' = a R$

    2.   Inversion $R' = \dfrac{1}{R}$

    3.   Addition $R' = R_1 + R_2$

These rules do not include multiplication. Multiplication is not allowed because squaring, for example, doubles the phase angle, and thus may destroy the positivity of the real part. Since these rules do not include multiplication, but only scaling, summation, and inversion, the impedance functions that occur in nature will often be represented mathematically as *continued fractions*.

The first two of Muir's rules are so obvious we will not prove them. The third rule deserves more careful attention. To prove any rule, we need to show three things about R', namely, it is causal, it is PR (the Fourier transform has a positive real part), and it has an inverse. This last part is the hard part with Muir's third rule, namely, that the sum of two impedances has a causal inverse. Proof of this fact will take about two pages, and introduce several additional concepts.

Impedance Defined from Reflectance

The size of the class of filters called *impedances* will be seen to be large, because impedances are derived by transformation from an easily specified family of filters called *reflectances,* say, c_t and its Fourier transform $C(\omega)$. To be a reflectance, the time function must be strictly causal, and the frequency function must be strictly less than unity. By *strictly causal* it is meant that the time function vanishes both at zero time and before. For example, take $-1 < \rho < +1$ and the reflectance c_t to be an impulse of size

----

†Personal communication with Francis Muir.

ρ after a time Δt. The Fourier transform is

$$C = \rho Z = \rho e^{i\omega\Delta t} \tag{18}$$

Obviously, the product of two reflectances is another reflectance.

An impedance has been defined to be a causal filter with a causal inverse and a Fourier transform whose real part is positive. It will be shown that from any reflectance C the expression

$$R = \frac{1 - C}{1 + C} \tag{19}$$

generates an impedance. There are three things to show: R is causal, has a causal inverse, and is PR. First because of the assumption that C has a magnitude strictly less than unity, $C\,\overline{C} < 1$, the denominator expands to a convergent $1 + C + C^2 + \cdots$. Second, the inverse of R is found by simply changing the sign of C. Third, multiply top and bottom by the complex conjugate:

$$\text{Re } R = \text{Re } \frac{(1 - C)(1 + \overline{C})}{\underset{positive}{\qquad}} \overset{?}{\geq} 0 \tag{20a}$$

$$\text{Re } R = \text{Re } \frac{(1 - C\,\overline{C}) + imaginary}{\underset{positive}{\qquad}} \geq 0 \tag{20b}$$

which shows that R has a real part that is positive.

The expression for $R(C)$ is easily back-solved for $C(R)$, but the converse theorem, that every R generates a reflectance, is harder to show. Nevertheless, it will be proved, along with a deeper theorem. A filter that is both causal and PR is said to be CPR. The deeper theorem is that *every CPR has an inverse and hence is an impedance*. This will be proved by showing that every CPR — say, $\hat{R}$, — can be used to construct a reflectance $\hat{C}$, which, since it is a reflectance, implies that the CPR $\hat{R}$ is an impedance R. Backsolving gives

$$\hat{C} = \frac{1 - \hat{R}}{1 + \hat{R}} \tag{21}$$

Proof requires that two things be shown — first, that the magnitude of $\hat{C}$ is less than unity. To show this, form the magnitude of the denominator and subtract the magnitude of the numerator. The result is four times the real part of $\hat{R}$, which is positive. Second, $\hat{C}$ must be proved causal. This is harder. $(1 + \hat{R})^{-1}$ can be expanded into a sum of positive powers of $\hat{R}$ and hence of positive powers of the delay operator. But the convergence of the series is not assured, because nothing requires $\hat{R}$ to be less than unity.

To prove that $\hat{C}$ is causal, we will take advantage of rule 1, namely, that an impedance can be scaled by any real positive number that you like, and it will still be an impedance. Consider a function that is similar to $\hat{C}$.

$$B \;=\; \frac{1 - \epsilon\,\hat{R}}{1 + \epsilon\,\hat{R}} \tag{22}$$

Choose ϵ small enough that for all ω, $\epsilon\,|\,\hat{R}\,| <1$. This ensures a convergent expansion for the denominator in positive powers of $\hat{R}$ and hence Z. The expansion contains only positive powers in the delay operator. Thus B is a reflectance, and its corresponding impedance is $\epsilon\,\hat{R}$. But an impedance can always be scaled by a positive number. Taking the number to be $1/\epsilon$ shows that $\hat{R}$ is an impedance. This completes the proof that every CPR is an impedance.

So impedances arise more easily than you might think. It is not necessary to have a reflectance C to insert into the relation $R = (1-C)/(1+C)$. We only need to have a CPR.

Functional Analysis

We will establish the following theorems about exponentials, logarithms, and powers of Fourier transforms of filters:

1. The exponential of a causal filter is causal.

2. The exponential of a causal filter is a minimum-phase filter.

3. The logarithm of a minimum-phase filter is causal.

4. The Fourier domain representation of a minimum-phase filter is a curve that does not enclose the origin of the complex plane.

5. Any power of a minimum-phase filter is minimum phase.

6. Any real fractional power $-1 \le \rho \le 1$ of an impedance function is an impedance function.

To establish theorem 1, define the Z-transform of an arbitrary causal function

$$U(Z) \;=\; u_0 + u_1 Z + u_2 Z^2 + \;\cdots \tag{23}$$

and substitute it into the familiar power series for the exponential function:

$$B(Z) \;=\; e^U \;=\; 1 + U + \frac{U^2}{2!} + \frac{U^3}{3!} + \cdots \qquad (\,|\,U\,| < \infty) \tag{24}$$

No negative powers of Z can be found in the right side of (24), so $B(Z)$ will have no negative powers of Z. Also, the factorials in the denominator assure us that (24) always converges, so b_t is always causal.

To establish theorem 2, that the exponential is not just causal but also minimum phase, consider

$$B_+ = e^{+U} \tag{25a}$$

$$B_- = e^{-U} \tag{25b}$$

Clearly both B_+ and B_- are causal, and they are inverses of one another. A minimum-phase filter is defined to be causal with a causal inverse. So B_+ and B_- are minimum phase.

Now we will establish the converse of theorem 2 — namely, theorem 3 — which states that the logarithm of a minimum-phase filter is causal. Take the logarithm of (24) and form the Z-derivative:

$$U = \ln B \tag{26a}$$

$$\frac{dU}{dZ} = u_1 + 2u_2 Z + 3u_3 Z^2 + \cdots \tag{26b}$$

$$\frac{dU}{dZ} = \frac{1}{B}\frac{dB}{dZ} \tag{26c}$$

Since B was assumed to be minimum phase, both $1/B$ and dB/dZ on the right of (26c) are causal. Since the product of two causals is causal, dU/dZ is causal. But dU/dZ cannot be causal unless U is causal. That proves theorem 3, disregarding the remote danger that B might converge while dB/dZ diverges.

Theorem 4 refers to the Fourier domain representation of the minimum-phase filter. In the complex plane, the filter gives parametric equations for a curve, say $[x(\omega), y(\omega)] = [Re\ B(Z), Im\ B(Z)]$. The phase angle $\phi(\omega)$ is defined by the arctangent of the ratio y/x. For example, the causal, non-minimum-phase filter $U(Z) = Z = e^{i\omega}$ gives the parametric equations $x = \cos\omega$ and $y = \sin\omega$ which define a circle surrounding the origin. Notice that the phase of $Z = e^{i\omega}$ is $\phi(\omega) = \omega$, which is a monotonically increasing function of ω. In the minimum phase case, $\phi(\omega{=}0) = \phi(\omega{=}2\pi)$. In the non-minimum-phase case, the curve loops the origin, so $\phi(\omega{=}0) = \phi(\omega{=}2\pi) + 2\pi$. Theorem 3 allows us to say that a general formula for minimum-phase filters is

$$B = e^{U(Z)} = \exp\left[\sum_{k=0}^{N} U_k \cos k\omega + i \sum_{k=0}^{N} U_k \sin k\omega \right] \tag{27a}$$

$$= \exp[r(\omega) + i\ \phi(\omega)] \tag{27b}$$

The phase $\phi(\omega)$, being a sum of periodic functions, is itself a periodic function of ω, which means that in the plane of $(\operatorname{Re} B, \operatorname{Im} B)$ the curve representing $B(\omega)$ does not enclose the origin.

On to theorem 5, which says that any power of a minimum-phase function is minimum phase. Consider

$$B^r = \left(e^{\ln B} \right)^r = e^{r \ln B} \tag{28}$$

Since B is assumed to be minimum phase, by theorem 3 $\ln B$ will be causal. Scaling by a real or complex constant r does not change causality. Exponentiating shows, by theorem 2, that B^r is minimum phase.

Finally the proof of theorem 6, that an impedance function can be raised to any real fractional power $-1 \leq \rho \leq +1$ and the result will still be an impedance function. An impedance function is defined to be a minimum-phase function with the additional property that the real part of its Fourier transform is positive. This means that the phase angle ϕ lies in the range $-\pi/2 < \phi < +\pi/2$. Raising the impedance function to the ρ power will compress the range to $-\pi\rho/2 < \phi < \pi\rho/2$. This will keep the real part of the impedance function positive. Theorem 5 states that any power of a minimum-phase function is causal, which is more than we need to be certain that a *fractional* real power of an impedance function will be causal.

Wide-Angle Wave Extrapolation

Let $s = -i\,\hat{\omega}$ denote the causal, positive, discrete representation of the differentiation operator, say,

$$s = -i\,\hat{\omega} = \frac{2}{\Delta t} \frac{1 - \rho Z}{1 + \rho Z} \tag{29}$$

Figure 2 compares hyperbolas constructed with ω to those constructed with $\hat{\omega}$. You see a pleasing drop in wraparound noise. It seems to work better than the ϵ in Section 4.1. As we will see, the introduction of complex-valued $\hat{\omega}$ leads to a more natural handling of the square root at the evanescent transition.

Consider the following recursion starting from $R_0 = s$:

$$R_{n+1} = s + \frac{X^2}{s + R_n} \tag{30}$$

This recursion produces continued fractions. Francis Muir introduced it as a means of developing wide-angle square-root approximations for migration (Section 2.1), and he developed his three rules to show that every R_n is an impedance function. To see why every R_n is an impedance function, first

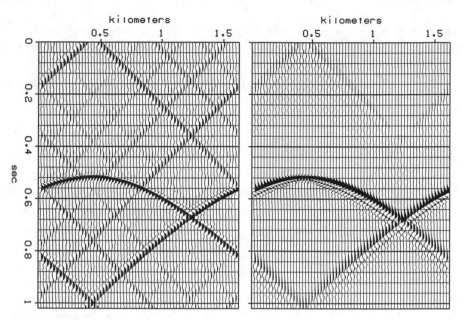

FIG. 4.6-2. Hyperbolas with real frequency (left) and complex frequency (right). (Plotting uses square root gain described in Section 4.1).

note that the denominator $s + R_n$ is, for $n = 0$, the sum of two impedance functions. Then its inverse is an impedance function, and multiplication by the real positive constant X^2 and addition of another s both preserve the properties of impedance functions. Recursively we see that all the R_n are impedances.

As N becomes large this recursion either converges or it does not. Supposing that it does, we can see what it will converge to by setting $R_{n+1} = R_n = R_\infty = R$. Thus,

$$R = s + \frac{X^2}{s + R} \tag{31a}$$

$$R(s + R) = s(s + R) + X^2 \tag{31b}$$

$$R^2 = s^2 + X^2 \tag{31c}$$

$$R = \sqrt{s^2 + X^2} \tag{31d}$$

In wave-extrapolation problems X^2 is $v^2 k_x^2$, where v is the wave velocity and k_x is the horizontal spatial frequency, namely, the Fourier dual to the horizontal x-axis. Performing these substitutions we have

$$R = \pm\sqrt{-\hat{\omega}^2 + v^2 k_z^2} \tag{32}$$

So R is like $\pm i k_z v$. Remember that R_0, the first approximation to R, is $-i\hat{\omega}$. So downgoing waves are

$$D(x, z, t) = D(x, 0, t)\, e^{ik_x x}\, e^{-R\, z/v}\, e^{-i\omega t} \tag{33}$$

To switch from downgoing to upcoming waves, we could either change the sign in front of R or we take the complex conjugate of R. The difference is what you want to do with the real part — do you want the wave to grow or not?

Consider the dissipation of waves in the exploding reflector model. They damp as they propagate from the explosion to the surface. This means that as we migrate them, they should be exponentially growing. But we don't really want that. We really want to assure that they are not growing, perhaps we even want them decaying as we extrapolate them back. So for migration we downward continue monochromatic waves with

$$U(x, z, t) = U(x, 0, t)\, e^{ik_x x}\, e^{-\bar{R}\, z/v}\, e^{-i\omega t} \tag{34}$$

although the real behavior of a wave from an exploding reflector wave would be

$$U(x, z, t) = U(x, 0, t)\, e^{ik_x x}\, e^{+R\, z/v}\, e^{-i\omega t} \tag{35}$$

To examine the phase of the complex quantity R, set $v = 1$ obtaining

$$R = \sqrt{(-i\hat{\omega})^2 + k_x^2} \tag{36}$$

First note that $(-i\hat{\omega})$ is causal because of its Z-transform representation. By squaring the Z-transform we see that $(-i\hat{\omega})^2$ is also causal. In the time domain, k_x^2 is a delta function at the time origin. Thus R^2 given by (36) is causal. Figure 3 shows how the phase of (36) is constructed from its constituents. To illustrate the behavior of $-i\hat{\omega}$ from zero to infinity, I include both an artist's conception and the function on itself, overlain at various magnifications. The function $-i\hat{\omega}$ is a periodic with ω and its real and imaginary parts plot to a closed curve. To show the rate of change of the function, I sampled ω at $2°$ intervals. From great distance the function is a circle. Close up it looks like a line parallel to the imaginary axis.

R^2 is causal and from figure 3 we can see that it has a "branch cut" property. That is, the phase of R has the positive real property. Theorem 5 forces R to be causal and minimum phase. That, with the phase defined by figure 3, proves that R, given by (36), is an impedance function.

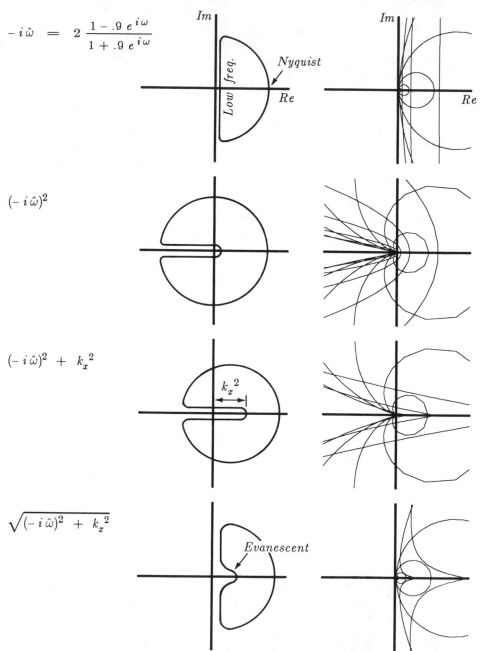

$$-i\,\hat{\omega} = 2\,\frac{1 - .9\,e^{i\omega}}{1 + .9\,e^{i\omega}}$$

$$(-i\,\hat{\omega})^2$$

$$(-i\,\hat{\omega})^2 + k_x^{\,2}$$

$$\sqrt{(-i\,\hat{\omega})^2 + k_x^{\,2}}$$

FIG. 4.6-3. Complex plane diagram of constituents of the extrapolation operator R given by (36). The center column shows an artist's conception. The right column shows the function at several magnifications simultaneously.

Fractional Integration and Constant Q

By equation (29) and theorem 6, fractional powers of integration and differentiation are also impedance functions. Kjartansson [1979] has advocated the fractional power as a stress-strain law for rocks. See also Madden [1976]. Classical studies in rock mechanics begin with a stress-strain law such as

$$stress \;=\; stiffness \;\times\; strain \;+\; viscosity \;\times\; strain\text{-}rate$$

which in the transform domain is

$$stress \;=\; [(-i\,\omega)^0 \;\times\; stiffness \;+$$

$$(-i\,\omega)^1 \;\times\; viscosity\,] \times strain \tag{37}$$

Experimentally, the viscoelastic law (37) does a poor job of describing real rocks. Let us try another mathematical form that is like (37) in its limiting behavior at high and low viscosity:

$$stress \;=\; const \;\times\; (-i\,\omega)^\epsilon \times strain \tag{38a}$$

$$=\; const \;\times\; (-i\,\omega)^{\epsilon-1} \times strain\text{-}rate \tag{38b}$$

Here ϵ close to zero gives elastic behavior and ϵ close to one gives viscous behavior. The fact that $(-i\,\omega)^{\epsilon-1}$ is an impedance function meshes nicely with the concepts that (1) stress may be determined from strain history and strain may be determined from stress history, and (2) stress times strain-rate is dissipated power. Kjartansson [1979] points out that $(-i\,\omega)^\gamma$ exhibits the mathematical property called *constant Q*, so that as a stress/strain law for fitting experimental data on rocks, it is far superior to (37). To see the constant Q property more clearly, express $(-i\,\omega)^\gamma$ in real and imaginary parts:

$$(-i\,\omega)^\gamma \;=\; |\,\omega\,|^\gamma \,[-i\; sgn\,(\omega)]^\gamma \tag{39a}$$

$$=\; |\,\omega\,|^\gamma \,[e^{-i\,\pi\, sgn\,(\omega)/2}]^\gamma \tag{39b}$$

$$=\; |\,\omega\,|^\gamma \left\{ \cos\left[\frac{\pi\,\gamma}{2}\, sgn\,(\omega)\right] - i\, \sin\left[\frac{\pi\,\gamma}{2}\, sgn\,(\omega)\right] \right\} \tag{39c}$$

$$=\; |\,\omega\,|^\gamma \left[\cos\left(\frac{\pi\,\gamma}{2}\right) - i\; sgn\,(\omega)\, \sin\left(\frac{\pi\,\gamma}{2}\right) \right] \tag{39d}$$

The constant Q property follows from the constant ratio between the real and imaginary parts of this function. Q itself is defined by

$$\frac{1}{Q} \;=\; \tan \pi\epsilon \;\approx\; \pi\epsilon \tag{40}$$

A pulse with a Q of about 10 is shown in figure 4.

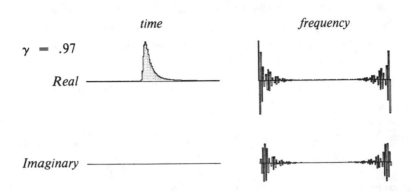

FIG. 4.6-4. The constant Q pulse given by $e^{-(-i\omega)^{.97}t_0}$. The frequency axis is represented by a discrete Fourier transform over 256 points. Zero time and zero frequency are on the left end of their respective axes.

EXERCISES

1. Take $\epsilon < 0$ and expand the integration operator for negative powers of Z. Explain the sign difference.

2. Let $\alpha > 0$ be a real, positive scaling constant, and let C be a reflectance function. Without using Muir's rules, prove that C' is a reflectance, where

$$\frac{1 - C'}{1 + C'} = \alpha \frac{1 - C}{1 + C}$$

   Note that you have proven Muir's first rule. Muir's third rule can also be proven in an analogous way, but with much more algebraic detail.

3. The word *isomorphism* means not only that any impedance R_1, R_2, R' can be mapped into a reflectance C_1, C_2, C', but also that Muir's three rules will be mapped into three rules for combining reflectances.

   a. What are these three rules?

   b. Although $C' = C_1 C_2$ does not turn out to be one of the three rules, it is obviously true. Either show that it is a consequence of the three rules or conclude that it is an independent rule that can be mapped back into the domain of the impedances to constitute a fourth rule.

4. Show that the log of the discrete causal integration operator, $\log[(1+Z)/(1-Z)]$, is one side of the discrete Hilbert transform. Show that the reflected pulse from a boundary between two media with the same velocity but slightly different Q is one side of the Hilbert transform.

5. Consider the fourth-order Taylor expansion for square root in an extrapolation equation

$$\frac{dP}{dz} = i\omega\left[1 - \frac{1}{2}\left(\frac{vk}{\omega}\right)^2 - \frac{1}{8}\left(\frac{vk}{\omega}\right)^4\right]P$$

   a. Will this equation be stable for the complex frequency $-i\omega = -i\omega_0 + \epsilon$? Why?

   b. Consider causal and anticausal time-domain calculations with the equation. Which, if any, is stable?

6. Consider a material velocity that may depend on the frequency ω and on the horizontal x-coordinate as well. Suppose that, luckily, the velocity can be expressed in the factored form $v(x,\omega) = v_1(x)\,v_2(\omega)$. Obtain a stable 45° wave-extrapolation equation. Hints: try

$$s = -\frac{i\omega}{v_2}$$

$$X^2 = positive\ eigenvalue\ of\ (v_1\partial_x)(v_1\partial_x)^T$$

7. Is the Levinson Recursion described in FGDP related to the rules in this section? If so, how? Hint: see Jones and Thron [1980].

8. Show the converse to theorem 4, namely, that if the phase curve of a causal function does not enclose the origin, then the inverse is causal.

4.7 Accuracy — the Contractor's View

A chain is no stronger than its weakest link. Economy dictates that all the links should be equally strong. Many broad questions merit study such as the errors associated with velocity uncertainty and with migration after, rather than before, stack. Having read this far, you are now qualified to attack these broad questions. Now we will narrow our focus and examine only the errors in downward continuation that result from familiar data processing approximations.

In the construction of a production program for wave-equation migration, weakness arises from approximations made in many different places. Economy dictates that funds to purchase accuracy should be distributed where they will do the most good. Geophysical contractors naturally become experts on accuracy/cost trade-offs in the migration of stacked data. Contractors will use the equations and program gathered below to obtain best results for the lowest cost. Users of reflection data are interested in learning to recognize imperfect migrations, so they may want to use the program to see the effect of various shortcuts.

In preparation for a big production job there are two general approaches to examining accuracy. The first approach, which gives the best insight into qualitative phenomena, is to make synthetic hyperbolas by the various methods. The synthetic hyperboloids can be be compared to the data at hand with a video movie system or by plotting on transparent paper. In the second approach you compute travel times of hyperboloids or spheroids of waves of different stepouts and frequencies for different mesh sizes, etc. Then an optimization program can be run to minimize the average error over the important range of parameters.

It is not necessary to write a time domain 45° finite difference migration program to see what its synthetic hyperboloids would look like. We can simply express all the formulas in the (ω, k_x)-domain and then do an inverse two-dimensional Fourier transformation. To facilitate comparisons between the many migration methods we will gather equations from different parts of the book in the order that they are needed. Then I'll present the program that makes the diffraction hyperbolas for many methods. Using the same equations you can compute travel times and solve the optimization problem as you wish.

Lateral Derivatives

First, k_x will range over $\pm\pi/\Delta x$. If the x-axis is going to be handled by finite differencing then we will need

$$\left(\frac{\hat{k}\ \Delta x}{2}\right)^2 = \frac{\sin^2 \dfrac{k\ \Delta x}{2}}{1 - b\ 4\sin^2 \dfrac{k\ \Delta x}{2}} \tag{4.3-7b}$$

So if the x-axis is going to be handled by finite differencing then subsequent reference to k_x should be replaced by $\hat{k}_x$. The finite differencing introduces the free parameter b. Likewise, you could also scale the whole expression by an adjustable parameter near unity. Also, Δx isn't necessarily fixed by the data collection. You could always interpolate the data before processing. A finite-difference method using interpolated data could be mandated by enough lateral velocity variation.

Viscosity and Causality

The frequency ω will range over $\pm\pi/\Delta t$. If the t-axis is going to be handled by finite differencing then we will need the Z-transform variable

$$Z = e^{i\ \omega\Delta t} \tag{4.6-2}$$

and the causal derivative

$$-i\,\hat{\omega} = \frac{2}{\Delta t}\frac{1 - \rho Z}{1 + \rho Z} \qquad -1 << \rho < 1 \tag{4.6-29}$$

The data can be subsampled or supersampled before processing, so Δt is an adjustable parameter. The causality parameter ρ should be a small amount less than unity, say $\rho = 1 - \epsilon$ where $\epsilon > 0$ is an adjustable parameter. You may want to introduce ρ even if you are migrating in the frequency domain because it reduces wraparound in the time domain — it is a kind of viscosity. The ϵ should be about inverse to the data length, say $1/N_t$ where N_t is the number of points on the time axis. (Because I made many plots of synthetic hyperbolas with square root gain, $\gamma = 1/2$, time wraparounds were larger than life. So I had the program default to ϵ four times larger). If you like to adjust free parameters, you could separately adjust numerator and denominator values of ρ. Subsequently, I'll distinguish between ω and $\hat{\omega}$, but you can take $\hat{\omega}$ to be ω if you don't care to introduce causality.

Retarded Muir Recurrence

The k_z square root may be computed with the square root function in your computer or by Muir's expansion. For Fourier domain calculations incorporating causality, you must use a complex square root function. This will also take care of the evanescent region automatically — you no longer have the discontinuity between evanescent and nonevanescent regions. The square root of a complex number is multivalued, so you better first check that your computer chooses the phase as described in figure 4.6-3. Mine did. But I found that limited numerical accuracy prevented me from achieving strict positivity of the real part of the impedance until I replaced the expression $\sqrt{s^2 + v^2 k^2} - s$ by its algebraic equivalent $v^2 k^2 / (\sqrt{s^2 + v^2 k^2} + s)$.

For finite differencing we will need the Muir recurrence. Let r_0 define the cosine of the angle that starts the Muir recurrence, often $0°$ or $45°$. This is another free parameter for optimization. See for example figure 2.1-1. This angle is also an angle of exact fit for all orders of the recurrence. Let

$$s = -i\,\hat{\omega} \tag{4.6-29}$$

Starting from $R_0 = r_0\,s$ the Muir recursion is

$$R_{n+1} = s + \frac{v^2\,k_x^2}{s + R_n} \tag{4.6-30}$$

For a diffraction program we will be evaluating $\exp(-Rz)$. Since R was proven in 4.6 to have a positive real part, the exponential should never grow. Finite difference calculations are normally done with retarded time. To retard time, $\exp(-Rz)$ is expressed as

$$e^{-R\,z/v} = e^{-(R-s)\,z/v}\ e^{-s\,z/v} \tag{1}$$

As discussed in Section 4.1, you probably don't want the time shift of retardation to be associated with viscous effects. So you will probably want to downward continue instead with

$$e^{-[R(-i\hat{\omega}) + i\hat{\omega}]\,z/v}\ e^{+i\omega z/v} \tag{2}$$

Notice the signs and distinction of ω from $\hat{\omega}$.

From equation (4.6-30) we see that $R - s$ should have a positive real part. I found that numerical roundoff sometimes prevented it. So the Muir recurrence was reorganized to incorporate the retardation. Let

$$R' = R - s \tag{3}$$

Equation (4.6-30) becomes

$$R'_{n+1} = \frac{v^2 k_x^2}{2s + R'_n} \tag{4}$$

From Muir's rules, you can see that R' will always have a positive real part if we start it that way, so we start it from

$$R'_1 = \frac{v^2 k_x^2}{s(1 + r_0)} \tag{5}$$

(Combining (5) and (3) gives the same 15° equation as does (4.6-30).) Mathematically (2) is identical to

$$e^{-R'z/v} \; e^{+i\omega z/v} \tag{6}$$

but numerically the exponential in (6) is assured to decay in z.

Stepping in Depth

If you are using finite differences on depth, or travel-time depth, then we have the relation

$$\frac{\hat{k}_z \, \Delta z}{2} = \tan \frac{k_z \, \Delta z}{2} \tag{4.3-14}$$

which can be used in $\exp ik_z z$. Since dissipation may be present, the quantities above may be complex. Let N_z be the number of depth layers, generally equal or less than N_t. Adapting (4.3-17) to (6) gives

$$e^{-R' N \Delta z/v} \approx \left[\frac{1 - R'\Delta z/2v}{1 + R'\Delta z/2v} \right]^N \tag{7}$$

Lightning Phase Shift Migration

Not only can the causality and viscosity features of time domain methods be incorporated in the frequency domain, but the square roots and complex exponentials of the phase shift method can be replaced by complex multiplies and divides. First note that the square root expansion need not begin from a starting guess or from a lower order iteration. It could begin from the square root previously obtained from the preceding ω or k. (I noticed this when an early version of the program had a bug that made all my 15° calculations look like 90° calculations)! Also, $\Delta z/v$ is necessarily small in the phase shift method to accommodate imaging at every time point. So the finite differencing form (7) is probably as good as the complex exponential. Why don't you try it?

Final Appearances

The output of an impulse-response program always looks awful. The main reason is the large area of (ω, k_x)-space which is near the Nyquist frequency or above the evanescent cutoff. Since we rarely sample data in time as coarsely as the Nyquist criterion permits, the program below defaults to final filtering with the filter $(1+Z)/(1-.8Z)$. This filter still passes a lot of energy outside the usual bandwidth of seismic data. Since all land data and most marine data do not have the zero frequency component, the program contains an option to filter further with $(1-Z)/(1-.8Z)$. I haven't displayed anything with this extra filter because I wanted this book to show all the artifacts you might encounter. Furthermore, I deliberately enhanced the visibility of artifacts on wiggle-trace, variable-area plots by plotting with the nonlinear $\gamma = 1/2$ gain described in Section 4.1. (Perspective hidden-line drawings always have linear gain). Since my plots are necessarily about 10 cm square in this book and in practice you will look at plots of about a hundred times the area, I plotted only one second of travel time.

Program

Many of the figures in the book were made with the program presented here. To enable you to reproduce them I am including the complete program. The parameter input and data output calls are site dependent, but I include them anyway to help clarify the defaults, increasing the odds that you can get exactly what I did.

```
# representations of  e^{-√(-iω/v)² + k² z}
integer output, outfd, fetch
integer iw,nw,ik,nk, omhat, kxhat, kzhat, degree, tfilt, xfilt
real v, dt, dx, dz, xf, x0, tf, tau0,  rho, bi, r0, eps,  pi, omega, k, vk2
complex cz, cs, cikz, cexp, cmplx, csqrt, cp(1024)
outfd = output()
                      call putch("esize","i", 8)      # complex numbers
nw = 256;             call putch( "n1","i",nw)        # inner index is ω
nk = 64;              call putch( "n2","i",nk)        # outer index is kₓ
                      call putch( "n3","i", 1)        # one frame movie

if( fetch(   "v", "f",   v) == 0 )   v  = 3.754       # rock vel
if( fetch(  "dt", "f",  dt) == 0 )   dt = .004        # Δt , sec
if( fetch(  "dx", "f",  dx) == 0 )   dx = .025        # Δx , km
if( fetch(  "dz", "f",  dz) == 0 )   dz = .004        # Δz , sec
if( fetch(  "xf", "f",  xf) == 0 )   xf = .25;        x0 =xf*nk*dx
if( fetch(  "tf", "f",  tf) == 0 )   tf = .5;         tau0=tf*nw*dt
```

```
if( fetch( "omhat", "i", omhat)==0)    omhat = 0        # ω̂
if( fetch( "kxhat", "i", kxhat) == 0 )  kxhat = 0        # k̂ₓ
if( fetch( "kzhat", "i", kzhat) == 0 )  kzhat = 0        # k̂_z
if( fetch("degree", "i",degree) == 0 )  degree= 90
if( fetch( "tfilt", "i", tfilt) == 0 )   tfilt = 1
if( fetch( "xfilt", "i", xfilt) == 0 )   xfilt = 1

if( fetch(  "rho", "f",   rho) == 0 )   rho  = 1-4./nw
if( fetch(   "bi", "f",    bi) == 0 )   bi   = 6.726       # b⁻¹
if( fetch(   "r0", "f",    r0) == 0 )   r0   = 0.7071
if( fetch(  "eps", "f",   eps) == 0 )   eps  = 0.

call putch("d1","f", dt);      call putch("label1","s","sec")
call putch("d2","f", dx);      call putch("label2","s","kilometers")
call hclose()                  # close data description file
pi = 3.14159265
do ik = 1, nk {                # loop over all kₓ
        k = 2*pi * (ik-1.) / nk
        if( k > pi )    k = k - 2*pi
        k = k / dx
        if( kxhat == 0 )
                vk2 = (v/2)**2 * k*k
        else
                vk2 = (v/2)**2 * (2/dx)**2 * sin(k*dx/2)**2  / (1 -
                                (4./bi) * sin(k*dx/2)**2 )

        do iw = 1, nw {                         # loop over all ω
                omega = 2*pi * (iw-1.) / nw
                if( omega > pi )   omega = omega - 2*pi
                omega = omega / dt
                cz = cexp( cmplx( 0., omega * dt ) )
                if( omhat == 0 )
                        cs = cmplx( 1.e-5 / dt, - omega)
                else
                        cs = (2./dt) * (1. - rho * cz) / (1. + rho * cz)
                if ( degree == 90 )
                        cikz = vk2 / ( csqrt( cs * cs + vk2 ) + cs )
                if ( degree == 15  |  degree == 45 )
                        cikz = vk2 / ( eps + (r0+1.) * cs )
                if (degree == 45 )
                        cikz =  vk2 / ( 2.*cs + cikz)
                if( real( cikz) < 0.) call erexit("cikz not positive real")

                if( kzhat == 0 )
                        cp(iw) = cexp( - tau0  * cikz )
                else
                        cp(iw) = ((1.-cikz * dz/2) / (1.+cikz * dz/2) ) ** (tau0/dz)
                cp(iw) = cp(iw) * cexp(cmplx(0., omega * tau0 ))        # unretard

                if( tfilt >= 1 ) cp(iw) = cp(iw) * (1+cz) / (1 - .8 * cz)
                if( tfilt >= 2 ) cp(iw) = cp(iw) * (1-cz) / (1 - .8 * cz)
                if( xfilt == 1 ) cp(iw) = cp(iw) * (1+cos(k*dx)) / (1+.85*cos(k*dx))
                cp(iw) = cp(iw) * cexp( cmplx(0.,k*x0) )
                }
        call rite( outfd, cp, 8*nw )    # write
        }
stop;   end
# Finally, you must 2-D Fourier Transform (Section 1.7), take real part, and plot.
```

Twenty-two plots displayed in this book were made with this program. Different input parameters for the different plots are in the table below.

Section and figure.	Default parameter overrides.
1.3-6a	tfilt=0
1.3-6b	
2.0-1a	
2.0-1b	$\hat{\omega}$
4.0-1a	$\hat{\omega}$, $v = 2.0$, xf=.5, xfilt=0
4.0-1b	$\hat{\omega}$, $v = 2.0$, xf=.5
4.1-4a	$15°$
4.1-4b	$15°$, $\epsilon = 1$
4.1-5a	$45°$, tf=.2
4.1-5b	$45°$, tf=.2, $\epsilon = 1$
4.2-4	$45°$, tf=.2, $\hat{\omega}$
4.3-4a	$v = 2.0$, $90°$
4.3-4b	$v = 2.0$, $15°$, $\hat{\omega}$, $\hat{k}_x$, $\hat{k}_z$
4.3-6a	$\hat{k}_x$, $\hat{\omega}$, $b^{-1} = 1000000.$
4.3-6b	$\hat{k}_x$, $\hat{\omega}$, $b^{-1} = 12.$
4.3-6c	$\hat{k}_x$, $\hat{\omega}$, $b^{-1} = 6.726$
4.3-6d	$\hat{k}_x$, $\hat{\omega}$, $b^{-1} = 6.$
4.3-6e	$\hat{k}_x$, $\hat{\omega}$, $b^{-1} = 5.$
4.6-2a	
4.6-2b	$\hat{\omega}$
4.7-1a	$\Delta z = .004$, $45°$, tf=.3, $\hat{\omega}$, $\hat{k}_x$, $\hat{k}_z$
4.7-1b	$\Delta z = .012$, $45°$, tf=.3, $\hat{\omega}$, $\hat{k}_x$, $\hat{k}_z$

It is rumored that accuracy can be improved by making the z mesh coarser. This couldn't happen if x and t were in the continuum, but since they may be discretized, there is the possibility of errors fortuitously canceling. To test this rumor, I tried tripling Δz (actually, tripling the increment in travel-time depth). The result is in figure 1. What do you think?

Complete as this analysis must seem, it is limited by the assumption of Fourier analysis that velocity is constant laterally. To handle this problem we turn now to the final lecture on techniques.

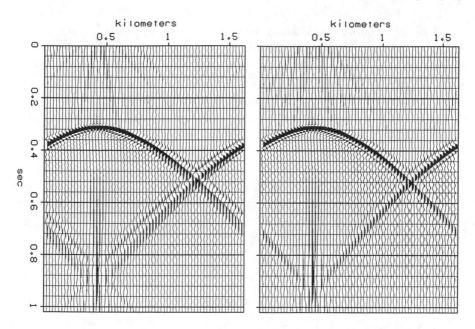

FIG. 4.7-1. Left is a 45° point diffraction in (x, z, t)-space with $\Delta z = v \, \Delta t$. At the right, with $\Delta z = 3 \, v \, \Delta t$.

4.8 The Bulletproofing of Muir and Godfrey

Stable extrapolation can be assured by preserving certain symmetries. It will be shown that stability is assured in both the differential equation

$$\frac{d\mathbf{q}}{dz} = -\mathbf{R}\,\mathbf{q} \tag{1}$$

and its Crank-Nicolson approximation

$$\frac{\mathbf{q}_{n+1} - \mathbf{q}_n}{\Delta z} = -\frac{\mathbf{R}}{2}(\mathbf{q}_{n+1} + \mathbf{q}_n) \tag{2}$$

provided that $\mathbf{R} + \mathbf{R}^*$ is a positive definite (actually, semidefinite) matrix. When stability was studied in the previous section the operator $\mathbf{R}$ was a scalar Z-transform. Because Z-transforms were used, the mathematics of that section was particularly suitable for time domain migrations. Because $\mathbf{R}$ was a scalar, the mathematics of that section was particularly suitable when data has been Fourier transformed over x. Here we will focus on the *matrix* character of $\mathbf{R}$. Thus we are concerning ourselves with migration in the x-domain. Our purpose in doing this theoretical work is to gain the ability to write a "bulletproof" program for migrating seismic data in the presence of lateral velocity variation. As an example, the familiar 45° extrapolation equation will be put in the bulletproof form. This section, combined with the previous one, gives a general theory for stable migration in (t, x)-space.

Stability of the Differential Equation

Let $\mathbf{q}^*$ denote the Hermitian conjugate of $\mathbf{q}$. For equation (1) to be stable the energy $\mathbf{q}^* \mathbf{q}$ must be either constant or decaying during depth extrapolation.

$$\frac{d}{dz} (\mathbf{q}^* \mathbf{q}) \leq 0$$

$$\mathbf{q}^* \mathbf{q}_z + \mathbf{q}_z^* \mathbf{q} \leq 0 \tag{3}$$

Substituting equation (1) into equation (3) gives

$$\mathbf{q}^* \mathbf{R} \mathbf{q} + \mathbf{q}^* \mathbf{R}^* \mathbf{q} \geq 0$$

$$\mathbf{q}^* (\mathbf{R} + \mathbf{R}^*) \mathbf{q} \geq 0 \tag{4}$$

Equation (4) shows that $\mathbf{R} + \mathbf{R}^*$ must be positive semidefinite for the differential equation to be stable.

Stability of the Difference Equation

The stability of the difference equation can be shown in a similar way, but with some extra clutter. First observe the identity

$$(\mathbf{a}^* \mathbf{a} - \mathbf{b}^* \mathbf{b}) \equiv \frac{1}{2} [(\mathbf{a} + \mathbf{b})^* (\mathbf{a} - \mathbf{b}) + (\mathbf{a} - \mathbf{b})^* (\mathbf{a} + \mathbf{b})] \tag{5}$$

Letting $\mathbf{a} = \mathbf{q}_{n+1}$ and $\mathbf{b} = \mathbf{q}_n$, equation (5) becomes

$$(\mathbf{q}_{n+1}^* \mathbf{q}_{n+1} - \mathbf{q}_n^* \mathbf{q}_n) =$$

$$= \frac{1}{2} [(\mathbf{q}_{n+1} + \mathbf{q}_n)^* (\mathbf{q}_{n+1} - \mathbf{q}_n) + (\mathbf{q}_{n+1} - \mathbf{q}_n)^* (\mathbf{q}_{n+1} + \mathbf{q}_n)] \tag{6}$$

Now, replace the $(\mathbf{q}_{n+1} - \mathbf{q}_n)$ terms by equation (2):

$$= -\frac{\Delta z}{4} [(\mathbf{q}_{n+1} + \mathbf{q}_n)^* \mathbf{R} (\mathbf{q}_{n+1} + \mathbf{q}_n) +$$

$$(\mathbf{q}_{n+1} + \mathbf{q}_n)^* \mathbf{R}^* (\mathbf{q}_{n+1} + \mathbf{q}_n)]$$

$$= -\frac{\Delta z}{4} [(\mathbf{q}_{n+1} + \mathbf{q}_n)^* (\mathbf{R} + \mathbf{R}^*) (\mathbf{q}_{n+1} + \mathbf{q}_n)] \tag{7}$$

This equation establishes the result: If the matrix $\mathbf{R} + \mathbf{R}^*$ is positive definite, then $\mathbf{q}_{n+1}^* \mathbf{q}_{n+1}$ is less than $\mathbf{q}_n^* \mathbf{q}_n$.

Application to 45° Wavefield Extrapolation

The scalar wave equation for the extrapolation of a downgoing wavefield is

$$\frac{d\mathbf{q}}{dz} = i\, k_z\, \mathbf{q} = -\mathbf{R}\, \mathbf{q} \tag{8}$$

where the $\mathbf{R}$ operator takes the usual form

$$\mathbf{R} = -i\, k_z = \frac{-i\,\omega}{v} \sqrt{1 - \frac{v^2 k_x^2}{\omega^2}} \tag{9}$$

Our plan is to approximate the square root by the usual continued fraction expansion and then identify $i\, k_x$ with ∂_x to obtain a space-domain equation. The main effort we must make stems from our refusal to make the usual assumption that $v(x, z)$ is independent of x. Since $\partial_x v\, q$ differs from $v\, \partial_x q$, the space representation does not seem to be unique, and we may wonder how the variable q relates to physical wave variables like pressure and displacement. Since (9) is purely imaginary, the depth-invariance of the quadratic $\mathbf{q}^*\mathbf{q}$ can be interpreted as the downward energy flux across the datum at depth z. Our main effort will be to assure that $\mathbf{q}^*\mathbf{q}$ does indeed remain depth-invariant when $v(x, z) \neq const.$ The task of determining the relation between the energy flux variable $\mathbf{q}$ and the physical variables will be left to the reader.

First $v^2 k_x^2$ must be represented in the space domain. Thinking of the x-derivative operator $\partial/\partial x = \partial_x$ as a large bidiagonal matrix with $(1, -1)/\Delta x$ along the diagonal and $\mathbf{V}(x)$ as a diagonal matrix, we are attracted to expressions like $(\mathbf{V}\,\partial_x)^{\mathrm{T}}(\mathbf{V}\,\partial_x)$ or $(\mathbf{V}\,\partial_x)(\mathbf{V}\,\partial_x)^{\mathrm{T}}$ because they are symmetric, positive, semidefinite matrices. In simplest form, such numerical representations are tridiagonal matrices that can be abbreviated as

$$\mathbf{T} = \begin{cases} (\mathbf{V}\,\partial_x)\,(\mathbf{V}\,\partial_x)^{\mathrm{T}} \\ (\mathbf{V}\,\partial_x)^{\mathrm{T}}\,(\mathbf{V}\,\partial_x) \end{cases} \tag{10a,b}$$

At a later time accuracy or some other consideration could determine the choice in (10). Even other expressions could be used, provided they are real, symmetric, and positive definite.

In the previous section the constant velocity, 45° expansion of (9) was shown to be

$$\mathbf{R} \;=\; \frac{1}{v} \; \frac{v^2 k_x^2}{-\,i\,\omega\,2 + \dfrac{v^2 k_x^2}{-\,i\,\omega\,2}} \tag{11}$$

This scalar $\mathbf{R}$ always has a positive real part because $-\,i\,\omega$ is always represented in an impedance form, and the whole expression is built up satisfying Muir's rules for combining impedance functions. In going to the x-domain notice that $(i\,k_x)^2 = -\partial_{xx}$ and $(\partial_x)^\mathsf{T} = -\,\partial_x$. So the positive scalar $v^2 k_x^2$ corresponds to the positive eigenvalues of (10).

The expression of the bulletproof, square-root operator $\mathbf{R}$ in the space domain will now be given as

$$\mathbf{M} \;=\; \frac{\mathbf{T}}{-\,i\,\omega\,2\,\mathbf{I} + \dfrac{\mathbf{T}}{-\,i\,\omega\,2}} \tag{12a}$$

$$\mathbf{R} \;=\; \mathbf{V}^{-1/2}\,\mathbf{M}\,\mathbf{V}^{-1/2} \tag{12b}$$

Use of the division sign in (12) is justifiable because the matrix $\mathbf{T}$ commutes with the identity matrix $\mathbf{I}$. (A hazard in this work is that $\mathbf{T}$ does not commute with the diagonal matrix $\mathbf{V}$). The matrix $\mathbf{M}$ has the properties required of $\mathbf{R}$ since a basic matrix theorem says that the eigenvalues of a polynomial of a real symmetric matrix are the polynomials of the eigenvalues. In other words, replacing $\mathbf{T}$ in (12) by one of its eigenvalues produces a complex $\mathbf{M}$ whose real part is positive, so that $\mathbf{M}^* + \mathbf{M}$ is positive as required. What is needed is to show that the following matrix is positive definite:

$$\mathbf{R} + \mathbf{R}^* \;=\; \mathbf{V}^{-1/2}\,(\mathbf{M} + \mathbf{M}^*)\,\mathbf{V}^{-1/2} \tag{13}$$

A matrix $\mathbf{A}$ is positive definite if for arbitrary d, the scalar $d^*\mathbf{A}\,d$ is positive. The diagonal matrix $\mathbf{V}^{-1/2}$ can certainly be absorbed into d and d will still be arbitrary, so the proof is complete.

In programming it is a nuisance to put $\mathbf{V}^{-1/2}$ on each side of the matrix $\mathbf{M}$. Actually you can put $\mathbf{V}^{-1}$ on either side. In general, some other quadratic form such as $\mathbf{q}^*\mathbf{U}\,\mathbf{q}$ where $\mathbf{U}$ is strictly positive definite will be decreasing if $\mathbf{R}^*\mathbf{U} + \mathbf{U}\,\mathbf{R}$ is positive definite.

5

Some Frontiers

In this final chapter of the book are gathered together imaging concepts that have been published, but have not yet come into routine industrial use. The first part of this chapter develops the mathematical concept of linear moveout and how it relates to velocity analysis. Data can be focused so that the interval velocity can be read directly. The latter part of the chapter is about multiple reflections. Here too linear moveout helps to define the problem. You will see basic mathematical tools that have the power to deal with multiple reflections and lateral velocity variations. This chapter has many data processing proposals. They are not descriptions of production processes!

Interpreting Seismic Data

Initially I regarded this chapter as one for specialists interested mainly in devising new processes. Then I realized that in dealing with things that don't seem to work as they are expected to, we are really, for the first time, struggling to contend with reality, not with what theory predicts. This can hold much interest for skilled interpreters.

The heart of petroleum prospecting is the interpretation of reflection seismic data. What is seismic interpretation? To be a "routine interpreter" you must know everything on which theory and practice generally agree. To be a *good* interpreter you must know the "noise level" of alternate phenomena with similar effects. Anomalies in seismic data can arise from the complexity of the earth itself, from seismic wave propagation in the earth (deep, near surface, or out of plane), or from imperfections in recording and imaging techniques. To make realistic judgements in so wide a realm, you must be a seismologist who is part geologist, part engineer, and part mathematician. This chapter will not teach you to be a good interpreter, but it will offer you a chance to observe some critical thinking about the relationship of seismic theory to seismic data.

Leaning

Echo delay is much like depth. We usually measure angles by their departure from the *vertical* ray, while in reality zero-offset data is rarely recorded. The best seismic data is usually far from vertical. In this chapter a pattern of thinking is developed that is oriented about a selected nonvertical ray. Rotation of coordinates does not solve the problem since after rotation, the plane on which measurements would be made would no longer be simply $z = 0$. Rotation would also make a mess of the simple seismic velocity function $v(z)$ by making it a strongly two-dimensional function $v'(x', z')$. The view of offset presented in Chapter 3 may have seemed rather complete, but in fact it was not very general because square roots were expanded about the vertical ray. The Stolt stretch development in Section 4.5 illustrated the advantage of leaving the hyperbola top and getting out on the flanks.

Linear moveout (LMO) is the way to reorient our thinking about non-vertical rays. While not widely incorporated in the modern production environment, this deeper view of offset is of special interest to researchers. It offers an understanding of multiple reflections, a subject untouched in Chapter 3. It also offers a better understanding of velocity estimation.

Stepout Review

In Section 1.5 a *Snell wave* was defined as a plane wave that has become nonplanar by moving into a velocity-stratified medium $v = v(z)$. A plane wave keeps its angle of propagation constant, while a Snell wave keeps its stepout dt/dx a constant function of z. Figure 1 shows a Snell wave incident on the earth's surface. The wavefronts at successive times are not parallel to each other; they are horizontal translations of one another. The *slowness* of horizontal motion is called the *stepout*. It is measured in units of inverse velocity and is given as milliseconds per meter or as seconds per kilometer. The slowness, denoted as p, is also called the *ray parameter* or the *Snell parameter:*

$$p = \frac{dt}{dx} = \frac{\sin\theta(z)}{v(z)} = const(z) \tag{1}$$

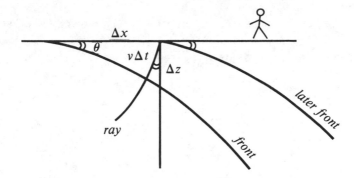

FIG. 5.0-1. Wavefront arrival at earth's surface, showing that observation of dt/dx gives the ratio $dt/dx = (\sin\theta)/v$.

5.1 Radial Traces

Radial trace sections were introduced in Section 3.6 as an alternative to constant-offset sections. In Section 3.6 the goal was to achieve a proper migration of nonzero-offset data. We also saw the definition of dip moveout (DMO). DMO simplifies further analysis because after DMO we can analyze gathers assuming that they come from a horizontally layered earth.

A *radial trace gather* is defined by a deformation of an ordinary gather. Let the ordinary gather be denoted by $P(x,t)$. Let the radial parameter be denoted by $r = x/t$. Then the radial trace gather $P'(r,t)$ is defined by the deformation $P'(r,t) = P(rt,t)$.

The horizontal location x of the tip of a ray moves according to $x = vt\sin\theta$. So in a constant-velocity medium, the radial trace with a fixed $r = x/t$ contains all the energy that propagates at angle θ.

The constancy of propagation angle within a radial trace should be helpful in the analysis of multiple reflections. It should also be helpful in compensation for the shot waveform, since the antenna effects of the shot and geophone arrays are time-invariant on each radial trace.

Assuming reflectors at depth z_j and constant velocity, hyperbolic travel-time curves are

$$t^2 v^2 = x^2 + z_j^2 \tag{1}$$

Let us see what happens to the hyperbola (1) when the offset x is transformed to the radial parameter $r = x/t$. We get an equation for a family of curves in the (r, t)-plane (plotted in figure 1).

$$z_j^2 = t^2(v^2 - r^2) \qquad (2)$$

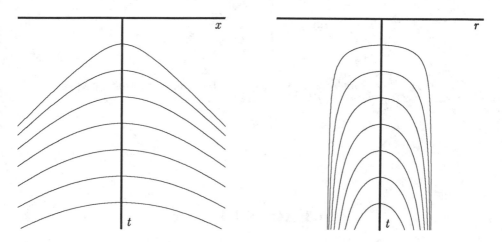

FIG. 5.1-1. Family of hyperbolas before and after transformation to radial space.

The asymptotes, instead of being along sloping lines $x^2 = \pm v^2 t^2$, are along vertical lines $r = \pm v$. The filled region of the (r, t)-plane is rectangular, while the filled region of the (x, t)-plane is triangular.

Figure 2 shows a field profile before and after transformation to radial space. Zero traces were interspersed between live ones to clarify the shape of the deformation. To understand this deformation, it helps to remember that a field trace is a curve of constant $x = rt$.

An interesting aspect of the radial-trace transformation is its effect on ground roll. A simple model of ground roll is a wave traveling horizontally at a constant rate. So on a radial-trace gather the ground roll is found as d.c. (zero frequency) on a few radial traces near $r = \pm v_{roll}$. Figure 3 shows an approximation to the idealization.

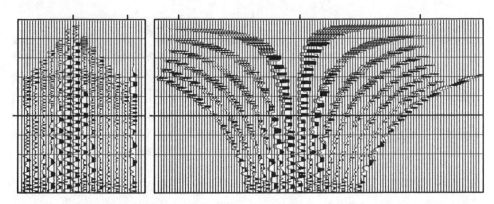

FIG. 5.1-2. Field profile from Alberta (Western Geophysical) interspersed with zero traces, shown before and after radial-trace deformation.

Moveout-Corrected Radial Traces

Moveout correction may be regarded as a transformation from time to depth. When the moveout correction is properly done, all traces should show the same depth-dependent reflectivity. In principle, radial moveout correction proceeds by introducing z and eliminating t with the substitution $tv = \sqrt{z^2 + x^2}$. In practice you would prefer a travel-time-depth axis to a depth axis. So the transformation equation becomes $t = \sqrt{\tau^2 + x^2/v^2}$. Eliminating x with rt we get

$$t = \frac{\tau}{\sqrt{1 - r^2/v^2}} \qquad (3)$$

Inspecting (3) we see that moveout correction in radial-trace coordinates is a *uniform compression* of the time t-axis into a τ-axis. The amount of compression is fixed when r is fixed. The amount of compression does not change with time. The uniformity of the compression is an aid to modeling and removing the effects of shot waveforms and multiple reflections. It is curious to note that moveout correction of radial traces *compresses* time, while moveout correction of constant-offset data *stretches* time nonuniformly. Figure 3 shows a field profile before and after the radial-trace transformation.

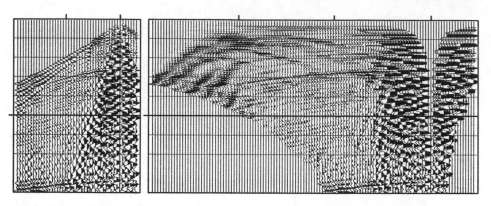

FIG. 5.1-3. Field profile from Alberta (Western Geophysical) shown before and after deformation into radial traces.

Snell Traces

The radial-trace coordinate system can be used no matter what the velocity of the earth. But the coordinate system has a special advantage when the velocity is constant, because then it gathers all the energy of a fixed propagation angle. The logical generalization to stratified media is to gather all the energy with a fixed Snell parameter. A *Snell trace* is defined (Ottolini) as a trajectory on the (x, t)-plane where the stepout $p = dt/dx$ would be constant if the velocity were $v(z)$. Where the velocity increases with depth, the Snell traces bend upward. The Snell trace trajectory is readily found by integrating the ray equations:

$$x = \int_0^z \tan \theta \; dz \tag{4a}$$

$$t = \int_0^z \frac{dz}{v \cos \theta} \tag{4b}$$

To do moveout correction on the Snell traces, introduce the vertical travel-time depth t such that $dz = v \, d\tau$. The radial-trace moveout-correction equations become

$$x(p, \tau) = \int_0^\tau \frac{p \, v^2}{\sqrt{1 - p^2 v^2}} \; d\tau \tag{5a}$$

$$t(p, \tau) = \int_0^\tau \frac{1}{\sqrt{1 - p^2 v^2}} \; d\tau \tag{5b}$$

Where the earth velocity is stratified, Snell traces have a theoretical advantage over radial traces. However they have the disadvantage that the curves could become multibranched, so that the transformation would not be one-to-one. So in practice you might use a simplified velocity model instead of your best estimate of the true velocity.

More philosophically, the transition from constant-offset traces to radial traces is a big one, whereas the transition from radial traces to Snell traces is not so large. Since the use of radial traces is not widespread, we can speculate that the practical usefulness of Snell traces may be further limited.

5.2 Slant Stack

Slant stack is a transformation of the offset axis. It is like steering a beam of seismic waves. I believe I introduced the term *slant stack* (Schultz and Claerbout [1978]) as a part of a migration method to be described next in Section 5.3. I certainly didn't invent the slant-stack concept! It has a long history in exploration seismology going back to Professor Rieber in the 1930s and to Professor Riabinkin in the Soviet Union. Mathematically, the slant-stack concept is found in the Radon [1917] transformation.

The slant-stack idea resembles the Snell trace method of organizing data around emergent angle. The Snell trace idea selects data based on a hypothetical velocity predicting the local stepout $p = dt/dx$. Slant stack does not *predict* the stepout, but *extracts* it by filtering. Thus slant stack does its job correctly whether or not the velocity is known. When the velocity of the medium is known, slant stack enables immediate downward continuation even when mixed apparent velocities are present as with diffractions and multiple reflections.

Slant Stacking and Linear Moveout

Looking on profiles or gathers for events of some particular stepout $p = dt/dx$ amounts to scanning hyperbolic events to find the places where they are tangent to a straight line of slope p. The search and analysis will be easier if the data is replotted with *linear moveout* — that is, if energy located at offset $x = g-s$ and time t in the (x, t)-plane is moved to time $\tau = t - px$ in the (x, τ)-plane. This process is depicted in figure 1. The linear moveout converts all events stepping out at a rate p in (x, t)-space to "horizontal" events in (x, τ)-space. The presence of horizontal timing lines facilitates the search for and the identification and measurement of the locations of the events.

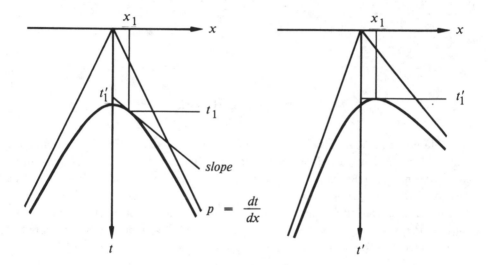

FIG. 5.2-1. Linear moveout converts the task of identifying tangencies to constructed parallel lines to the task of locating the tops of convex events.

After linear moveout $\tau = t - px$, the components in the data that have Snell parameters near p are slowly variable along the x-axis. To extract them, apply a low-pass filter on the x-axis, and do so for each value of τ. The limiting case of low-frequency filtering is extracting the mean. This leads to the idea of slant stack.

To *slant stack*, do linear moveout with $\tau = t - px$, then sum over x. This is the same as summing along slanted lines in (t, x)-space. In either case, the entire gather $P(x, t)$ gets converted to a single trace that is a function of τ.

Slant stack assumes that the sum over observed offsets is an adequate representation of integration over all offset. The (slanted) integral over offset will receive its major contribution from the zone in which the path of integration becomes tangent to the hyperboloidal arrivals. On the other hand, the contribution to the integral is vanishingly small when the arrival-time curve crosses the integration curve. The reason is that propagating waves have no zero-frequency component.

The strength of an arrival depends on the length of the zone of tangency. The Fresnel definition of the length of the zone of tangency is based on a half-wavelength condition. In an earth of constant velocity (but many flat layers) the width of the tangency zone would broaden with time as the hyperbolas flatten. This increase goes as $\sqrt{t}$, which accounts for half the spherical-divergence correction. In other words, slant stacking takes us from two dimensions to one, but a $\sqrt{t}$ remains to correct the conical wavefront of three dimensions to the plane wave of two.

Slant-Stack Gathers are Ellipses.

A slant stack of a data gather yields a single trace characterized by the slant parameter p. Slant stacking at many p-values yields a *slant-stack gather*. (Those with a strong mathematical-physics background will note that slant stacking transforms travel-time curves by the Legendre transformation. Especially clear background reading is found in *Thermodynamics,* by H.B. Callen, Wiley, 1960, pp. 90-95).

Let us see what happens to the familiar family of hyperbolas $t^2 v^2 = z_j^2 + x^2$ when we slant stack. It will be convenient to consider the circle and hyperbola equations in *parametric* form, that is, instead of $t^2 v^2 = x^2 + z^2$, we use $z = vt \cos \theta$ and $x = vt \sin \theta$ or $x = z \tan \theta$. Take the equation for linear moveout

$$\tau \;=\; t \,-\, p \, x \tag{1}$$

and eliminate t and x with the parametric equations.

$$\tau \;=\; \frac{z}{v \, \cos \theta} \,-\, \frac{\sin \theta}{v} \, z \tan \theta \;=\; \frac{z}{v} \cos \theta \tag{2}$$

$$\tau \;=\; \frac{z}{v} \sqrt{1 - p^2 v^2} \tag{3}$$

Squaring gives the familiar ellipse equation

$$\left(\frac{\tau}{z} \right)^2 \;+\; p^2 \;=\; \frac{1}{v^2} \tag{4}$$

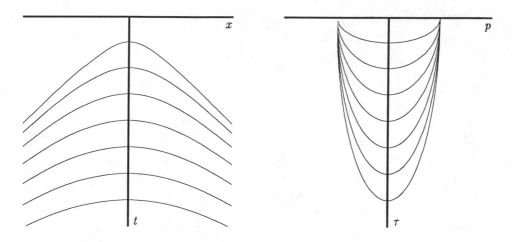

FIG. 5.2-2. Travel-time curves for a data gather on a multilayer earth model of constant velocity before and after slant stacking.

Equation (4) is plotted in figure 2 for various reflector depths z_j.

Two-Layer Model

Figure 3 shows the travel times of waves in a two-layer model. As is the usual case, the velocity is higher in the deeper layer. At the left are the familiar hyperboloidal curves. Strictly, the top curve is exactly a hyperbola whereas the lower curve is merely hyperboloidal. The straight line through the origin represents energy traveling horizontally along the earth's surface. The lower straight line is the head wave. (In seismology it is often called the *refracted* wave, but this name can cause confusion). It represents a ray that hits the deeper layer at critical angle and then propagates horizontally along the interface.

The right side of the figure shows the travel-time curves after slant stacking. Note that curves cross one another in the (x, t)-space but they do not cross one another in the (p, τ)-space. The horizontal axis $p = dt/dx$ has physical dimensions inverse to velocity. Indeed, the velocity of each layer may be read from its travel-time curve as the maximum p-value on its ellipse. The head waves — which are straight lines in (x, t)-space — are points in (p, τ)-space located where the ellipsoids touch. The top curve in (p, τ)-space is exactly an ellipse, and the lower curve is merely ellipsoidal.

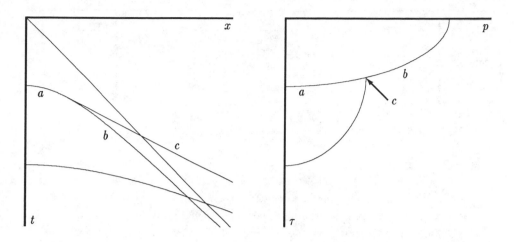

FIG. 5.2-3. Identification of precritical reflection (a), postcritical reflection (b), and head wave (c).

Interval Velocities from Slant Stacks

Section 1.5 showed that downward continuation of Snell waves is purely a matter of time shift. The amount of time shift depends only on the angle of the waves. For example, a frequency domain equation for the shifting is

$$P(\omega, p, z_2) \;=\; P(\omega, p, z_1)\, e^{-i\,\frac{\omega}{v}\sqrt{1 - p^2 v^2}\,(z_2 - z_1)} \tag{5}$$

Downward continuing to the first reflector, we find that the first reflections should arrive at zero time. In migration it is customary to retard time with respect to the zero-dip ray. So downward continuation in retarded time flattens the first reflection without changing the zero-dip ray. Time shifting the data to align on the first-layer reflection is illustrated by the third panel in figure 4. The first panel shows the velocity model, and the second panel shows the slant stacks at the surface. After the first reflector is time aligned, we have the data that should be observed at the bottom of the first layer. Now the next deeper curve is an exact ellipse. Estimate the next deeper velocity from that next deeper ellipse. Continue the procedure to all depths. This method of velocity estimation was proposed and tested by P. Schultz [1982].

Figure 4 illustrates the difficulty caused by a shallow, high-velocity layer. Reflection from the bottom of any deeper, lower-velocity layer gives an incomplete ellipse. It does not connect to the ellipse above because it seems to want

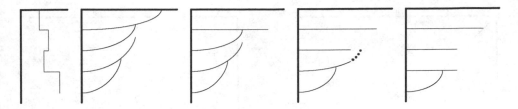

FIG. 5.2-4. Schultz flattening on successive layers.

to extend beyond. The large p-values (dotted in the figure) are missing because they are blocked by the high-velocity (low p) layer above. The cutoff in p happens where waves in the high-velocity layer go horizontally. So there are no head waves on deeper, lower-velocity layer bottoms.

Schultz's method of estimating velocity from an ellipse proceeds by summing on scanning ellipses of various velocities and selecting the one with the most power. So his method should not be troubled by shallow high-velocity layers. It is interesting to note that when the velocity does increase continuously with depth, the velocity-depth curve can be read directly from the rightmost panel of figure 4. The velocity-depth curve would be the line connecting the ends (maximum p) of the reflections, i.e. the head waves.

Interface Velocity from Head Waves

The determination of earth velocity from head waves is an old subject in seismology. Velocity measurement from head waves, where it is possible, refers to a specific depth —the depth of the interface— so it has even better depth-resolving power than an *interval* velocity (the velocity of a depth interval between two reflections).

Traditionally, head-wave velocity analysis involved identification (picking) of travel times. Travel times are hard to pick out on noisy data. Clayton and McMechan [1981] introduced a new method based on the wavefield itself, instead of on picked travel times. They did for the velocity analysis of head waves what wave-equation migration did for reflections.

The same idea for getting velocity from back-scattered head waves on sections (Section 3.5) can be used on ordinary head waves on common-midpoint gathers. On gathers you have the extra information not on a section that downward continuation focuses energy on zero offset. The focus is not a featureless point. Take original data to consist of a head wave only, with no reflection. Downward continuation yields a focus at zero offset. The

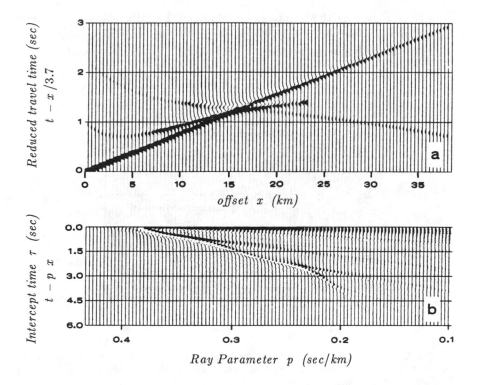

FIG. 5.2-5. The upper figure (a) contains a synthetic head-wave profile (plotted with linear moveout). The data is transformed by slant stack to the lower half of the figure (b). The result of downward continuation of this slant-stacked wavefield (b) is shown in figure 6. (Clayton & McMechan)

focus is a concentrated patch of energy oriented with the same stepout dt/dh as the original unfocused head wave. Summing through the focus at all possible orientations (slant stack) transforms the data $u(h,\tau)$ to dip space, say $\bar{u}(p,\tau)$. The velocity of the earth at travel-time depth τ is found where the seismic energy has concentrated on the (p,τ)-plane. The velocity is given directly by $v(\tau)=1/p(\tau)$. Given $v(\tau)$, $v(z)$ is readily found. Or the entire calculation could be done in depth z directly instead of in travel-time depth τ.

Clayton and McMechan actually do the downward continuation and the slant stack in the opposite order. They slant stack first and then downward continue. In principle these processes can be done in either order. Remember that we are bootstrapping to the correct earth velocity. Slant stacking does not depend on the earth's velocity, but downward continuation does. Slant stacking need be done only once if it is done first, which is why Clayton and McMechan do it that way. Figures 5 and 6 show one of their examples.

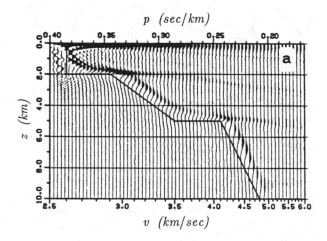

FIG. 5.2-6. The result of downward continuation of the slant-stacked wavefield in figure 5b with the correct velocity-depth function (the solid line). (Clayton & McMechan)

Compare the method of Clayton and McMechan to that of Schultz. Schultz flattens the reflections by a method that is sensitive to the large p parts of the ellipse. Clayton and McMechan look only at the largest p part of the ellipse. Schultz has the advantage that a method based on reflection is not troubled by high-velocity layers, but the disadvantage that decision making is required during the descent. Clayton and McMechan present the interpreter with a plane of information from which the interpreter selects the velocity. Clayton and McMechan's velocity space is a linear, invertible function of the data. Section 5.4 will describe a linear, invertible transformation of *reflection* data (not head waves) to velocity space.

Slant Stack and Fourier Transform

Let $u(x, t)$ be a wavefield. The *slant stack* $\bar{u}(p, \tau)$ of the wavefield is defined mathematically by

$$\boxed{\bar{u}(p, \tau) = \int u(x, \tau + px)\, dx}\tag{6}$$

The integral across x in (6) is done at constant τ, which is a slanting line in the (x, t)-plane.

Slant stack is readily expressed in Fourier space. The definition of the two-dimensional Fourier transformation of the wavefield $u(x, t)$ is

$$U(k, \omega) = \iint e^{i\omega t - ikx}\, u(x, t)\, dx\, dt\tag{7}$$

Recall the definition of Snell's parameter in Fourier space $p = k/\omega$ and use it to eliminate k from the 2-D Fourier transform (7).

$$U(\omega p, \omega) = \iint e^{i\omega(t - px)}\, u(x, t)\, dx\, dt\tag{8}$$

Change the integration variable from t to $\tau = t - px$.

$$U(\omega p, \omega) = \int e^{i\omega\tau}\left[\int u(x, \tau + px)\, dx\right] d\tau\tag{9}$$

Insert the definition (6) into (9).

$$U(\omega p, \omega) = \int e^{i\omega\tau}\, \bar{u}(p, \tau)\, d\tau\tag{10}$$

Think of $U(\omega p, \omega)$ as a one-dimensional function of ω that is extracted from the (k, ω)-plane along the line $k = \omega p$.

The inverse Fourier transform of (10) is

$$\boxed{\bar{u}(p, \tau) = \int e^{-i\omega\tau}\, U(\omega p, \omega)\, d\omega}\tag{11}$$

The result (11) states that a slant stack can be created by Fourier-domain operations. First you transform $u(x, t)$ to $U(k, \omega)$. Then extract $U(\omega p, \omega)$ from $U(k, \omega)$. Finally, inverse transform from ω to τ and repeat the process for all interesting values of p.

Getting $U(\omega p, \omega)$ from $U(k, \omega)$ seems easy, but this turns out to be the hard part. The line $k = \omega p$ will not pass nicely through all the mesh points (unless $p = \Delta t / \Delta x$) so some interpolation must be done. As we have seen from the computational artifacts of Stolt migration, Fourier-domain interpolation should not be done casually. Interpolation advice is found in

Section 4.5.

Both (6) and (11) are used in practice. In (6) you have better control of truncation and aliasing. For large datasets, (11) is much faster.

Inverse Slant Stack

Tomography in medical imaging is based on the same mathematics as inverse slant stack. Simply stated, (two-dimensional) tomography or inverse slant stacking is the reconstruction of a function given line integrals through it. The inverse slant-stack formula will follow from the definition of two-dimensional Fourier integration:

$$u(x,t) = \int e^{-i\omega t} \left[\int e^{ikx} U(k,\omega) \, dk \right] d\omega \tag{12}$$

Substitute $k = \omega p$ and $dk = \omega \, dp$ into (12). Notice that when ω is negative the integration with dp runs from positive to negative instead of the reverse. To keep the integration in the conventional sense of negative to positive, introduce the absolute value $|\omega|$. (More generally, a change of variable of volume integrals introduces the *Jacobian* of the transformation). Thus,

$$u(x,t) = \int e^{-i\omega t} \left[\int e^{i\omega px} U(\omega p,\omega) \, |\omega| \, dp \right] d\omega \tag{13}$$

$$u(x,t) = \int \left\{ \int e^{-i\omega t} \left[U(\omega p,\omega) e^{i\omega px} \, |\omega| \right] d\omega \right\} dp \tag{14}$$

Observe that the { } in (14) contain an inverse Fourier transform of a product of three functions of frequency. The product of three functions in the ω-domain is a convolution in the time domain. The three functions are first $U(\omega p,\omega)$, which by (11) is the FT of the slant stack. Second is a delay operator $e^{i\omega px}$, i.e an impulse function of time at time px. Third is an $|\omega|$ filter. The $|\omega|$ filter is called a *rho* filter. The *rho* filter does not depend on p so we may separate it from the integration over p. Let "*" denote convolution. Introduce the delay px as an argument shift. Finally we have the inverse slant-stack equation we have been seeking:

$$\boxed{u(x,t) = rho(t) * \int \bar{u}(p, t-px) \, dp} \tag{15}$$

It is curious that the inverse to the slant-stack operation (6) is basically another slant-stacking operation (15) with a sign change.

Plane-Wave Superposition

Equation (15) can be simply interpreted as plane-wave superposition. To make this clear, we first dispose of the *rho* filter by means of a definition.

$$\tilde{u}(p,\tau) = rho(\tau) * \bar{u}(p,\tau) \tag{16}$$

Equation (16) will be seen to be more than a definition. We will see that $\tilde{u}(p,\tau)$ can be interpreted as the *plane-wave spectrum*. Substituting the definition (16) into both (15) and (6) gives another transform pair:

$$u(x,t) = \int \tilde{u}(p, t-px)\,dp \tag{17}$$

$$\tilde{u}(p,\tau) = rho(\tau) * \int u(x, \tau+px)\,dx \tag{18}$$

To confirm that $\tilde{u}(p,\tau)$ may be interpreted as the plane-wave spectrum, we take $\tilde{u}(p,\tau)$ to be the impulse function $\delta(p - p_0)\,\delta(\tau - \tau_0)$ and substitute it into (17). The result $u(x,t) = \delta(t - p_0 x - \tau_0)$ is an impulsive plane wave, as expected.

Reflection Coefficients — Spherical versus Planar

The amplitudes that you see on the reflected waves on a field profile are affected by many things. Assume that corrections can be made for the spherical divergence of the wave, the transmission coefficients through the layers, inner bed multiples, etc. What remains are the *spherical-wave reflection coefficients*. Spherical-wave reflection strengths are not the same as the plane-wave reflection coefficients calculated in FGDP or by means of Zoeppritz [1919] equations. Theoretical analyses of reflection coefficient strengths are always based on Fourier analysis. Equations (17) and (18) provide a link between plane-wave reflection coefficients and cylindrical-wave reflection coefficients. See page 196 for going from cylinders to spheres.

The Rho Filter

In practical work, the *rho* filter is often ignored because it can be absorbed into the rest of the filtering effects of the overall data recording and processing activity. However, the *rho* filter is not inconsequential. The integrations in the slant stack enhance low frequencies, and the *rho* filter pushes them back to their appropriate level. Let us inspect this filter. The *rho* filter has the same spectrum as does the time derivative, but their time functions are very different. The finite-difference representation of a time derivative is short, only Δt in time duration. Because of the sharp corner in the absolute-value function, the *rho* filter has a long time duration. The

Hilbert kernel $-1/t$ has a Fourier transform $i\ sgn(\omega)$. Notice that $|\omega| = (-i\omega) \times i\ sgn(\omega)$. In the time domain this means that $d/dt\,(-1/t) = 1/t^2$, so $rho(t) = 1/t^2$.

An alternate view is that the *rho* filter should be divided into two parts, with half going into the forward slant stack and the other half into the inverse. Then slant stacking would not cause the power spectrum of the data to change. An interesting way to divide the $|\omega|$ is $|\omega| = \sqrt{-i\omega}\,\sqrt{i\omega}$. It was shown in Section 4.6 that $\sqrt{-i\omega}$ is a causal time function and $\sqrt{i\omega}$ is anticausal. More details about slant stacks are found in Phinney et al. [1981]

In practice, slant stack is not so cleanly invertible as 2-D FT, so various iteration and optimization techniques are often used.

EXERCISES

1. Assume that $v(z) = const$ and prove that the width of a Fresnel zone increases in proportion to $\sqrt{t}$.

2. Given $v(z)$, derive the width of the Fresnel zone as a function of t .

5.3 Snell Waves and Skewed Coordinates

Slant stacks are closely related to Snell waves. But there is more to it than that. Three different types of gathers (CSP, CGP, and CMP) can be slant stacked, and the meaning is different in each case.

A Snell wave can be synthesized by slant stacking ordinary reflection data. Snell waves are described by wave-propagation theory. You can expect to be able to write a wave equation that really describes the Snell wave despite complexities of lateral velocity variation, multiple reflections, shear waves, or all these complications at once. Contrast this to a CDP stack where downward continuation is already an approximation even when velocity is constant. Of course we can always return to data analysis in shot-geophone space. But the slant stack is a stack, and that means there is already some noise reduction and data compression.

Snell Wave Information in Field Data

The superposition principle allows us to create an impulse function by a superposition of sinusoids of all frequencies. A three-dimensional generalization is the creation of a point source by the superposition of plane waves going in all directions. Likewise, a plane wave can be a superposition of many Huygens secondary point sources. A Snell wave can be simulated by an appropriate superposition, called a *slant stack,* of the point-source data recorded in exploration.

Imagine that all the shots in a seismic survey were shot off at the same time. The downgoing wave would be approximately a plane wave. (Let us ignore the reality that the world is 3-D and not 2-D). The data recorded from such an experiment could readily be simulated from conventional data simply by summing the data field $P(s, g, t)$ over all s. In each common-geophone profile the traces would be summed with no moveout correction.

To simulate a nonvertical Snell wave, successive shots must be delayed (to correspond to a supersonic airplane), according to some prescribed $p_s = dt/ds$.

What happens if data is summed over the geophone axis instead of the shot axis? The result is point-source experiments recorded by receiver antennas that have been highly tuned to receive vertically propagating waves. Time shifting the geophones before summation simulates a receiver antenna that records a Snell wave, say, $p_g = dt/dg$ upcoming at an angle $\sin \theta = p_g\, v$.

Integration over an axis is an extreme case of low-pass filtering over an axis. Between the two extremes of the point-source case and the plane-wave case is the case of directional senders and receivers.

The simple process of propagation spreads out a point disturbance to a place where, from a distance, the waves appear to be nearly plane waves or Snell waves. Little patches of data where arrivals appear to be planar can be analyzed as though they were Snell waves.

In summary, a *downgoing* Snell wave is achieved by dip filtering in *shot* space, whereas an *upcoming* Snell wave is achieved by dip filtering in *geophone* space.

Muting and Data Recording

The basic goal of muting is to remove horizontally moving energy. Such energy is unrelated to a deeper image. Typically muting is performed as described in Section 3.5 — that is, a weighting function zeroes data generally beyond some value of $(g-s)/t$. There is no question that muting removes much horizontally moving energy, but more can be done. Because of backscattering, horizontally moving energy can often be found inside the mute zone. The way to get rid of it is to use a dip filter instead of a weighting function. Before modern high-density recording, slow moving noises were often aliased on the geophone cable, so dip filtering wasn't feasible. If the emergent angle isn't close enough to vertical, that is, if dt/dg isn't small enough, then the waves can't have come from the exploration target. On explosion data, filtering is not so easily applied in shot space as it is in geophone space because data is not very densely recorded in shot space. Don't fall into the trap of thinking that this dip filtering can be done on a common-*mid* point gather. Back-scattered ground roll has no moveout on a common-midpoint gather (see Section 3.2).

Marine water-bottom scatter is frequently so strong that it is poorly suppressed by conventional processing. In Section 3.2 we saw the reason: point scatterers imply hyperbolic arrivals, which have steep dip, hence they have the stacking velocities of sediment rather than water. What is needed are two dip filters — one to reject waves leaving the shots at nonpenetrating angles, and the other to reject waves arriving at the geophones at non-penetrating angles.

Present-day field arrays filter on the basis of spatial frequency k_x. More high-frequency energy would be left in the data if the recording equipment used dip (k/ω) filters instead of spatial-frequency k filters. The causal dip filters described in Section 2.5 might work nicely.

Synthesizing the Snell Wave Experiment

Let us synthesize a downgoing Snell wave with field data, then imagine how the upcoming wave will look and how it will carry to us information about the subsurface.

Slant stacking will take the survey line data $P(s,g,t)$, which is a function of shot location s, geophone location g, and travel time t, and sum over the shot dimension, thereby synthesizing the upcoming wave $U(g,t)$, which should have been recorded from a downgoing Snell wave. This should be the case even though there may be lateral velocity variation and multiple reflections.

The summation process is confusing because three different kinds of time are involved:

t = travel time in the point-source field experiments.

t' = $t - p(g - s)$ = interpretation time. The shallowest reflectors are seen just after $t' = 0$.

t_{pseudo} = time in the Snell pseudoexperiment with a moving source.

Time in the pseudoexperiment in a horizontally layered earth has the peculiar characteristic that the further you move out the geophone axis, the later the echoes will arrive. Transform directly from the field experiment time t to interpretation time t' by

$$t' = t_{pseudo} - p\,x = t - p(g - s) \tag{1}$$

Figure 1 depicts a downgoing Snell wave.

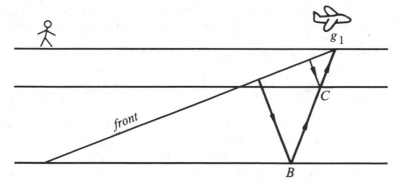

FIG. 5.3-1. Wavefront of a Snell wave that reflects from two layers, carrying information back up to g_1.

Figure 2 shows a hypothetical common-geophone gather, which could be summed to simulate the Snell wave seen at location g_1 in figure 1. The lateral offset of B from C is identical in figure 1 with that in figure 2 (at two places in figure 2). Repeating the summation for all geophones synthesizes an upcoming wave from a downgoing Snell wave.

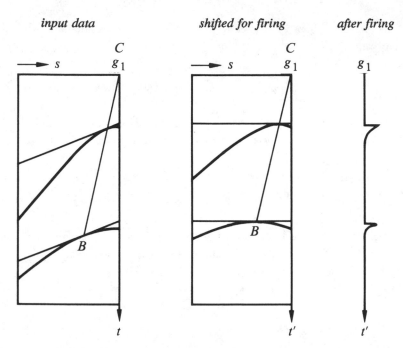

FIG. 5.3-2. On the left is a common-geophone gather at g_1 over two flat reflectors. In the center the data is shifted by linear moveout in preparation for the generation of the synthetic Snell wave by summation over shots. On the right is shown the Snell wave trace recorded at geophone g_1. A Snell wave seismic section consists of many side-by-side traces like g_1.

The variable t' may be called an interpretation coordinate, because shallow reflectors are seen just after $t' = 0$, and horizontal beds give echoes that arrive with no horizontal stepout, unlike the pseudo-Snell wave. For horizontal beds, the detection of lateral location depends upon lateral change in the reflection coefficient. In figure 1, the information about the reflection strength at B is recorded rightward at g_1 instead of being seen above B, where it would be on conventional stack. The moving of received data to an appropriate lateral location is thus an additional requirement for full interpretation.

Figure 3 shows the same two flat layers as figures 1 and 2, but there are also anomalous reflection coefficients at points A, B, and C. Point A is directly above point B. The path of the wave reflected at B leads directly to C and thence to g_1. Subsequent frames show the diffraction hyperbolas associated with these three points. Notice that the pseudo-Snell waves

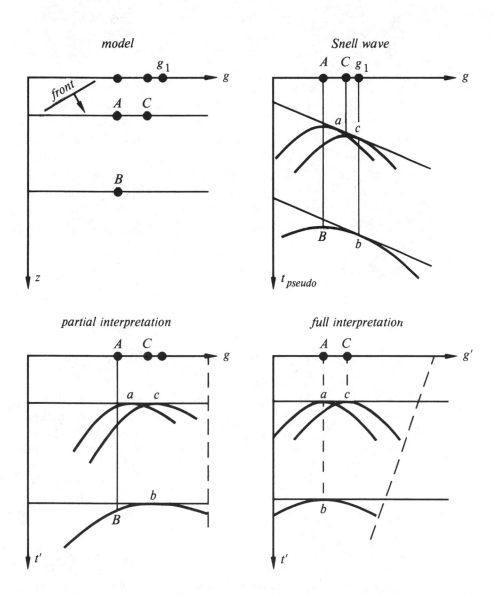

FIG. 5.3-3. Top left is three point scatterers on two reflectors. Top right is the expected Snell wave. Bottom left is the Snell wave after linear moveout. Bottom right is after transform to full interpretation coordinates. At last a, b, and c are located where A, B, and C began.

reflecting from the flat layers step out at a rate p. Hyperbolas from the scatters A, B, and C come tangent to the Snell waves at points a, b, and c. Notice that b and c lie directly under g_1 because all are aligned along a raypath with Snell parameter p. The points A, B, and C locate the tops of the hyperbolas since the earliest arrival must be directly above the point scatterer, no matter what the incident wavefield. Converting to the interpretation coordinate t' in the next frame offers the major advantage that arrivals from horizontal layers become horizontal. But notice that the hyperboloids have become skewed. Limiting our attention to the arrivals with little stepout, we find information about the anomalous reflection coefficients entirely in the vicinities of a, b, and c, which points originally lay on hyperbola flanks. These points will not have the correct geometrical location, namely that of A, B and C, until the data is laterally shifted to the left, to, say, $g' = g - f(t')$. Then a will lie above b. The correct amount of shift $f(t')$ is a subject that relates to velocity analysis. The velocity analysis that pertains to this problem will be worked out in the next section.

What's Wrong with Snell Waves?

Before the DSR was developed, I thought that the only proper way to analyze seismic data was to decompose it into Snell waves. Since a Fresnel zone seems to be about 10° wide, not many Snell waves should be required. The small number of required sections was important because of the limited power of computers in the 1970's. I knew that each Snell wave is analyzable by a single square-root equation, and that even multiple reflections can be handled by methods described in FGDP and here in Section 5.6. Theoretically this approach was a big improvement over CDP stack, which is hardly analyzable at all. A practical problem for downgoing Snell waves, however, is that they may become complicated early if they encounter lateral velocity inhomogeneity shortly after they depart the earth's surface. I no longer believe that Snell waves are a panacea, although I am unsure what their ultimate role will be. But many waves behave a little like they are Snell waves. This motivates the development of a coordinate system that is ideal for Snell waves, and good for waves that are not far from being Snell waves.

Lateral Invariance

The nice thing about a vertically incident source of plane waves $p = 0$ in a horizontally stratified medium is that the ensuing wavefield is laterally invariant. In other words, an observation or a theory for a wavefield would in this case be of the form $P(t) \times const(x)$. Snell waves for any particular nonzero p-value are also laterally invariant. That is, with

$$t' \ = \ t - p \, x \qquad\qquad (2a)$$

$$x' \ = \ x \qquad\qquad (2b)$$

lateral invariance is given by the statement

$$P(x, t) \ = \ P'(t') \times const(x') \qquad\qquad (3)$$

Obviously, when an apparently two-dimensional problem can be reduced to one dimension, great conceptual advantages result, to say nothing of sampling and computational advantages. Before proceeding, study equation (3) until you realize why the wavefield can vary with x but be a constant function of x' when (2b) says $x = x'$.

The coordinate system (2) is a retarded coordinate system, not a moving coordinate system. Moving coordinate systems work out badly in solid-earth geophysics. The velocity function is never time-variable in the earth, but it becomes time-variable in a moving coordinate system. This adds a whole dimension to computational complexity.

The goal is to create images from data using a *model* velocity that is a function of all space dimensions. But the coordinate system used will have a *reference* velocity that is a function of depth only.

Snell Wave Coordinates

A Snell wave has three intrinsic planes, which suggests a coordinate system. First are the layer planes of constant z, which include the earth's surface. Second is the plane of rays. Third is the moving plane of the wavefront. The planes become curved when velocity varies with depth.

The following equations define *Snell wave coordinates:*

$$z'(z, x, t) \ = \ z \, \frac{\cos \theta}{v} \qquad\qquad (4a)$$

$$x'(z, x, t) \ = \ z \tan \theta \ + \ x \qquad\qquad (4b)$$

$$t'(z, x, t) \ = \ z \, \frac{\cos \theta}{v} \ - \ x \, \frac{\sin \theta}{v} \ + \ t \qquad\qquad (4c)$$

Equation (4a) simply defines a travel-time depth using the vertical phase velocity seen in a borehole. Interfaces within the earth are just planes of constant z'.

Setting x' as defined by equation (4b) equal to a constant, say, x_0, gives the equation of a ray, namely, $(x - x_0)/z = -\tan\theta$. Different values of x_0 are different rays.

Setting t' as defined by equation (4c) equal to a constant gives the equation for a moving wavefront. To see this, set $t' = t_0$ and note that at constant x you see the borehole speed, and at constant z you see the airplane speed.

Mathematically, one equation in three unknowns defines a plane. So, setting the left side of any of the equations (4a,b,c) to a constant gives an equation defining a plane in (z, x, t)-space. To get some practice, we will look at the intersection of two planes. Staying on a wavefront requires $dt' = 0$. Using equation (4c) gives

$$dt' = 0 = \frac{\cos\theta}{v}\, dz - \frac{\sin\theta}{v}\, dx + dt \qquad (5)$$

Combining the constant wavefront equation $dt' = 0$ with the constant depth equation $dz' = dz = 0$ gives the familiar relationship

$$\frac{dt}{dx} = p \qquad (6)$$

When coordinate planes are nonorthogonal, the coordinate system is said to be *affine*. With affine coordinates, such as (4), we have no problem with computational tractability, but we often do have a problem with our own confusion. For example, when we display movies of marine field data, we see a sequence of (h, t)-planes. Successive planes are successive shot points. So the data is displayed in (s, h) when we tend to think in the orthogonal coordinates (y, h) or (s, g). With affine coordinates I find it easiest to forget about the coordinate axis, and think instead about the perpendicular plane. The shot axis **s** can be thought of as a plane of constant geophone, say, cg. So I think of the marine-data movie as being in (cs, ch, ct)-space. In this movie, another plane, really a family of planes, the planes of constant midpoints cy, sweep across the screen, along with the "texture" of the data (Section 3.0).

To define Snell coordinates when the velocity is depth-variable, it is only necessary to interpret (4) carefully. First, all angles must be expressed in terms of p by the Snell substitution $\sin\theta = p\, v(z)$. Then z must everywhere be replaced by the integral with respect to z.

Snell Waves in Fourier Space

The chain rule for partial differentiation says that

$$
\begin{bmatrix} \partial_t \\ \partial_x \\ \partial_z \end{bmatrix}
=
\begin{bmatrix} t'_t & x'_t & z'_t \\ t'_x & x'_x & z'_x \\ t'_z & x'_z & z'_z \end{bmatrix}
\begin{bmatrix} \partial_{t'} \\ \partial_{x'} \\ \partial_{z'} \end{bmatrix}
\qquad \text{(7a,b,c)}
$$

In Fourier space, equations (7a) and (7b) may be interpreted as

$$ -i\,\omega \;=\; -i\,\omega' \qquad\qquad (8a) $$

$$ i\,k_x \;=\; +p\,\omega' + i\,k'_x \qquad\qquad (8b) $$

Of particular interest is the energy that is flat after linear moveout (constant with x'). For such energy $\partial/\partial x' = i\,k'_x = 0$. Combining (8a) and (8b) gives the familiar equation

$$ p \;=\; \frac{k}{\omega} \qquad\qquad (9) $$

EXERCISES

1. Explain the choice of sign of the s-axis in figure 1.

2. Equation (4) is for *upgoing* Snell waves. What coordinate system would be appropriate for *downgoing* Snell waves?

3. Express the scalar wave equation in the coordinate system (4). Neglect first derivatives.

4. Express the dispersion relation of the scalar wave equation in terms of the Fourier variables (ω', k'_x, k'_z).

5.4 Interval Velocity by Linear Moveout

Linear moveout forms the basis for a simple, graphical method for finding seismic velocity. The method is particularly useful for the analysis of data that is no longer in a computer, but just exists on a piece of paper. Additionally, the method offers insights beyond those offered by the usual computerized hyperbola scan. Using it will help us rid ourselves of the notion that angles should be measured from the vertical ray. Non-zero Snell parameter can be the "default."

Ultimately this method leads to a definition of *velocity spectrum,* a plane in which the layout of data, after a linear invertible transformation, shows the seismic velocity.

Graphical Method for Interval Velocity Measurement

A wave of velocity v from a point source at location $(x, z) = (0, z_s)$ passes any point (x, z) at time t where

$$v^2 t^2 = x^2 + (z - z_s)^2 \tag{1}$$

In equation (1) x should be replaced by either half-offset h or midpoint y. Then t is two-way travel time; the velocity v is half the rock velocity; and $(z - z_s)$ is the distance to an image source.

Differentiating (1) with respect to t (at constant z) gives

$$v^2 2 t = 2 x \frac{dx}{dt} \tag{2}$$

$$v^2 = \frac{x}{t} \frac{dx}{dt} \tag{3}$$

Figure 1 shows that the three parameters required by (3) to compute the material velocity are readily measured on a common-midpoint gather.

Equation (3) can be used to estimate a velocity whether or not the earth really has a constant velocity. When the earth velocity is stratified, say, $v(z)$, it is easy to establish that the estimate (3) is exactly the root-mean-square (RMS) velocity. First recall that the bit of energy arriving at the point of tangency propagates throughout its entire trip with a constant Snell parameter $p = dt/dx$.

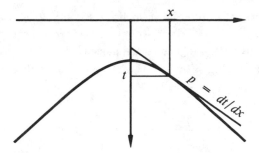

FIG. 5.4-1. A straight line, drawn tangent to hyperbolic observations. The slope p of the line is arbitrary and may be chosen so that the tangency occurs at a place where signal-to-noise ratio is good. (Gonzalez)

The best way to specify velocity in a stratified earth is to give it as some function $v'(z)$. Another way is to pick a Snell parameter p and start descending into the earth on a ray with this p . As the ray goes into the earth from the surface $z = 0$ at $t = 0$, the ray will be moving with a speed of, say, $v(p, t)$. It is an elementary exercise to compute $v(p, t)$ from $v'(z)$ and vice versa. The horizontal distance x which a ray will travel in time t is given by the time integral of the horizontal component of velocity, namely,

$$x = \int_0^t v(p, t) \sin \theta \, dt \qquad (4)$$

Replacing $\sin \theta$ by pv and taking the constant p out of the integral yields

$$x = p \int_0^t v(p, t)^2 \, dt \qquad (5)$$

Recalling that $p = dt / dx$, insert (5) into (3):

$$v^2_{measured} = \frac{x}{t} \frac{dx}{dt} \qquad (6)$$

$$v^2_{measured} = \frac{1}{t} \int_0^t v(p, t)^2 \, dt \qquad (7)$$

which justifies the assertion that

$$v_{measured} = v_{root\text{-}mean\text{-}square} = v_{RMS} \qquad (8)$$

Equation (7) is exact. It does not involve a "small offset" assumption or a "straight ray" assumption.

Next compute the *interval* velocity. Figure 2 shows hyperboloidal arrivals from two flat layers. Two straight lines are constructed to have the same slope p. Then the tangencies are measured to have locations (x_1, t_1) and (x_2, t_2). Combining (6) with (4), and using the subscript j to denote the jth tangency (x_j, t_j), gives

$$x_j \frac{dx}{dt} = \int_0^{t_j} v(p, t)^2 \, dt \qquad (9)$$

Assume that the velocity between successive events is a constant $v_{interval}$, and subtract (9) with $j+1$ from (9) with j to get

$$(x_{j+1} - x_j) \frac{dx}{dt} = (t_{j+1} - t_j) \, v_{interval}^2 \qquad (10)$$

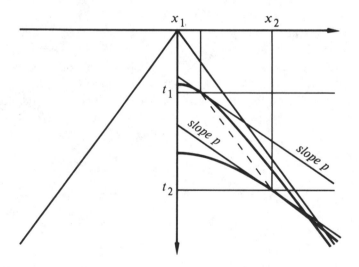

FIG. 5.4-2. Construction of two parallel lines on a common-midpoint gather which are tangent to reflections from two plane layers. (Gonzalez)

Solving for the interval velocity gives

$$v_{interval}^2 = \frac{x_{j+1} - x_j}{t_{j+1} - t_j} \frac{dx}{dt} \qquad (11)$$

So the velocity of the material between the jth and the $j+1^{\text{st}}$ reflectors can be measured directly using the square root of the product of the two slopes in (11), which are the dashed and solid straight lines in figure 2. The advantage of manually placing straight lines on the data, over automated analysis, is that you can graphically visualize the noise sensitivity of the measurement, and you can select on the data the best offsets at which to make the measurement.

If you do this routinely you quickly discover that the major part of the effort is in accurately constructing two lines that are tangent to the events. When you run into difficulty, you will find it convenient to replot the data with linear moveout $t' = t - px$. After replotting, the lines are no longer sloped but horizontal, so that any of the many timing lines can be used. Locating tangencies is now a question of finding the tops of convex events. This is shown in figure 3.

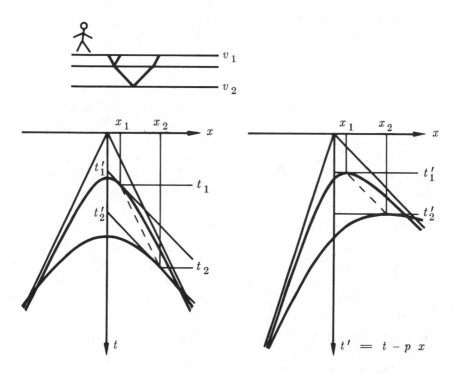

FIG. 5.4-3. Measurement of interval velocity by linear moveout. (Gonzalez)

In terms of the time t', equation (11) is

$$v_{interval}^2 = \frac{1}{\dfrac{\Delta t}{\Delta x}}\frac{1}{p} = \frac{1}{\dfrac{\Delta t'}{\Delta x} + p}\frac{1}{p} \tag{12}$$

Earth velocity is measured on the right side of figure 3 by measuring the slope of the dashed line, namely $\Delta t'/\Delta x$, and inserting it into equation (12). (The value of p is already known by the amount of linear moveout that was used to make the plot).

Common-Midpoint Snell Coordinates

Common-*midpoint* slanted wave analysis is a more conservative approach to seismic data analysis than the Snell wave approach. The advantage of common-midpoint analysis is that the effects of earth dip tend to show up mainly on the midpoint axis, and the effect of seismic velocity shows up mainly on the offset axis. Our immediate goal is to define an invertible, wave-equation approach to determination of interval velocity.

The disadvantage of common-midpoint analysis is that it is nonphysical. A slant stack at common geophone simulates a downgoing Snell wave, and you expect to be able to write a differential equation to describe it, no matter what ensues, be it multiple reflection or lateral velocity variation. A common-*midpoint* slant stack does not model anything that is physically realizable. Nothing says that a partial differential equation exists to extrapolate such a stack. This doesn't mean that there is necessarily anything wrong with a common-midpoint *coordinate* system. But it does make us respect the Snell wave approach even though its use in the industrial world is not exactly growing by leaps and bounds.

(Someone implementing common-midpoint slant stack would immediately notice that it is easier than slant stack on common-geophone data. This is because at a common midpoint, the tops of hyperboloids must be at zero offset, the location of the Fresnel zone is more predictable, and interpolation and missing data problems are much alleviated).

Seismic data is collected in time, geophone, shot, and depth coordinates (t, g, s, z). A new four-component system will now be defined. Midpoint is defined in the usual way:

$$y(t, g, s, z) = \frac{g + s}{2} \tag{13}$$

Travel-time depth is defined using the vertical phase velocity in a borehole. Two-way travel times are used, in order to be as conventional as possible:

$$\tau(t, g, s, z) \;=\; 2\,z\,\frac{\cos\theta}{v} \tag{14}$$

Next the surface offset h' is defined. We will not use the old definition of offset. For this method, shots and geophones should not go straight down, but along a ray. This can be so if h' is defined as follows:

$$h'(t, g, s, z) \;=\; \frac{g - s}{2} + z\,\tan\theta \tag{15}$$

With this new definition of h' the separation of the shot and geophone decreases with depth for constant h'.

Define the *LMO time* as the travel time in the point-source experiment less the linear moveout. So, at any depth, the LMO time is $t - p(g - s)$. As h' was defined to be the *surface* half-offset, t' is defined to be the *surface* LMO time. From the LMO time of a buried experiment, the LMO time at the surface is defined by adding in the travel-time depth of the experiment:

$$t' \;=\; t - p(g - s) + \tau \tag{16}$$

You may like to think of (16) as a "slant" on time retardation for upcoming waves, say, $t' = t_{LMO} + z_{slant}/v$. Formally,

$$t'(t, g, s, z) \;=\; t - p(g - s) + 2\,z\,\frac{\cos\theta}{v} \tag{17}$$

Figure 4 is a geometrical representation of these concepts.

From the geometry of figure 4 it will be deduced that a measurement of a reflection at some particular value of (h', t') directly determines the velocity. Write an equation for the reflector depth:

$$v\left[\frac{t'}{2} + p\,h'\right]\cos\theta \;=\; \text{reflector depth} \;=\; \frac{h'}{\tan\theta} \tag{18}$$

Using Snell's law to eliminate angles and solving for velocity gives

$$v^2 \;=\; \frac{1}{p}\,\frac{1}{p + \dfrac{t'}{2h'}} \tag{19}$$

This is consistent with equation (12).

Gathering the above definitions into a group, and allowing for depth-

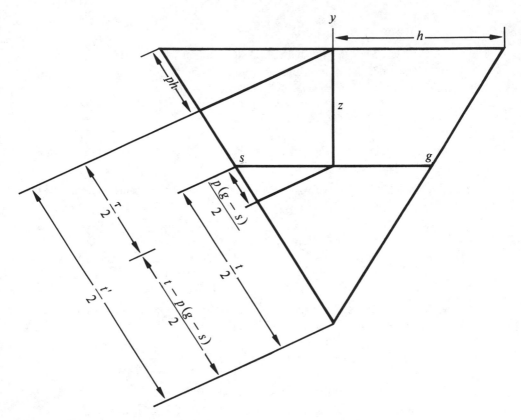

FIG. 5.4-4. The CMP-LMO coordinate frame geometry. This is a natural coordinate system for describing waves that resemble a reference Snell wave.

variable velocity by replacing z by the integral over z, we get

$$t'(t,g,s,z) \;=\; t - p(g-s) + 2\int_0^z \frac{\cos\theta}{v}\,dz \qquad (20\text{a})$$

$$y(t,g,s,z) \;=\; \frac{g+s}{2} \qquad (20\text{b})$$

$$h'(t,g,s,z) \;=\; \frac{g-s}{2} + \int_0^z \tan\theta\,dz \qquad (20\text{c})$$

$$\tau(t,g,s,z) \;=\; 2\int_0^z \frac{\cos\theta}{v}\,dz \qquad (20\text{d})$$

Before these equations are used, all the trigonometric functions must be eliminated by Snell's law for stratified media, $\sin \theta(z) = p\, v(z)$. Snell's parameter p is a numerical constant throughout the analysis.

The equation for interval velocity determination (12) again arises when dt'/dz from (20a) and dh'/dz from (20c) are combined:

$$\frac{dt'}{dh'} = \frac{2\cos\theta}{v\tan\theta} \tag{21}$$

Eliminating the trig functions with $p\, v = \sin\theta$ allows us to solve for the interval velocity:

$$v^2 = \frac{1}{p}\,\frac{1}{p + \dfrac{1}{2}\dfrac{dt'}{dh'}} \tag{22}$$

At the earth's surface $z = 0$, seismic survey data can be put into the coordinate frame (20) merely by making a numerical choice of p and doing the linear moveout. No knowledge of velocity $v(z)$ is required so far. Then we look at the data for some tops of the skewed hyperbolas. Finding some, we use equation (12), (19) or (22) to get a velocity with which to begin downward continuation.

Waves can be described in either the (t, g, s, z) physical coordinates or the newly defined coordinates (t', y, h', τ). In physical coordinates the image is found at

$$t = 0 \qquad \text{and} \qquad g = s \tag{23a,b}$$

To express these conditions in the Snell coordinates, insert (23) into (20a) and (20d). The result is what programmers call the stopping condition:

$$t' = \tau \tag{24}$$

This is the depth at which the velocity information should be best focused in the (h', t')-plane. Next some downward-continuation equations.

Differential Equations and Fourier Transforms

The chain rule for partial differentiation gives

$$\begin{bmatrix} \partial_t \\ \partial_g \\ \partial_s \\ \partial_z \end{bmatrix} = \begin{bmatrix} t'_t & y_t & h'_t & \tau_t \\ t'_g & y_g & h'_g & \tau_g \\ t'_s & y_s & h'_s & \tau_s \\ t'_z & y_z & h'_z & \tau_z \end{bmatrix} \begin{bmatrix} \partial_{t'} \\ \partial_y \\ \partial_{h'} \\ \partial_\tau \end{bmatrix} \tag{25}$$

In our usual notation the Fourier representation of the time derivative ∂_t is $-i\omega$. Likewise, $\partial_{t'}$ and the spatial derivatives $(\partial_y, \partial_{h'}, \partial_r, \partial_g, \partial_s, \partial_z)$ are associated with $i(k_y, k_{h'}, k_r, k_g, k_s, k_z)$. Using these Fourier variables in the vectors of (25) and differentiating (20) to find the indicated elements in the matrix of (25), we get

$$
\begin{bmatrix} -\omega \\ k_g \\ k_s \\ k_z \end{bmatrix}
=
\begin{bmatrix}
1 & 0 & 0 & 0 \\
-p & 1/2 & 1/2 & 0 \\
p & 1/2 & -1/2 & 0 \\
\dfrac{2\cos\theta}{v} & 0 & \tan\theta & \dfrac{2\cos\theta}{v}
\end{bmatrix}
\begin{bmatrix} -\omega' \\ k_y \\ k_{h'} \\ k_r \end{bmatrix}
\qquad (26a,b,c,d)
$$

Let S be the sine of the takeoff angle at the source and let G be the sine of the emergent angle at the geophone. If the velocity v is known, then these angles will be directly measurable as stepouts on common-geophone gathers and common-shot gathers. Likewise, on a constant-offset section or a slant stack, observed stepouts relate to an apparent dip Y, and on a linearly moved-out common-midpoint gather, stepouts measure the apparent stepout H'. The precise definitions are

$$
S = \frac{v\,k_s}{\omega} \qquad\qquad G = \frac{v\,k_g}{\omega} \qquad (27a,b)
$$

$$
Y = \frac{v\,k_y}{2\omega} \qquad\qquad H' = \frac{v\,k_{h'}}{2\omega} \qquad (27c,d)
$$

With these definitions (26b) and (26c) become

$$
G = p\,v + Y + H' = Y + (H' + p\,v) \qquad (28a)
$$

$$
S = -p\,v + Y - H' = Y - (H' + p\,v) \qquad (28b)
$$

The familiar offset stepout angle H is related to the LMO residual stepout angle H' by $H' = H - pv$. Setting H' equal to zero means setting $k_{h'}$ equal to zero, thereby indicating integration over h', which in turn indicates slant stacking data with slant angle p. Small values of H'/v or $k_{h'}/\omega$ refer to stepouts near to p.

Processing Possibilities

The double-square-root equation is

$$
\frac{k_z}{\omega} = -\frac{1}{v}\left(\sqrt{1 - S^2} + \sqrt{1 - G^2} \right) \qquad (29)
$$

With substitutions (26a,d), and (27a,b) the DSR equation becomes

$$\frac{k_\tau}{\omega} = 1 - \frac{pv}{1 - p^2 v^2} H' - \frac{1}{2} \left\{ \left[1 - \frac{2pv(H' - Y) + (H' - Y)^2}{1 - p^2 v^2} \right]^{1/2} \right.$$

$$\left. + \left[1 - \frac{2pv(H' + Y) + (H' + Y)^2}{1 - p^2 v^2} \right]^{1/2} \right\} \qquad (30)$$

Equation (30) is an exact representation of the double-square-root equation in what is called *retarded Snell midpoint coordinates.*

The coordinate system (20) can describe any wavefield in any medium. Equation (20) is particularly advantageous, however, only in stratified media of velocity near $v(z)$ for rays that are roughly parallel to any ray with the chosen Snell parameter p. There is little reason to use these coordinates unless they "fit" the wave being studied. Waves that fit are those that are near the chosen p value. This means that H' doesn't get too big. A variety of simplifying expansions of (30) are possible. There are many permutations of magnitude inequalities among the three ingredients pv, H', and Y. You will choose the expansion according to the circumstances. The appropriate expansions and production considerations, however, have not yet been fully delineated. But let us take a look at two possibilities.

First, any dataset can be decomposed by stepout into many datasets, each with a narrow bandwidth in stepout space — CMP slant stacks, for example. For any of these datasets, H' could be ignored altogether. Then (30) would reduce to

$$\frac{k_\tau}{\omega} = 1 - \frac{1}{2} \left\{ \left[1 - \frac{-2pvY + Y^2}{1 - p^2 v^2} \right]^{1/2} + \left[1 - \frac{+2pvY + Y^2}{1 - p^2 v^2} \right]^{1/2} \right\} \qquad (31a)$$

or

$$\frac{k_\tau}{\omega} = 1 - \frac{1}{2 \sqrt{1 - p^2 v^2}} \left[\sqrt{1 - (Y - pv)^2} + \sqrt{1 - (Y + pv)^2} \right] \qquad (31b)$$

The above approach is similar to the one employed by Richard Ottolini in his dissertation.

Next, let us make up an approximation to (30) which is separable in Y and H'. We will be using separation methodology introduced in Section 3.4. Equation (31b) provides the first part. Then take $Y = 0$ and keep *all* terms up to quadratics in H':

$$\frac{k_\tau}{\omega} = \frac{H'^2}{2(1 - p^2 v^2)^2} \qquad (32)$$

A separable approximation of (30) is (31b) plus (32). It is no accident that there are no linear powers of H' in (32). The coordinate system was designed so that energy near the chosen model $Y = 0$ and $H = pv$ should not drift in the (h', t')-plane as the downward continuation proceeds.

The velocity spectrum idea represented by equation (32) is to use the H' term to focus the data on the (h', t')-plane. After focusing, it should be possible to read interval velocities directly as slopes connecting events on the gathers. This approach was used in the dissertation of Alfonso Gonzalez [1982].

EXERCISE

1. A hyperbola is identified on a zero-offset section. The top is obscured but you can measure (p, x, t) at two places. What is the earth velocity? Given the same measurements on a field profile (constant s) what is the earth velocity?

5.5 Multiple Reflection, Current Practice

Near the earth's surface are a variety of unconsolidated materials such as water, soil, and the so-called weathered zone. The contrast between these near-surface materials and the petroleum reservoir rocks below is often severe enough to produce a bewildering variety of near-surface resonances. These resonance phenomena are not predicted and cannot be explained by the methods described in previous chapters.

Hard Sea Floor Example

Figure 1 shows textbook-quality multiple reflections from the sea floor. Hyperbolas $v^2 t^2 - x^2 = z_j^2$ appear at uniform intervals $z_j = j \, \Delta Z$, $j = 0, 1, 2, \cdots$. The data is unprocessed other than by multiplication by a spherical divergence correction t. Air is slower and lighter than water while sea-floor sediment is almost always faster and denser. This means that

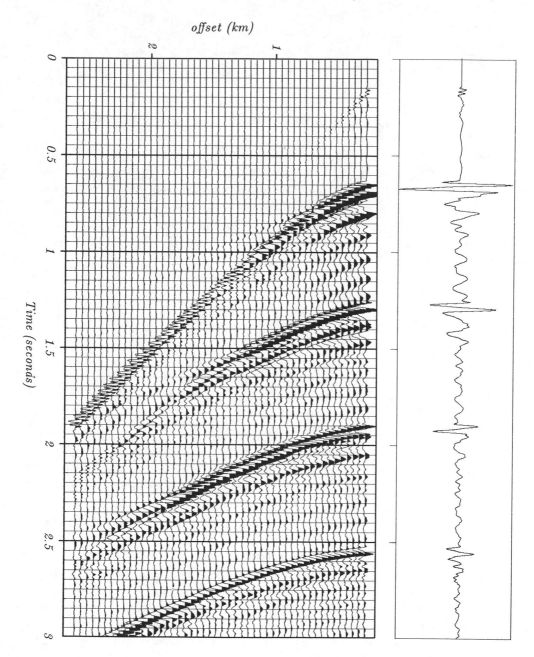

FIG. 5.5-1. Marine profile of multiple reflections from Norway. At the right, the near trace is expanded. (GECO)

successive multiple reflections almost always have alternating polarity. The polarity of a seismic arrival is usually ambiguous, but here the waveform is distinctive and it clearly alternates in polarity from bounce to bounce. The ratio of amplitudes of successive multiple reflections is the reflection coefficient. In figure 1, the reflection coefficient seems to be about -0.7. Multiply reflected head waves are also apparent, as are alternating polarities on them. Since the head-wave multiple reflections occur at critical angle, they should have a -1.0 reflection coefficient. We see them actually increasing from bounce to bounce. The reason for the increase is that the spherical-divergence correction is based on three-dimensional propagation, while the head waves are really spreading out in two dimensions.

Multiple reflections are fun for wave theorists, but they are a serious impediment to geophysicists who would like to see the information-bearing primary reflections that they mask.

Deconvolution in Routine Data Processing

The water depth in figure 1 is deeper than typical of petroleum prospecting. Figures 2 and 3 are more typical. In figure 2, the depth is so shallow it is impossible to discern bounces. With land data the base of the weathered zone is usually so shallow and indistinct that it is generally impossible to discern individual reflections. The word *shallow* as applied to multiple reflections is *defined* to mean that the reflections reoccur with such rapidity that they are not obviously distinguished from one another.

Statisticians have produced a rich literature on the subject of deconvolution. For them the problem is really one of estimating a source waveform, not of removing multiple reflections. There is a certain mathematical limit in which the multiple-reflection problem becomes equivalent to the source-waveform problem. This limit holds when the reverberation is confined to a small physical volume surrounding the shot or the geophone, such as the soil layer. The reason that the source-waveform and multiple-reflection problems are equivalent in this limit is that the downgoing wave from a shot is not simply intrinsic to the shot itself but also includes the local soil resonances. The word *ghost* in reflection seismology refers to the reflection of the source pulse from the surface (or sometimes from the base of the weathered layer). Because the source is so near to these reflectors, we often regard the ghost as part of the source waveform too.

An extensive literature exists on the vertical-incidence model of multiple reflections. Among wave-propagation theorists, the removal of all multiples is called *inversion*. It seems that for inversion theory to be applicable to the real problem, the theory must include a way to deal with an unknown, spectrally incomplete, shot waveform.

Routine work today typically ignores inversion theory and presumes the mathematical limit within which multiples may be handled as a shot waveform. The basic method was first developed for the industry by Schneider, Larner, Burg, and Backus [1964] of GSI (figure 2). Despite many further theoretical developments and the continuing active interest of many practical workers, routine deconvolution is little changed.

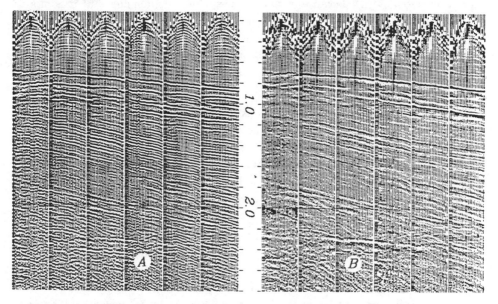

FIG. 5.5-2. Field profiles before (left) and after (right) deconvolution. (distributed by GSI, circa 1965)

Conventional industrial deconvolution (figure 2) has many derivations and interpretations. I will state in simple terms what I believe to be the essence of deconvolution. Every seismogram has a spectrum. The spectrum is a product of many causes. Some causes are of fundamental interest. Others are extraneous. It is annoying when a seismogram is resonant just because of some near-surface phenomena. Deconvolution is basically a process in which strong resonances are measured, and then a filter is designed to suppress them. The filter is designed to have a spectrum that is roughly inverse to the spectrum of the raw data. Thus the output of the filter is roughly white (equal amounts of all frequencies). From the earliest times, seismologists have found that reflection seismic data rarely makes sense much outside the frequency band 10-100 Hz, so as a final step, frequencies outside the band are removed. (The assumption that the output spectrum should be white seems to most seismologists to be a weak assumption, but practice usually shows it better than interpreting earth images from raw data).

Another nonmathematical explanation of why deconvolution is a success in practice is that it equalizes the spectrum from trace to trace. It *balances* the spectra (Tufekcic et al [1981]). Not only is it annoying when a seismogram is resonant just because of some near-surface phenomena, but it is more annoying when the wave spectrum varies from trace to trace as the near surface varies from place to place. A variable spectrum makes it hard to measure stepouts. Notice that the conventional industrial deconvolution described above includes *spectral balancing* as a byproduct. Figure 3 shows data that needs spectral balancing.

The above interpretation of deconvolution and why it works is different from what is found in most of the geophysical literature. Deconvolution is often interpreted in terms of the predictability of multiple reflections and the nonpredictability of primary reflections. It is shown in FGDP how multiple reflections are predicted. They are predicted, not by a strictly convolutional model, but approximately so. Prediction by convolution works best when the reverberation is all in shallow layers. Then it is like a source waveform.

Cardiovascular research is well integrated with routine practice, whereas pulmonary research is not. I compare this to migration and velocity theory being a good guide to industrial practice, whereas deconvolution theory is less so. The larger gap between theory and practice is something to be aware of. Some fields are more resistant to direct attack. In them you progress by more indirect routes. This is confusing for the student and demoralizing for the impatient. But that is the way it is. For more details, see Ziolkowski [1984].

The next few pages show land data with buried geophones confirming that source waveforms are mainly near-surface reverberation. Then we turn to departures from the convolutional model.

A Vertical Seismic Profile (VSP)

Seismologists always welcome the additional information from a *vertical seismic profile* (VSP). A VSP is some collection of seismograms recorded from the surface to a borehole. Routine well-based measurements such as rock cuttings and electric logs record *local* information, often just centimeters from the well. It is nice to think of the earth as horizontal strata, but this idealization fails at some unknown distance from the well. Surface reflection

FIG. 5.5-3. Profile from the North Sea. (Western Geophysical) Observe strong reverberations with a period of about 90 ms. These are multiple reflections from the sea floor. Note that the strongest signal occurs at increasing offset with increasing time. This is because the strongest multiples are often at critical angle. The strength of the reverberation diminishes abruptly 1.8 km behind the ship. This implies that the sea floor changes abruptly at that point.

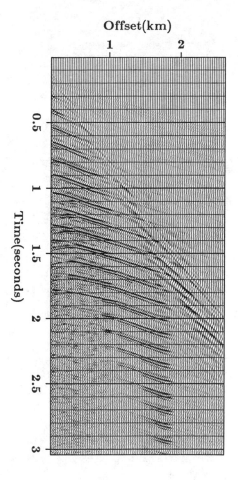

seismology, although it is further from the "ground truth" of well-bore measurements, provides the needed information about lateral continuity. But surface reflection data has resolving-power limitations as well as other uncertainties. The VSP provides information at an intermediate scale and also provides a calibration of the surface seismic method. Unfortunately, VSPs are costly and we rarely have them.

The subject of VSP occupies several books and many research papers. (See Gal'perin [1974] and Balch et al [1982]). Here we will just look at a single VSP to get some idea of source waveforms and multiple reflections. The VSP shown in figure 4 is from a typical land area. The multiple reflections are not so severe as with the marine data shown elsewhere in this chapter. The earliest arrival in figure 4 is the primary downgoing wave. Downgoing waves increase their travel time with depth, the slope of the arrival curve giving the downward component of velocity. After the first downgoing waves

Arco VSP - P-wave detector

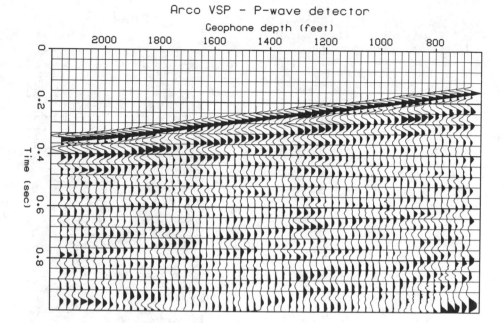

FIG. 5.5-4. Vertical seismic profile. The source is at the earth's surface near the borehole. The horizontal axis is the receiver depth. The vertical axis is travel time from zero to one second. Amplitudes are scaled by $t^{1.5}$. (ARCO)

arrive, you can see more downgoing waves with the same velocity. Upgoing waves have the opposite slope of the downgoing waves. These are also visible in figure 4.

Since late echoes are weaker than early ones, seismic data is normally scaled upwards with time before being displayed. There is no universal agreement in either theory or practice of what scaling is best. I have usually found t^2 scaling to be satisfactory for reflection data. (See Section 4.1). Figure 4 shows that $t^{1.5}$ scaling keeps the first arrival at about a constant amplitude on the VSP.

Viewing figure 4 from the side shows that the downgoing pulse is followed by a waveform that is somewhat consistent from depth to depth. The degree of consistency is not easy to see because of interference with the upcoming wave. As far as I can tell from the figure, the downgoing wave at the greatest depth is equal to that at the shallowest depth.

Figure 5 shows the same data augmented by some shallow receivers. You will notice that the downgoing wave no longer seems to be independent of

depth. So we can conclude that, as a practical matter, the downgoing
waveform seems to be mainly a result of near-surface reverberation.

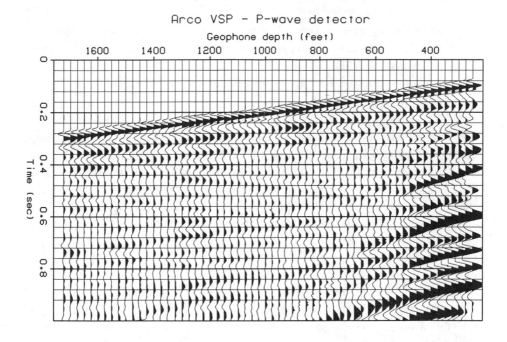

FIG. 5.5-5. The data of figure 4 augmented with shallower receivers. Amplitudes are scaled by $t^{1.5}$. (ARCO)

The energy in the first burst in figure 4 is roughly comparable to the
remaining energy. The remaining energy would be less if the VSP were
displayed without $t^{1.5}$ scaling, but since the surface reflection data is normally displayed with some such scaling (often t^2), it makes more sense statistically to speak of the energy on the scaled data. So the reverberating energy
is roughly comparable to the first arriving energy.

Below the near-surface region, the downgoing wave changes slowly with
depth. Now we should ask how much the downgoing wave would change if
the experiment were moved laterally. Obviously the borehole will not move
laterally and we will be limited to data where only the surface source moves
laterally. Since near-surface variations often change rapidly in the lateral
direction, we may fear that the downgoing waveform also changes rapidly
with shot location. The reverberation near a shot is repeated similarly near

any surface receiver. The resulting composite reverberation is the convolution of near-shot reverberation and near-geophone reverberation. So to get the information needed to deconvolve surface seismic data, the VSP should be recorded with many surface source locations.

Unfortunately such *offset VSP* data is rarely available. When petroleum production declines and expensive secondary recovery methods are contemplated, the cost of VSP will not seem so high. The production lost during VSP acquisition may be more easily weighed against future gains.

Again we should think about the meaning of "bad" data. Seismic data is generally repeatable whenever it is above the level of the ambient microseismic noise. But often the signals make no sense. The spatial correlations mean nothing to us. Most data at late times fits this description. Perhaps what is happening is this: (1) The downgoing waveform is getting a long trail; (2) the trail is a chaotic function of the surface location; and (3) the energy in the trail exceeds the energy in the first pulse. So, with so much randomness in the downgoing wave, the upcoming wave is necessarily incomprehensible.

Deep Marine Multiples, a Phenomenon of Polar Latitudes

It has frequently been noted that sea-floor multiple reflection seems to be a problem largely in the polar latitudes only — rarely in equatorial regions. This observation might be dismissed as being based on the statistics of small numbers, but two reasons can be given why the observation may be true. Each of these is of interest whether or not the statistics are adequate.

It happens that natural gas is soluble in water and raises the temperature of freezing, particularly at high pressure. Ice formed when natural gas is present is called *gas hydrate*. Thus there can be, under the liquid ocean, trapped in the sediments, solid gas hydrate. The gas hydrate stiffens the sediment and enhances multiple reflections.

A second reason for high multiple reflections at polar latitudes has to do with glacial erosion. Ordinarily ocean bottoms are places of slow deposition of fine-grained material. Such freshly deposited rocks are soft and generate weak multiple reflections. But in polar latitudes the scouring action of glaciers removes sediment. Where erosion is taking place the freshly exposed rock is stronger and stiffer than newly forming sediments. Thus, stronger sea-floor reflections.

Continents erode and deposit at all latitudes. However, one might speculate that on balance, continental shelves are created by deposition in low and middle latitudes, and then drift to high latitudes where they erode. While highly speculative, this theory does provide an explanation for the association of multiple reflections with polar latitudes.

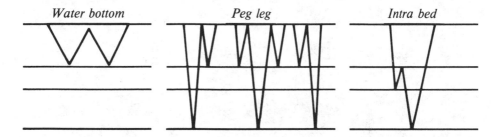

FIG. 5.5-6. Raypaths are displayed for (a) a water-bottom multiple, (b) a pegleg multiple family, and (c) a short-path multiple.

Pegleg and Intrabed Multiple Reflections

Multiple reflections fall into one of three basic categories — see figure 6.

Water-bottom multiples are those multiples whose raypaths lie entirely within the water layer (figure 6a). Since the sea floor usually has a higher reflectivity than deeper geological horizons, water-bottom multiples often have strong amplitudes. In deep water these multiples can be very clear and distinct. A textbook-quality example is shown in figure 1.

Pegleg multiple reflections are variously defined by different authors. Here pegleg multiples (figure 6b) are defined to be those multiples that undergo one reflection in the sedimentary sequence and other reflections in the near surface.

To facilitate interpretation of seismic data, let us review the timing and amplitude relations of vertical-incidence multiple reflections in layered media. Take the sea-floor two-way travel time to be t_1 with reflection coefficient c_1. Then the nth multiple reflection comes at time $n\, t_1$ with reflection strength c_1^n. Presume also a deeper primary reflection at travel-time depth t_2 with reflection coefficient c_2. The sea-floor peglegs arrive at times $t_2 + n\, t_1$. Note that peglegs come in families. For example, the time $t_2 + 2t_1$ could arise from three paths, $t_2 + 2t_1$, $t_1 + t_2 + t_1$, or $2t_1 + t_2$. So the nth order pegleg multiple echo is really a summation of $n+1$ rays, and thus its strength is proportional to $(n+1)\, c_2\, c_1^n$. The sea-floor reverberation is c_1^n, which is not the same function of n as the function that describes reverberation on sediments, $(n+1)\, c_1^n$. Ignoring the sea-floor

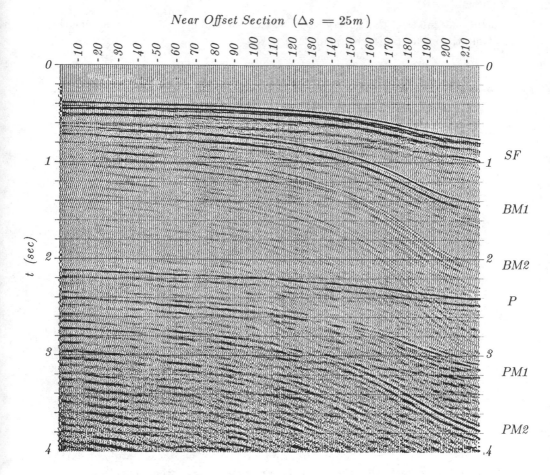

FIG. 5.5-7. Near-offset section — offshore Labrador (Flemish Cap). The offset distance is about 9 shotpoints. SF = sea floor, BM1 = first bottom multiple, BM2 = second bottom multiple, P = primary, PM1 = first pegleg multiple, and PM2 = second pegleg multiple. (AMOCO-Canada, Morley)

reverberation itself you can just think of $(n + 1) c_1^n$ as a shot waveform.

Every multiple must have a "turn-around" where an upcoming wave becomes a downgoing wave. Almost all readily recognized multiples are *surface* multiples, that is, they have their turn-arounds at the earth's surface. Figure 7 shows some clear examples. In land data the turn-around can be at the base of the soil layer, which is almost the same as being at the earth's surface.

A raypath that is representative of yet another class of multiples, called *short-path* or *intrabed* multiples, is shown in figure 6c. Their turn-around is *not* at or near the earth's surface. These multiples are rarely evident in field data, although figure 8 shows a clear case in which they are. When they are identified, it is often because the seismic data is being interpreted using some accompanying well logs. The reason that short-path multiples are so rarely observed compared to peglegs is that the reflection coefficients within the sedimentary sequence are so much lower than on the free surface. The weakness of individual short-path multiples may be compensated for, however, by the *very large* numbers in which they can occur. Any time a seismic section becomes incomprehensible, we can hypothesize that the data has become overwhelmed by short-path multiples.

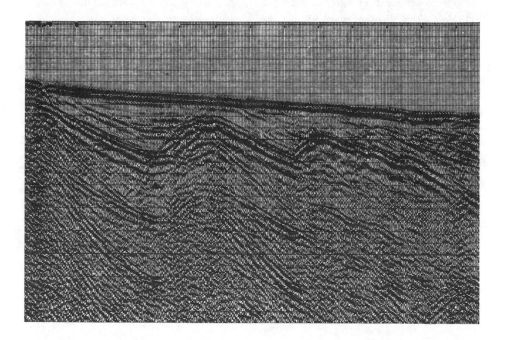

FIG. 5.5-8. A rare case of unambiguous intrabed multiple reflections. The data was recorded near Puerto Rico. The inner-bed multiple is between the sea floor and the basement. Thus its travel time is $t_{base} + (t_{base} - t_{floor})$. Do you see it? (Western Geophysical)

The Need to Distinguish between Types of Sections

By 1974, wave-equation methods had established themselves as a successful way to migrate CDP-stacked sections. Bolstered by this success, Don Riley and I set out to apply the wave equation to the problem of predictive suppression of deep-water multiple reflections. Hypothesizing that diffraction effects were the reason for all the difficulty that was being experienced then with deep-water multiple reflection, we developed a method for the modeling and predictive removal of diffracted multiple reflections (see FGDP, Chapter 11-4). We didn't realize that in practice the multiple reflection problem would be so much more difficult than the primary reflection problem. For primaries, the same basic migration method works on zero-offset sections, CDP-stacked sections, or vertical-incidence plane-wave sections. Our multiple-suppression method turned out to be applicable only to vertical plane-wave stacks. Don Riley prepared figure 9, which shows some comparisons.

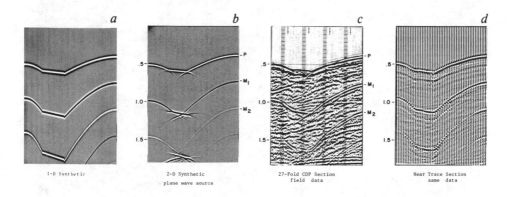

FIG. 5.5-9. Diffracted multiple-reflection examples: (*a*) 1-D synthetic, (*b*) 2-D synthetic, vertical plane-wave source, (*c*) 27-fold CDP data section (GSI), (*d*) near-trace section. (Riley)

One thing to keep in mind while studying the comparisons in figure 9 is that on the field data there are likely to be aspects of propagation in three dimensions that may go unrecognized. The third dimension is always a "skeleton in the closet." It doesn't usually spoil two-dimensional migration, but that doesn't assure us that it won't spoil 2-D wave-equation multiple suppression.

Examples of Shallow-Water Multiples with Focusing

The exploding-reflector concept does not apply to multiple reflections, so there is no simple wave-theoretic means of predicting the focusing behavior of multiples on a near-trace section. Luckily multiples on vertical plane-wave stacks are analyzable. They may give us some idea about the focusing behavior of multiple reflections on other seismic sections. A vertically downgoing plane wave is simulated by a common-geophone stack without moveout. This isn't the same as the familiar CDP stack, but it is analyzable with the techniques described in Chapters 1 and 2.

Consider a multiple reflection that has undergone several surface bounces. The seismic energy started out as a downgoing plane wave. It remained unchanged until its first reflection from the sea floor. The sea-floor bounce imposed the sea-floor topography onto the plane wave. In a computer simulation the topography would be impressed upon the plane wave by a step with the lens equation. Then the wave diffracted its way up to the surface and back down to the sea floor. In a computer another topographic lens shift would be applied. The process of alternating diffraction and lensing would be repeated as often as you would care to keep track of things. Figure 10 shows such a simulation. A striking feature of the high-order multiple reflections in figure 10 is the concentration of energy into localized regions. It is easy to see how bounces from concave portions of the sea floor can overcome the tendency of acoustic energy to spread out. These regions of highly concentrated energy that occur late on the time axis do not resemble primaries at all. With primaries a localized disturbance tends to be spread out into a broad hyperbola. Primary migration of the highly concentrated bursts of energy seen on figure 10 must lead to semicircles. Such semicircles are most unlikely geological models — and are all too often predicted by the industry's best migration programs.

The most important thing to learn from the synthetic multiple reflections of figure 10 is that multiples need not resemble primaries. Semicircles that occur on migrated stacks could be residual multiple reflections. Unfortunately, there is no simple theory that says whether or not focused multiples on vertical wave stacks should resemble those on zero-offset sections or CDP stacks. Luckily some data exists that provides an answer. Figure 11 is a zero-offset section which establishes that such focusing phenomena are indeed found in qualitative, if not quantitative, form on reflection survey data.

The marine data exhibited in figure 11 clearly displays the focusing phenomena in the synthetic calculations of figure 10. This suggests that we should utilize our understanding in a quantitative way to predict and suppress the multiple reflections in order to get a clearer picture of the earth's

FIG. 5.5-10. Simulated sea-floor multiple reflections. The vertical exaggeration is 5. Little focusing is evident on the gentle sea-floor topography, but much focusing is evident on high-order multiple reflections. At late times there is a lack of lateral continuity, really unlike primary reflection data.

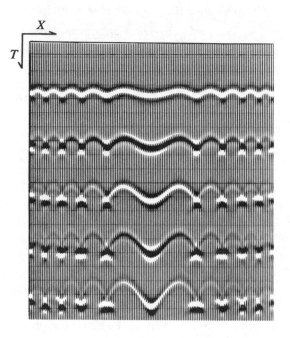

subsurface. There are several reasons why this would not be easy to do. First, the Riley theory applies to vertical wave stacks. These are quantitatively different from common-midpoint stacks. Second, the effective seismic sea-floor depth is not a known input: it must somehow be determined from the data itself. Third, the water depth in figure 11 is so shallow that individual bounces cannot be distinguished.

Why Deconvolution Fails in Deep Water

It has been widely observed that deconvolution generally fails in deep water. A possible reason for this is that deep water is not the mathematical limit at which the multiple-reflection problem is equivalent to the shot-waveform problem. But that is not all. Theory predicts that under ordinary circumstances multiples should alternate in polarity. The examples of figure 1 and figure 2 confirm it. You will have trouble, however, if you look for alternating polarity on CDP stacks. The reason for the trouble also indicates why deconvolution tends to fail to remove deep multiples from CDP stacks.

Recall the timing relationships for multiples at zero offset. The *reverberation period* is a constant function of time. Because of moveout, this is not the case at any other offset. Normal-moveout correction will succeed in

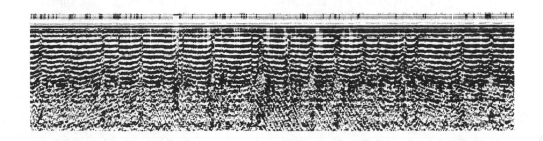

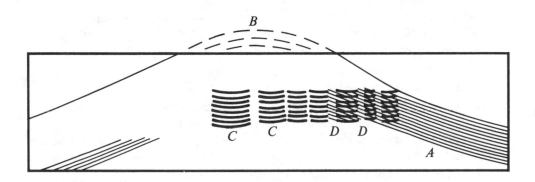

FIG. 5.5-11. Example of focusing effects on multiple reflections in near-trace section at Chukchi Sea. These effects are obscured by stacking. (U.S. Geological Survey)

A. Existing structure.

B. Former structure unevenly eroded away leaving localities of sea floor convex or concave.

C. High order multiple reflections focusing where the sea floor is concave.

D. Existing structural dip exposed in windows where the multiples are weak (i.e., where convex sea floor causes multiple to spread rapidly).

restoring zero-offset timing relationships in a constant-velocity earth, but when the velocity increases with depth, the multiples will have a slower RMS velocity than the primaries. So the question is what velocity to use, and whether, in typical land and marine survey situations, the residual time shifts are greater than a half-wavelength. No equations are needed to answer this question. All that is needed is the general observation that conventional common-midpoint stacking suppresses multiples because they have lower velocities than primaries. This observation implies that normal moveout routinely time shifts multiple reflections a half-wavelength or more out of their natural zero-offset relationships.

To make matters worse, the amplitude relationships that we expect at zero offset are messed up. Reflection coefficient is a function of angle. But on a seismogram from some particular offset, each multiple reflection will have reflected at a different angle.

Vertical incidence timing relationships are *approximately* displayed on CDP stacks. The practical difficulty is that the CDP stack does not mimic the vertical-incidence situation well enough to enable satisfactory prediction of multiples from primaries.

Before stack, on marine data, moveout could be done with water velocity, but then any peglegs would not fit the normal-incidence timing relationship. Since peglegs are often the worst part of the multiple-reflection problem, moveout should perhaps be done with pegleg velocity. No matter how you look at it, all the timing relationships for deep multiple reflections cannot be properly adjusted by moveout correction.

EXERCISE

1. On some land data it was noticed that a deep multiple reflection arrived a short time earlier than predicted by theory. What could be the explanation?

5.6 Multiple Reflection — Prospects

To improve our ability to suppress multiples, we try to better characterize them. The trouble is that a realistic model has many ingredients. Few of
the theories that abound in the literature have had much influence on routine
industrial practice. I would put these unsuccessful theories into two
categories:

1. Those that try to achieve everything with statistics, oversimplifying
    the complexity of the spatial relations

2. Those that try to achieve everything with mathematical physics,
    oversimplifying the noisy and incomplete nature of the data

Multiple reflection is a good subject for nuclear physicists, astrophysi-
cists, and mathematicians who enter our field. Those who are willing to take
up the challenge of trying to carry theory through to industrial practice are
rewarded by learning some humility. I'll caution you now that I haven't
pulled it all together in this section either!

Here two approaches will be proposed, both of which attend to geometry
and statistics. Both approaches are new and little tested. Regardless of how
well they may work, I think you will find that they illuminate the task.

The first approach, called *CMP slant stack,* is a simple one. It
transforms data into a form in which all offsets mimic the simple, one-
dimensional, zero-offset model. The literature about that model in both
statistics and mathematical physics is extensive.

The second approach is based on a *replacement impedance* concept. It is
designed to accommodate rapid lateral variations in the near surface. It is
easiest to explain for a hypothetical marine environment where the sole
difficulty arises from lateral variation in the sea-floor reflectivity. The basic
idea is downward continuation of directional shots and directional geophones
to just beneath the sea floor, but no further. This is followed by upward continuation through a replacement medium that has a zero sea-floor reflection
coefficient. This process won't eliminate all the multiple reflections, but it
should eliminate the most troublesome ones.

Transformation to One Dimension by Slant Stack

A rich literature (c.f. FGDP) exists on the one-dimensional model of multiple reflections. Some authors develop many facets of wave-propagation theory. Others begin from a simplified propagation model and develop many facets of information theory. These one-dimensional theories are often regarded as applicable only at zero offset. However, we will see that all other offsets can be brought into the domain of one-dimensional theory by means of slant stacking.

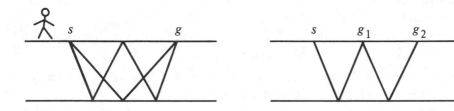

FIG. 5.6-1. Rays at constant-offset (left) arrive with various angles and hence various Snell parameters. Rays with constant Snell parameter (right) arrive with various offsets. At constant p all paths have identical travel times.

The way to get the timing and amplitudes of multiples to work out like vertical incidence is to stop thinking of seismograms as time functions at constant offset, and start thinking of constant Snell parameter. In a layered earth the complete raypath is constructed by summing the path in each layer. At vertical incidence $p = 0$, it is obvious that when a ray is in layer j its travel time t_j for that layer is independent of any other layers which may also be traversed on other legs of the total journey. This independence of travel time is also true for any other fixed p. But, as shown in figure 1, it is not true for a ray whose total offset $\sum f_j$, instead of its p, is fixed. Likewise, for fixed p, the horizontal distance f_j which a ray travels while in layer j is independent of other legs of the journey. Thus, in addition, $t_j + const\ f_j$ for any layer j is independent of other legs of the journey. So $t_j' = t_j - pf_j$ is a property of the jth layer and has nothing to do with any other layers which may be in the total path. Given the layers that a ray crosses, you add up the t_j and the f_j for each layer, just as you would in the vertical-incidence case. Some paths are shown in figure 2.

To see how to relate field data to slant stacks, begin by searching on a common-midpoint gather for all those patches of energy (tangency zones)

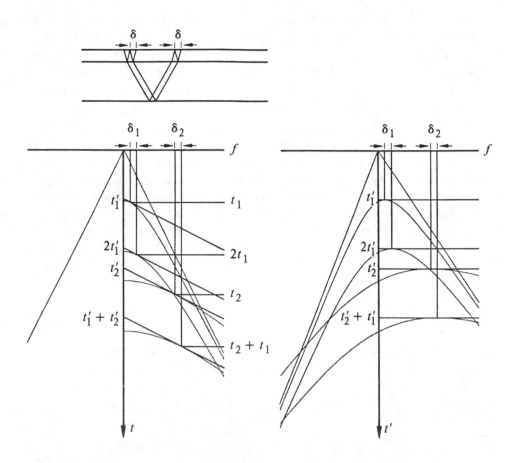

FIG. 5.6-2. A two-layer model showing the events $(t_1, 2t_1, t_2, t_2+t_1)$. On the top is a ray trace. On the left is the usual data gather. On the right the gather is replotted with linear moveout $t' = t - pf$. Plots were calculated with $(v_1, v_2, 1/p)$ in the proportion $(1,2,3)$. Fixing our attention on the patches where data is tangent to lines of slope p, we see that the arrival times have the vertical-incidence relationships — that is, the reverberation period is fixed, and it is the same for simple multiples as it is for peglegs. This must be so because the ray trace at the top of the figure applies precisely to those patches of the data where $dt/dx = p$. Furthermore, since $\delta_1 = \delta_2$, the times $(t_1', 2t_1', t_2'+t_2')$ also follow the familiar vertical-incidence pattern. (Gonzalez)

where the hyperboloidal arrivals attain some particular numerical value of slope $p = dt/df$. These patches of energy seen on the surface observations each tell us where and when some ray of Snell's parameter p has hit the surface. Typical geometries and synthetic data are shown in figures 2 and 3.

Both the t_j and the t_j' behave like the times of normal-incident multiple reflections. While the lateral location of any patch unfortunately depends on the velocity model $v(z)$, slant stacking makes the lateral location irrelevant. In principle, slant stacking could be done for many separate values of p so that the (f, t)-space would get mapped into a (p, t)-space. The nice thing about (p, t)-space is that the multiple-suppression problem decouples into many separate one-dimensional problems, one for each p-value. Not only that, but the material velocity is not needed to solve these problems. It is up to you to select from the many published methods. After suppressing the multiples you inverse slant stack. Once back in (f, t)-space you could estimate velocity and further suppress multiples using your favorite stacking method.

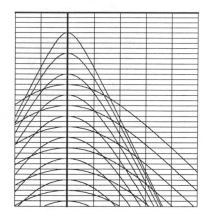 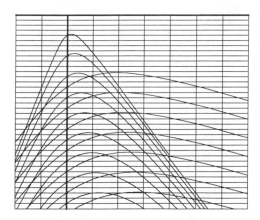

FIG. 5.6-3. The same geometry as figure 2 but with more multiple reflections.

Figure 3 is a "workbook" exercise. By picking the tops of all events on the right-hand frame and then connecting the picks with dashed lines, you should be able to verify that sea-bottom peglegs have the same interval velocity as the simple bottom multiples. The interval velocity of the sediment can be measured from the primaries. The sediment velocity can also be measured by connecting the nth simple multiple with the nth pegleg multiple.

Transformation to one dimension by slant stack for deconvolution is a process that lies on the border between experimental work and industrial practice. See for example Treitel et al [1982]. Its strength is that it correctly handles the angle-dependences that arise from the source-receiver geometry as well as the intrinsic angle-dependence of reflection coefficient. One of its weaknesses is that it assumes lateral homogeneity in the reverberating layer. Water is extremely homogeneous, but sediments at the water bottom can be quite inhomogeneous.

Near-Surface Inhomogeneity

Soils have strange acoustic behavior. Their seismic velocities are usually less than or equal to the speed of sound in water (1500 meters/sec). It is not uncommon for soil velocity to be five times slower than the speed of sound in water, or as slow as the speed of sound in air (300 meters/sec). Where practical, seismic sources are buried under the weathered zone, but the receivers are almost always on the surface. About the only time you may encounter buried receivers is in a marshy area. There field operations are so difficult that you will have many fewer receivers than normal.

A source of much difficulty is that soils are severely laterally inhomogeneous. It is not rare for two geophones separated by 10 meters to record quite different seismograms. In particular, the uphole transit time (the seismic travel time from the bottom of a shot hole to the surface near the top of the hole) can easily exhibit time anomalies of a full wavelength. All this despite a flat level surface. How can such severe, unpredictable, travel-time anomalies in the weathered zone be understood? By river meanders, tiny shallow gas pockets, pocketed carbonates, glacial tills, etc. All these irregularities can be found at depth too, but they are worse at the surface before saturation and the pressure of burial reduce the acoustic inhomogeneity. See also Section 3.7.

The shallow marine case is somewhat better. Ample opportunities for lateral variations still exist — there are buried submarine channels as well as buried fossil river channels. But the dominant aspect of the shallow marine case becomes the resonance in the water layer. The power spectrum of the observed data will be controlled by this resonance.

Likewise, with land data, the power spectrum often varies rapidly from one recording station to the next. These changes in spectrum may be interpreted as changes in the multiple reflections which stem from changes in the effective depth or character of the weathered zone.

Modeling Regimes

Downward-continuation equations contain four main ingredients: the slowness of the medium at the geophone $v(g)^{-1}$; likewise at the shot $v(s)^{-1}$; the stepout in offset space k_h/ω; and the dip in midpoint space k_y/ω. These four ingredients all have the same physical dimensions, and modeling procedures can be categorized according to the numerical inequalities that are presumed to exist among the ingredients. One-dimensional work ignores three of the four — namely, dip, stepout, and the difference $v(g)^{-1} - v(s)^{-1}$. CMP slant stack includes the stepout k_h/ω. Now we have a choice as to whether to include the dip or the lateral velocity variation. The lateral velocity variation is often severe near the earth's surface where the peglegs live. Recall the simple idea that typical rays in the deep subsurface emerge steeply at a low-velocity surface. When using continuation equations in the near surface, we are particularly justified in neglecting dip, that is $v^{-1} \gg k_y/\omega$. It is nice to find this excuse to neglect dip since our field experiments are so poorly controlled in dip out of the plane of the experiment. Offset stepout, on the other hand, is probably always much larger in the plane of the survey line than out of it.

Another important ingredient for modeling or processing multiple reflections is the coupling of upcoming and downgoing waves. This coupling introduces the reflectivity beneath the shot $c(s)$ and the receiver $c(g)$. An important possibility, to which we will return, is that $c(s)$ may be different from $c(g)$, even though all the angles may be neglected.

Subtractive Removal of Multiple Reflections

Stacking may be thought of as a multiplicative process. Modeling leads to subtractive processes. The subtractive processes are a supplement to stacking, not an alternative: After subtracting, you can stack.

First we try to model the multiple reflections, then we try to subtract them from the data. In general, removal by subtraction is more hazardous than removal by multiplication. To be successful, subtraction requires a correct amplitude as well as a timing error of less than a quarter-wavelength.

Statistically determined empirical constants may be introduced to account for discrepancies between the modeling and reality. In statistics this is known as *regression*. For example, knowing that a collection of data points should fit a straight line, we can use the method of least-sum-squared-residuals to determine the best parameters for the line. A careful study of the data points might begin by removing the straight line, much as we intend to remove multiple reflections. Naturally an adjustable parameter can help account for the difficulty expected in calculating the precise amplitude for the multiples. An unknown timing error is much harder to model. Because of the nonlinearity of the mathematics, a slightly different, more tractable approach is to take as adjustable parameters the coefficients in a convolution filter. Such a filter could represent any scale factor and time shift. It is tempting to use a time-variable filter to account for time-variable modeling errors. An inescapable difficulty with this is that a filter can represent a lot more than just scaling and amplitude. And the more adjustable parameters you use, the more the model will be able to fit the data, whether or not the model is genuinely related to the data.

The difficulty of subtracting multiple reflections is really just this: If an inadequate job is done of modeling the multiples — say, for example, of modeling the geometry or velocity — then you need many adjustable parameters in the regression. With many adjustable parameters, primary reflections get subtracted as well as multiples. Out goes the baby with the bath water.

Slanted Deconvolution and Inversion

Because of the wide offsets used in practice, it has become clear that seismologists must pay attention to differences in the sea floor from bounce to bounce. A straightforward and appealing method of doing so was introduced by Taner [1980] — that was his *radial-trace* method. A radial trace is a line cutting through a common-shot profile along some line of constant $r = h/t$. Instead of deconvolving a seismogram at constant offset, we deconvolve on a radial trace. The deconvolution can be generalized to a downward-continuation process. Downward continuation of a radial trace may be approximated by time shifting. Unfortunately, there is a problem when the data on the line consists of both sea-floor multiples and peglegs, because these require different trajectories. The problem is resolved, at least in principle, by means of Snell waves. Estevez, in his dissertation [1977], showed theoretically how Snell waves could also be used to resolve other difficulties, such as diffraction and lateral velocity variation (if known). An example illustrating the relevance of the differing depths of the sea floor on different bounces, is shown in figure 4.

Incompleteness of the data causes us to have problems with most inversion methods. Data can be incomplete in time, space, or in its spectrum. Any recursive method must be analyzed to ensure that an error made at shallow depths will not compound uncontrollably during descent. All data is spectrally incomplete. Also, with all data there is uncertainty about the shot waveform. At the p-values for which pegleg multiples are a problem, the first sea-floor bounce usually occurs too close to the ship to be properly recorded.

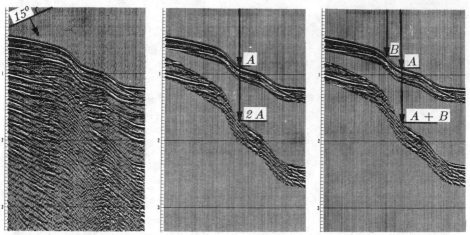

FIG. 5.6-4. Time of multiple depends on sum of all times. (Estevez [1977])

To solve this problem, Taner built a special auxiliary recording system.

It is an advantage for Snell wave methods that slant stacking creates some signal-to-noise enhancement from the raw field data, but it is a disadvantage that the downward continuation must continue to all depths. The methods to be discussed next are before-stack methods, but they do not require downward continuation much below the sea floor.

The Split Backus Filter

We are preparing a general strategy, *impedance replacement,* for dealing with surface multiple reflections. This strategy will require heavy artillery drawn from both regression theory and wave-extrapolation theory. So as not to lose sight of the goals, we will begin with an example drawn from an idealized geometry. That reality is not too far from this idealization was demonstrated by Larry Morley, whose doctoral dissertation [1982] illustrates a successful test of this method and describes the impedance-replacement strategy in more detail.

Imagine that the sea floor is flat. Near the shot the sea-floor reflection coefficient is taken as c_s. Near the geophone it is taken to be c_g. Near the geophone the reverberation pattern is

$$\frac{1}{1 + c_g\, Z} \;=\; 1 - c_g Z + c_g{}^2 Z^2 - c_g{}^3 Z^3 + c_g{}^4 Z^4 + \cdots \qquad (1)$$

where Z is the two-way delay operator for travel to the water bottom. (See Section 4.6 or FGDP for Z-transform background). Near the shot there is a similar reverberation sequence:

$$\frac{1}{1 + c_s\, Z} \;=\; 1 - c_s\, Z + c_s{}^2 Z^2 - c_s{}^3 Z^3 + c_s{}^4 Z^4 + \cdots \qquad (2).$$

Ignoring the difference between c_s and c_g leads to the Backus [1959] reverberation sequence, which is the product of (1) and (2):

$$\frac{1}{1 + c\, Z}\; \frac{1}{1 + c\, Z} \;=\; 1 - 2cZ + 3c^2 Z^2 - 4c^3 Z^3 + 5c^4 Z^4 + \cdots \qquad (3)$$

The denominator in (3) is the Backus filter. Applying this filter should remove the reverberation sequence. Morley called the filter which results from explicitly including the difference at the shot and geophone a *split Backus* filter. The depth as well as the reflection coefficient may vary laterally. (The effect of dip is second order). Thus the split Backus operator can be taken to be

$$\left(1 + c_s\, e^{\,i\,\omega\tau(s)}\right)\; \left(1 + c_g\, e^{\,i\,\omega\tau(g)}\right) \qquad (4)$$

Inverting (4) into an expression like (3), you will find that the nth term splits into n terms. This just means that paths with sea-floor bounces near the shot can have different travel times than paths with bounces near the geophone.

Figure 5, taken from Morley's dissertation, shows that split pegleg multiples are an observable phenomenon. His interpretation of the figure follows:

[The figure] is a constant-offset section (COS) from the same line for an offset halfway down the cable (a separation of 45 shot points with this geometry). The first-order pegleg multiple starting at 2.5 seconds on the left and running across to 3 seconds on the right is "degenerate" (unsplit) on the near-trace section but is split on the COS due to the sea-floor topography. The maximum split is some 200 mils around shot points 180-200. This occurs, as one might expect, where the sea floor has maximum dip; i.e., where the difference between sea-floor depths at the shot and geophone positions is greatest.

Most present processing ignores the Backus filter altogether and solves for an independent deconvolution filter for each seismic trace. This introduces a great number of free parameters. By comparison, a split Backus approach should do a better job of preserving primaries.

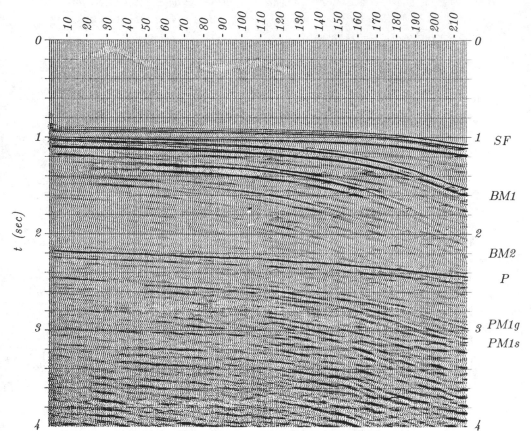

Constant Offset Section $(\Delta s = 25m)$

FIG. 5.6-5. Constant-offset section (COS) from the same line as figure 7 in Section 5.5. Offset distance is about 46 shotpoints. Notice that the first-order pegleg multiple is now split into two distinct arrivals, $PM1s$ and $PM1g$. (AMOCO Canada, Morley)

In practice we would expect that any method based on the split Backus concept would need to include the effect of moveout. Luckily, velocity contrast would reduce the emerging angle for peglegs. Of course, residual moveout problems would be much more troublesome with water-bottom multiples. Presumably the process should be applied after normal moveout in that case. Let us take a look at the task of estimating a split Backus operator.

Sea-Floor Consistent Multiple Suppression

Erratic time shifts from trace to trace have long been dealt with by the so-called surface-consistent statics model. Using this model you fit the observed time shifts, say, $t(s, g)$, to a regression model $t(s, g) \approx t_s(s) + t_g(g)$. The statistically determined functions $t_s(s)$ and $t_g(g)$ can be interpreted as being derived from altitude or velocity variations directly under the shot and geophone. Taner and Coburn [1980] introduced the closely related idea of a surface-consistent frequency response model that is part of the statics problem. We will be interpreting and generalizing that approach. Our intuitive model for the data $P(s, g, \omega)$ is

$$P(s, g, \omega) \approx \frac{1}{1 + c_s\, e^{\,i\,\omega\tau(s)}}\, \frac{1}{1 + c_g\, e^{\,i\,\omega\tau(g)}} \times \tag{5}$$

$$e^{\,i\frac{\omega}{v}\sqrt{z^2 + 4h^2}}\, H(h, \omega)\, Y(y, \omega)\, F(\omega)$$

The first two factors represent the split Backus filter. The next factor is the normal moveout. The factor $H(h, \omega)$ is the residual moveout. The factor $Y(y, \omega)$ is the depth-dependent earth model beneath the midpoint y. The last factor $F(\omega)$ is some average filter that results from both the earth and the recording system.

One problem with the split Backus filter is a familiar one — that the time delays $\tau(s)$ and $\tau(g)$ enter the model in a nonlinear way. So to linearize it the model is generalized to

$$P'(s, g, \omega) \approx S(s, \omega)\, G(g, \omega)\, H(h, \omega)\, Y(y, \omega)\, F(\omega) \tag{6}$$

Now S contains all water reverberation effects characteristic of the shot location, including any erratic behavior of the gun itself. Likewise, receiver effects are embedded in G. Moveout correction was done to P, thereby defining P'.

Theoretically, taking logarithms gives a linear, additive model:

$$\ln P'(s, g, \omega) \approx \tag{7}$$

$$\ln S(s, \omega) + \ln G(g, \omega) + \ln H(h, \omega) + \ln Y(y, \omega) + \ln F(\omega)$$

The phase of P', which is the imaginary part of the logarithm, contains the *travel-time* information in the data. This information begins to lose meaning as the data consists of more than one arrival. The phase function becomes discontinuous, even though the data is well behaved. In practice, therefore, attention is restricted to the real part of (7), which is really a statement about power spectra. The decomposition (7) is a linear problem, perhaps best solved by iteration because of the high dimensionality involved. In reconstructing S and G from power spectra, Morley used the Wiener-Levinson technique, explicitly forcing time-domain zeroes in the filters S and G to account for the water path. He omitted the explicit moveout correction in (5), which may account for the fact that he only used the inner half of the cable.

Replacement-Medium Concept of Multiple Suppression

In seismology wavelengths are so long that we tend to forget it is physically possible to have a directional wave source and a directional receiver. Suppose we had, or were somehow able to simulate, a source that radiated only down and a receiver that received only waves coming up. Then suppose that we were somehow able to downward continue this source and receiver beneath the sea floor. This would eliminate a wide class of multiple reflections. Sea-floor multiples and peglegs would be gone. That would be a major achievement. One minor problem would remain, however. The data might now lie along a line that would not be flat, but would follow the sea floor. So there would be a final step, an easy one, which would be to upward continue through a replacement medium that did not have the strong disruptive sea-floor reflection coefficient. The process just described would be called impedance replacement. It is analogous to using a replacement medium in gravity data reduction. It is also analogous to time shifting seismograms for some *replacement velocity*. (See Section 3.7).

The migration operation downward continues an upcoming wave. This is like downward continuing a geophone line in which the geophones can receive only upcoming waves. In reality, buried geophones see both upcoming and downgoing waves. The directionality of the source or receiver is built into the sign chosen for the square-root equation that is used to extrapolate the wavefield. With the reciprocal theorem, the shots could also be downward continued. Likewise shots physically radiate both up and down, but we can imagine shots that radiate either up or down, and mathematically the choice is a sign. So the results of four possible experiments at the sea floor, all possibilities of upward and downward directed shots and receivers, can be deduced.

Extrapolating all this information across the sea-floor boundary requires an estimate of the sea-floor reflection coefficient. This coefficient enters the calculation as a scaling factor in forming linear combinations of the waves above the sea floor. The idea behind the reflection-coefficient estimation can be expressed in two ways that are mathematically equivalent:

1.   The waves impinging on the boundary from above and below should have a cross-correlation that vanishes at zero lag.

2.   There should be minimum power in the wave that impinges on the boundary from below.

After the geophones are below, you must start to think about getting the shots below. To invoke reciprocity, it is necessary to invert the directionality of the shots and receivers. This is why it was necessary to include the auxiliary experiment of upward-directed shots and receivers.

EXERCISES

1.   Refer to figure 3.

   a.   What graphical measurement shows that the interval velocity for simple sea-floor multiples equals the interval velocity for peglegs?

   b.   What graphical measurements determine the sediment velocity?

   c.   With respect to the velocity of water, deduce the numerical value of the (inverse) Snell parameter p.

   d.   Deduce the numerical ratio of the sediment velocity to the water velocity.

2.   Consider the upcoming wave U observed over a layered medium of layer impedances given by $(I_1, I_2, I_3, \cdots)$, and the upcoming wave U' at the surface of the medium $(I_2, I_2, I_3, \cdots)$. Note that the top layer is changed.

   a.   Draw raypaths for some multiple reflections that are present in the first medium, but not in the second.

   b.   Presuming that you can find a mathematical process to convert the wave U to the wave U', what multiples are removed from U' that would not be removed by the Backus operator?

   c.   Utilizing techniques in FGDP, chapter 8, derive an equation for U' in terms of U, I_1, and I_2 that does not involve $I_3, I_4 \cdots$.

5.7 Profile Imaging

A field profile consists of the seismograms of one shot and many receivers along a line. Migration of a single profile, or of many widely separated profiles, demands a conceptual basis that is far removed from anything discussed so far in this book, namely, exploding-reflector and survey-sinking concepts. Such a conceptual basis exists, predates (Claerbout [1970]), and seems more basic than that of exploding reflectors or survey sinking. I call this older imaging concept the U/D imaging concept.

The sinking concept seems to demand complete coverage in shot-geophone space. Exploding reflectors requires many closely spaced shots. On the other hand, profile imaging with the U/D concept has no requirement for density along the shot axis. An example of a dataset that could only be handled by the older concept is a sonobuoy. A sonobuoy is a hydrophone with a radio transmitter. It is thrown overboard, and a ship with an air gun sails away, repeatedly firing until the range is too great. The principle of reciprocity says that the data is equivalent to a single source with a *very* long line of geophones.

While improving technology is leading to greater sampling density on the geophone axis, we are unlikely to see increasing density in shot space. There are only twenty-four hours in a day, and we must wait ten seconds between shots for the echoes to die down. So, given a certain area to survey and a certain number of months to work, we end out with an irreducible shotpoint density. Indeed, with three-dimensional geometries proving their worth, we may see *less* spatial sampling density. Poor sample density in shot space is a small impediment to profile methods.

Unlike the exploding-reflector method and the survey-sinking method, U/D concepts readily incorporate modeling and analysis of multiple reflections. Indeed, an ingenious algorithm for simultaneous migration and de-reverberation is found in FGDP. In principle it can be applied to either field profiles or slant stacks.

Wave equation methods have been suggestive of new ways of making weathering-layer corrections. Yet none have yet become widely accepted in practice, and it is too early to tell whether a DSR approach or a profile approach will work better.

All these considerations warrant a review of the profile migration method and the U/D imaging concept. We could easily see a revival of these in one form or another.

The U/D Imaging Concept

The U/D imaging concept says that *reflectors exist in the earth at places where the onset of the downgoing wave is time-coincident with an upcoming wave.* Figure 1 illustrates the concept.

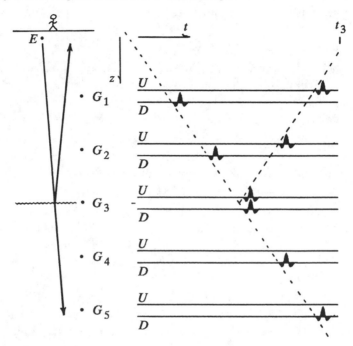

FIG. 5.7-1. Upcoming and downgoing waves observed with buried receivers. A disturbance leaves the surface at $t=0$ and is observed passing the buried receivers $G_1...G_5$ at progressively later times. At the depth of a reflector, z_3, the G_3 receiver records both the upcoming and downgoing waves in time coincidence. Shallower receivers also record both waves. Deeper receivers record only D. The fundamental principle of reflector mapping states that reflectors exist where U and D are time-coincident. (Riley)

It is easy to confuse the *survey-sinking* concept with the U/D concept because of the similarity of the phrases used to describe them: "downward continue the shots" sounds like "downward continue the downgoing wave." The first concept refers to computations involving only an upcoming wavefield $U(s,g,z,t)$. The second concept refers to computations involving both upcoming $U(x, z, t)$ and downgoing $D(x, z, t)$ waves. No particular source location enters the U/D concept; the source could be a downgoing plane wave.

In profile migration methods, the downgoing wave is usually handled theoretically, typically as an impulse whose travel time is known analytically or by ray tracing. But this is not important: the downgoing wave could be handled the same way as the upcoming wave, by the Fourier or finite-difference methods described in previous chapters. The upcoming wave could be expressed in Cartesian coordinates, or in the moveout coordinate system to be described below.

The time coincidence of the downgoing and upcoming waves can be quantified in several ways. The most straightforward seems to be to look at the zero lag of the cross-correlation of the two waves. The image is created by displaying the zero-lagged cross-correlation everywhere in (x, z)-space.

The time coincidence of the upcoming wave and the earliest arrival of a downgoing wave gives evidence of the existence of a reflector, but in principle, more can be learned from the two waves. The amplitude ratio of the upcoming to the downgoing wave gives the reflection coefficient.

In the Fourier domain, the product $U(\omega, x, z) \overline{D}(\omega, x, z)$ represents the zero lag of the cross-correlation. The reflection coefficient ratio is given by $U(\omega, x, z)/D(\omega, x, z)$. This ratio has many difficulties. Not only may the denominator be zero, but it may have zeroes in the wrong part of the complex plane. This happens when the downgoing wave is causal but not minimum phase. (See Section 4.6 and FGDP). The phase of the complex conjugate of a complex number equals the phase of the inverse of the number. Thus the ratio U/D and the product $U \overline{D}$ both have the same phase. It seems you can invent other functional forms that compromise the theoretical appeal of U/D with the stability of $U \overline{D}$.

Don C. Riley [1974] proposed another form of the U/D principle, namely, that the upcoming waves must vanish for all time before the first arrival of the downgoing wave. Riley's form found use in wave-equation dereverberation.

Migration with Moveout Correction

If the earth were truly inhomogeneous in all three dimensions, we could hardly expect the data of a single seismic line to make any sense at all. But reflection seismology usually seems to work, even when it is restricted to a single line. This indicates that the layered model of the earth is a reasonable starting point. Thus normal-moveout correction is usually a good starting process. Mathematically, NMO is an excellent tool for dealing with depth variation in velocity, but its utility drops in the presence of steep dip or a wide dip spectrum.

My early migration programs were based on concepts derived from single profiles. The data and the wave equation were transformed to a moveout-corrected coordinate system. This approach to migration is well suited to data that is sparsely sampled on the geophone axis. When steepness of dip becomes the ground on which migration is evaluated, then moveout correction offers little advantage; indeed, it introduces unneeded complexity. Whatever its merits or drawbacks, NMO commands our attention by its nearly universal use in the industrial world.

Moveout/Radial Coordinates in Geophone Space

Our theoretical analysis will abandon the geophone axis g in favor of a radial-like axis characterized by a Snell parameter p. (This really says nothing about the implied data processing itself, since it would be simple enough to transform final equations back to offset). The coordinate system being defined will be called a retarded, moveout-corrected, Snell trace, coordinate system. Ideal data in this coordinate system in a zero-dip earth is unchanged as it is downward continued. Hence the amount of work the differential equations have to do is proportional to the departure of the data from the ideal. Likewise the necessity for spatial sampling of the data increases in proportion to the departure of data from the ideal. Define

p	$(sin\ \theta)/v$, the Snell ray parameter
t_p	any one-way time from the surface along a ray with parameter p
g	the surface separation of the shot from the geophone
t'	one-way time, surface to reflector, along a ray
τ	travel-time depth of buried geophones, one-way time along a ray
t	travel time seen by buried geophones
$v(p, t_p)$	a stratified velocity function $v'(z)$, in the new coordinates

The coordinate system is based on the following simple statements: (1) travel time from shot to geophone is twice the travel time from shot to reflector, less the time-depth of the geophone; and (2) the horizontal distance traveled by a ray is the time integral of $v \sin \theta = pv^2$; (3) the vertical distance traveled by a ray is computed the same way as the horizontal distance, but with a cosine instead of a sine.

$$t(t', p, \tau) = 2t' - \tau \tag{1}$$

$$g(t', p, \tau) = 2p \int_0^{t'} v(p, t_p)^2 \, dt_p - p \int_0^\tau v(p, t_p)^2 \, dt_p \tag{2}$$

$$z(t', p, \tau) = \int_0^\tau v(p, t_p) \sqrt{1 - p^2 v^2} \, dt_p \tag{3}$$

Surfaces of constant t' are reflections. Surfaces of constant p are rays. Surfaces of constant τ are datum levels. Unfortunately, it is impossible to invert the above system explicitly to get (t', p, τ) as a function of (t, g, z). It is possible, however, to proceed analytically with the differentials. Form the Jacobian matrix

$$\begin{bmatrix} \partial_{t'} \\ \partial_p \\ \partial_\tau \end{bmatrix} = \begin{bmatrix} t_{t'} & g_{t'} & z_{t'} \\ t_p & g_p & z_p \\ t_\tau & g_\tau & z_\tau \end{bmatrix} \begin{bmatrix} \partial_t \\ \partial_g \\ \partial_z \end{bmatrix} \tag{4}$$

Performing differentiations only where they lead to obvious simplifications gives the transformation equation for Fourier variables:

$$\begin{bmatrix} -\omega' \\ k_p \\ k_\tau \end{bmatrix} = \begin{bmatrix} 2 & g_{t'} & 0 \\ 0 & g_p & z_p \\ -1 & g_\tau & z_\tau \end{bmatrix} \begin{bmatrix} -\omega \\ k_g \\ k_z \end{bmatrix} \tag{5}$$

It should be noted that (5) is a linear relation involving the Fourier variables, but the coefficients involve the original time and space variables. So (5) is in both domains at once. This is useful and valid so long as it is assumed that second derivatives neglect the derivatives of the coordinate frame itself. This assumption is often benign, amounting to something like spherical divergence correction.

Here we could get bogged down in detail, were we to continue to attack the nonzero offset case. Specializing to zero offset, namely, $p = 0$, we get

$$\begin{bmatrix} -\omega' \\ k_p \\ k_\tau \end{bmatrix} = \begin{bmatrix} 2 & 0 & 0 \\ 0 & 2t' - \tau & 0 \\ -1 & 0 & v \end{bmatrix} \begin{bmatrix} -\omega \\ k_g \\ k_z \end{bmatrix} \tag{6}$$

Equation (6) may be substituted into the single-square-root equation for downward continuing geophones, thereby transforming it to a retarded equation in the new coordinate system.

Historical Notes on a Mysterious Scale Factor

My first migrations of reflection seismic data with the wave equation were based on the U/D concept. The first wave-equation migration program was in the frequency domain and worked on synthetic profiles. Since people generally ignored such work I resolved to complete a realistic test on field data. Frequency-domain methods were deemed "academic." I found I could use the bilinear transformation of Z-transform analysis to convert the 15° wave equation to the time domain. As a practical matter, it was apparent that a profile migration program could be used on a section. But the theoretical justification was not easy. At that time I thought of the exploding-reflector concept as a curious analogy, not as a foundation for the derivation.

The actual procedure by which the first zero-offset section was migrated with finite differences was more circuitous and complicated than the procedure later introduced by Sherwood (Loewenthal et al [1976]) and adopted generally. The equation for profile migration in moveout-corrected coordinates has many terms. Neglecting all those with offset as a coefficient (since you are trying to migrate a zero-offset section), you are left with an equation that resembles the retarded, 15° extrapolation equation. But there is one difference. The $v\,\partial_{gg}$ term is scaled by a mysterious coefficient, $[t'/(2t'-\tau)]^2$. This is the equation I used. As the travel-time depth τ increases from zero to the stopping depth t', the mystery coefficient increases slowly from 1/4 to 1.

Unfortunately my derivation was so complicated that few people followed it. (You notice that I do not fully include it here). My 1972 paper includes the derivation but by way of introduction it takes you through a conceptually simpler case, namely, the seismic section that results from a downgoing plane-wave source. This simpler case brings you quickly to the migration equation. But the mystery coefficient is absent. Averaged over depth the mystery coefficient averages to a half. (The coefficient multiplies the second x-derivative and arises from Δx decreasing as geophones descend along a coordinate ray path toward the shot). Sherwood telephoned me one day and challenged me to explain why the coefficient could not be replaced by its average value, 1/2. I could give no practical reason, nor can I today. So he abandoned my convoluted derivation and adopted the exploding-reflector model as an assumption, thereby easily obtaining the required 1/2. I felt more comfortable about the mystery coefficient later when the survey-sinking concept emerged from my work with Doherty, Muir, and Clayton.

My first book, FGDP, describes how the U/D concept can be used to deal with the three problems of migration, velocity analysis, and multiple suppression. In only one of these three applications, namely, zero-offset migration (really CDP-stack migration), has the wave-equation methodology become a part of routine practice. None-the-less, the U/D concept has been generally forgotten and replaced by Sherwood's exploding-reflector concept.

5.8 Predictions for the Next Decade

In the 1960s seismologists learned how to apply time-series optimization theory to seismic data — see FGDP for that. Eventually time series reached the point of diminishing returns because its approach to spatial relations was oversimplified. In the 1970s seismologists learned to apply the wave equation. That's what this book has been about. You can see that the job of applying the wave equation is not yet complete, but we have come a long way. Perhaps we have solved most of our "first-order" problems, and the problems that remain are mainly "second order." For second-order effects to be significant, all the first-order phenomena must be reasonably accounted for.

Some first-order effects that this book has touched on only lightly relate to obvious, as well as subtle, imperfections in seismic data.

Problems in the Database

We often have a problem of *truncation*. The recording cable is of course finite in length, and perceptible waves generally travel well beyond it. The seismic survey itself has finite dimensions. We also have the problem of *gaps*. Gaps in seismic data may occur unpredictably, as when a gun misfires or surveyors are denied access to parcels of land in the midst of their survey. In addition we have the problem of spatial *aliasing*. Because of improving technology, we can expect a substantial reduction in aliasing on the geophone axis, but aliasing on the shot axis will remain. There are only twenty-four hours in a day, and we must wait ten seconds between shots for the echoes to die down. So, given a certain area to survey and a certain number of months to survey it in, we end out with a certain number of shotpoints per square kilometer. With marine data, the spacing in the line of the path of the ship presents no problems compared to the problems presented by data spacing off the line.

Migration provides a mapping from a data space to a model space. This transformation is invertible (in the nonevanescent subspace). When data is missing, the transformation matrix gets broken into two parts. One part operates on the known data values, and the other part operates on the missing values. Except for a bit of Section 3.5, this book ignores the missing part. Although a strategy is presented in 3.5 for handling the missing part, it is very costly, and I believe it will ultimately be superseded or much improved.

Noisy data can be defined as data that doesn't fit our model. If the missing data were replaced by zeroes, for example, the data would be regarded as complete, but noisy. Data is missing where the signal-to-noise ratio is known to be zero. More general noise models are also relevant, but statistical treatment of partially coherent multidimensional wave *fields* is poorly developed in both theory and practice.

My prediction is that a major research activity of the next decade will be to try to learn to simultaneously handle both the physics and the statistics of wavefields.

Reuniting Optimization Theory and Wave Theory

Let's take a quick peek beyond this book into the future. A seismic image is typically a 1000×1000 plane, derived from a volume of about 1000^3 interrelated data points. There are unknowns present everywhere, not only in the earth model, but also in the data, as noise, as gaps, and as insufficient spatial density and extent of data recording. To assemble an interpretation we must combine principles from physics with principles from statistics. Presumably this could be done in some monster optimization formulation. A look at the theory of optimization shows that solution techniques converge in a number of iterations that is greater than the number of unknowns. Thus the solution to the problem, once we learn how to pose the problem properly, seems to require about a million times as much computing power as is available. What a problem!

But the more you look at the problem, the more interesting it becomes. First we have an optimization problem. Since we are constrained to make only a few iterations, say, three, we must go as far as we can in those three steps. Now, not only do we have the original optimization problem, but we also have the new problem of solving it in an optimum way. First we have correlated randomness in the raw data. Then, during optimization, the earth model changes in a correlated random way from one iteration to the next. Not only is the second optimization problem the practical one — it is deeper at the theoretical level.

Throw Away Your Paper Sections.

Current seismic interpretation often amounts to taking colored pencils and enhancing aspects of a computer-generated image. Seismic interpretation is entering an era in which the interpretation will all be done on a video screen. The basic reason is that a sheet of paper is only two-dimensional, while most reflection data is three-dimensional. Modern 3-D surveys really record four-dimensional data. A video screen can show a movie. The operator/interpreter can interact with the movie. There are things I would like to show you, but I cannot show you in a book. Seismic data or even a blank sheet of paper has *texture*. When a textured object moves, you immediately recognize it. But I couldn't show it to you with pictures in this book. (Imagine a sequence of pictures of a blank sheet of paper, each one shifted some way from the previous one). The perception of small changes is blocked by any eye movement between pictures. Astronomers look for changes in the sky by rapidly blinking between looking at photographs taken at different times. Our eyes are special computers. Movies often show "where something comes from," enabling us to notice the unexpected in the general ambiance.

Most seismic interpretation is done on *stacked sections*. The original data is three-dimensional, but one dimension is removed by summation. Theoretically, the summation removes only redundancy while it enhances the signal-to-noise ratio. In reality, things are much more complicated. And much more will be perceptible when summations are done by the human eye (just by increasing the speed of a movie). There will be two generations of seismic interpreters — those who can interpret the prestacked data they see on their video screens — and those who interpret only stacked sections.

STANFORD EXPLORATION PROJECT MEMBERSHIP SUMMARY

Sponsor	73-4	74-5	75-6	76-7	77-8	78-9	79-80	80-1	81-2	82-3	83-4
AGIP (Italy)	$-^1$	-	-	-	-	$+^2$	+	+	+	+	+
Amoco	+	+	+	+	+	+	+	+	+	+	+
ARAMCO (Saudi Arabia)	-	-	-	-	-	-	-	+	+	+	+
Arco	+	+	+	+	+	+	+	+	+	+	+
British Petroleum	-	-	-	-	-	+	+	+	+	+	+
Canterra (Canada)	-	-	-	-	-	-	-	-	+	+	+
CGG (France)	-	-	-	-	+	+	+	+	+	+	+
Chevron	+	+	+	+	+	+	+	+	+	+	+
Cities Service	-	-	-	+	+	+	+	+	+	-	-
Conoco	+	+	+	+	+	+	+	+	+	+	+
Digicon	+	+	+	+	+	+	+	+	+	+	+
Digitech (Canada)	-	-	+	-	-	-	-	-	-	-	-
Dome (Canada)	-	-	-	-	-	-	-	-	-	-	+
Exxon	+	+	+	+	+	+	+	+	+	+	+
GECO (Norway)	-	-	-	-	+	+	+	+	+	+	+
General Crude	-	-	-	+	-	-	-	-	-	-	-
Geosource (Petty–Ray)	+	+	+	+	+	+	+	+	+	+	+
Getty Oil	-	-	-	-	+	+	+	+	+	+	+
GSI	+	+	+	+	+	+	+	+	+	+	+
Gulf	-	-	-	+	+	+	+	+	+	+	+
IBM	-	-	-	+	+	+	+	+	+	+	+
INA (Yugoslavia)	+	+	+	+	-	-	-	-	-	-	-
Japex Geoscience	-	-	-	-	-	-	+	+	+	+	+
Koninklijke Shell (Holland)	-	+	+	+	+	+	+	+	+	+	+
Marathon	-	-	-	+	+	+	+	+	-	-	+
Mobil	+	+	+	+	+	+	+	+	+	+	+
Norsk Hydro (Norway)	-	-	-	-	-	-	-	-	+	+	+
Occidental	-	-	+	+	+	+	+	+	+	+	+
Petrofina (Belgium)	-	+	-	-	-	-	-	-	-	-	-
Phillips Petroleum	-	-	-	-	+	+	+	+	+	+	+
Prakla–Seismos (W. Germany)	-	-	-	-	+	+	+	+	+	+	+
Preusag (Germany)	+	+	+	-	-	-	-	-	-	-	-
Santa Fe Minerals	-	-	-	-	-	-	-	-	-	-	+
Schlumberger	-	-	-	-	-	-	-	-	+	+	+
Seiscom–Delta	+	+	+	-	-	-	-	-	-	-	-
Seismograph Service	+	+	+	+	+	+	+	+	+	+	+
Shell	+	+	+	-	-	-	-	-	-	-	-
SNEA (P) (France)	+	+	+	+	+	+	+	+	+	+	-
Sun Exploration	+	+	+	+	+	+	+	+	+	+	+
Superior Oil	-	-	-	-	-	-	+	+	+	+	+
Teledyne	+	+	+	-	-	+	+	+	-	-	-
Texaco	+	+	+	+	+	+	+	+	+	+	+
TOTAL (France)	+	+	+	+	+	+	+	+	+	+	+
Union	+	+	+	+	+	+	+	+	+	+	+
USGS	+	+	+	+	+	+	+	+	+	+	+
United Geophysical	+	+	-	-	-	+	+	-	-	-	-
Veritas Seismic (Canada)	-	-	-	-	-	-	-	-	-	-	+
Western Geophysical	+	+	+	+	+	+	+	+	+	+	+

---

[1]Dash (–) represents a nonmember
[2]Plus (+) represents a current member.

References

Aki, K., and Richards, P.G., 1980, *Quantitative seismology, theory and methods:* vol. I and II, Freeman, San Francisco.

Anstey, N.A., 1980, *Seismic exploration for sandstone reservoirs*, International Human Resources Development Corporation, Boston.

Backus, M.M., 1959, Water reverberations — their nature and elimination: *Geophysics*, vol. 24, pp. 233-261.

Backus, M.M., and Chen, R.L., 1975, Flat spot exploration: *Geophys. Prosp.*, vol. 23, pp. 533-577.

Balch, A.H., Lee, M.W., Miller, J.J., and Taylor, R.T., 1982, The use of vertical seismic profiles in seismic investigations of the earth: *Geophysics*, vol. 47, pp. 906-918.

Bally, A.W., 1983, Seismic expression of structural styles — A picture and work atlas: AAPG studies in geology #15 - vol. 1, 2, and 3, AAPG, P.O. Box 979, Tulsa, OK 74101

Baysal, E., Kosloff, D.D., and Sherwood, J.W.C., 1983, Reverse time migration: *Geophysics*, vol. 48, pp. 1514-1524.

Beresford-Smith, G., and Mason, I.M., 1980, A parametric approach to the compression of seismic signals by frequency transformation: *Geophys. Prosp.*, vol. 28, pp. 551-571

Berkhout, A.J., 1979, Steep dip finite-difference migration: *Geophys. Prosp.*, vol. 27, pp. 196-213.

Berkhout, A.J., and van Wulfften Palthe, D.W., 1979, Migration in terms of spatial deconvolution: *Geophys. Prosp.*, vol. 27, pp. 261-291.

Berkhout, A.J., 1980, *Seismic migration — Imaging of acoustic energy by wave field extrapolation:* Amsterdam/New York, Elsevier/North Holland Publishing Co.

Berkhout, A.J., and van Wulfften Palthe, D.W., 1980, Migration in the presence of noise: *Geophys. Prosp.*, vol. 28, pp. 372-383.

Berkhout, A.J., 1981, Wave field extrapolation techniques in seismic migration, a tutorial: *Geophysics*, vol. 46, pp. 1638-1656.

Berkhout, A.J., and De Jong, B.A., 1981, Recursive migration in three dimensions: *Geophys. Prosp.*, vol. 29, pp. 758-781.

Berryhill, R.T., 1979, Wave equation datuming: *Geophysics*, vol. 44, pp. 1329-1344.

Bleistein, N., 1984, *Mathematical methods for wave phenomena*, Academic Press

Bleistein, N., and Cohen, J.K., 1979, Direct inversion procedure for Claerbout's equations: *Geophysics*, vol. 44, pp. 1034-1040.

Bloxsom, H., 1979, *Migration and interpretation of deep crustal seismic reflection data:* Ph.D. thesis,[†] Stanford University, SEP-22.

---

† Ph.D. theses included in this bibliography are available from University Microfilms International, 300 N. Zeeb Road, Ann Arbor, MI 48106, USA.

Bolondi, G., Rocca, F., and Savelli, S., 1978, A frequency domain approach to two-dimensional migration: *Geophys. Prosp.*, vol. 26, pp. 750-772.

Bolondi, G., Loinger, E., and Rocca, F., 1982, Offset continuation of seismic sections: *Geophys. Prosp.*, vol. 30, pp. 813-828.

Brown, D.L., 1983, Applications of operator separation in reflection seismology: *Geophysics*, vol. 48, pp. 288-294.

Buhl, P., Diebold, J.B., and Stoffa, P.L., 1982, Array length magnification through the use of multiple sources and receiving arrays: *Geophysics*, vol. 47, pp. 311-315.

Burg, J.P., 1975, *Maximum entropy spectral analysis:* Ph.D. thesis, Stanford University, SEP-6.

Cherry, J.T., and Waters, K.H., 1968, Shear-wave recording using continuous signal methods, Part I - Early development: *Geophysics*, vol. 33, pp. 229-239.

Cerveny, V., I.A. Molotkov, I. Psencik, 1977, *Ray Method in Seismology*, Univerzita Karlova, Prague

Chun, J.H., and Jacewitz, C.A., 1981, Fundamentals of frequency domain migration: *Geophysics*, vol. 46, pp. 717-733.

Claerbout, J.F., 1970, Coarse grid calculations of waves in inhomogeneous media with application to delineation of complicated seismic structure: *Geophysics*, vol. 35, pp. 407-418.

Claerbout, J.F., 1971, Toward a unified theory of reflector mapping: *Geophysics*, vol. 36, pp. 467-481.

Claerbout, J.F., and Johnson, A.G., 1971, Extrapolation of time-dependent waveforms along their path of propagation: *Geophysical Journal of the Royal Astronomical Society*, vol. 26, pp. 285-293.

Claerbout, J.F., 1971, Numerical holography: *Acoustical Holography*, vol. 3, pp. 273-283.

Claerbout, J.F., and Doherty, S.M., 1972, Downward continuation of moveout corrected seismograms: *Geophysics*, vol. 37, pp. 741-768.

Claerbout, J.F., and Muir, F., 1973, Robust modeling with erratic data: *Geophysics*, vol. 38, pp. 826-844.

Claerbout, J.F., 1976, *Fundamentals of geophysical data processing:* McGraw-Hill (sometimes referred to in the text of this book as FGDP).

Clayton, R.W., and Engquist, B., 1980, Absorbing side boundary conditions for wave-equation migration: *Geophysics*, vol. 45, pp. 895-904.

Clayton, R.W., 1981, *Wavefield inversion methods for refraction and reflection data:* Ph.D. thesis, Stanford University, SEP-27.

Clayton, R.W., and McMechan, G.A., 1981, Inversion of refraction data by wave field continuation: *Geophysics*, vol. 46, pp. 860-868.

Cohen, J.K., and Bleistein, N., 1979, Velocity inversion procedure for acoustic waves: *Geophysics*, vol. 44, pp. 1077-1087.

Crank, J., and Nicolson, P., 1947, A Practical method for numerical evaluation of solutions of partial differential equations of the heat-conduction type: Proc. Cambridge Philos. Soc., vol. 43, p. 50.

Dahm, C.G., and Graebner, R.J., 1982, Field development with three-dimensional seismic methods in the Gulf of Thailand - a case history: *Geophysics*, vol. 47, pp. 149-176.

Deans, S.R., 1983, The Radon transform and some of its applications: John Wiley, pp. 204-217.

Dent, B., 1983, Compensation of marine seismic data for the effects of highly variable water depth using ray-trace modeling — A case history: *Geophysics*, vol. 48, pp. 910-933.

Diebold, J.D., and Stoffa, P.L., 1981, The travel time equation, tau-p mapping and inversion of common midpoint data: *Geophysics*, vol. 46, pp. 238-254.

Diet, J.D., and Fourmann, J.M., 1979, Determination of the shape parameter W in Stolt's frequency domain migration: presented at 49th Annual International Society of Exploration Geophysicists Meeting, New Orleans.

Doherty, S.M., and Claerbout, J.F., 1976, Structure independent seismic velocity estimation: *Geophysics*, vol. 41, pp. 850-881.

Doherty, S.M., 1975, *Structure independent seismic velocity estimation:* Ph.D. thesis, Stanford University, SEP-4.

References

Dohr, G.P., and Stiller, P.K., 1975, Migration velocity determination: Part II. Applications: *Geophysics*, vol. 40, pp. 6-16.

Embree, P., Burg, J.P., and Backus, M.M., 1963, Wide-band velocity filtering — the pie-slice process: *Geophysics*, vol. 28, pp. 948-974.

Engquist, B., and Majda, A., 1979, Radiation boundary conditions for acoustic and elastic waves: *Communications on Pure and Applied Mathematics*, vol. 32, pp. 313-320.

Erickson, E.L., Miller, D.E., and Waters, K.H., 1968, Shear-wave recording using continuous signal methods, Part II - Later experimentation: *Geophysics*, vol. 33, pp. 240-254.

Estevez, R., 1977, *Wide-angle diffracted multiple reflections:* Ph.D. thesis, Stanford University, SEP-12.

Estevez, R., and Claerbout, J.F., 1982, Wide-angle diffracted multiple reflections: *Geophysics*, vol. 47, pp. 1255-1272.

Fenati, D. and Rocca, F. 1984, Seismic reciprocity field tests from the Italian Peninsula: *Geophysics* vol. 49, pp. 1690-1700

FGDP: see Claerbout, J.F., 1976.

French, W.S., 1974, Two-dimensional and three-dimensional migration of model-experiment reflection profiles: *Geophysics*, vol. 39, pp. 265-277.

French, W.S., 1975, Computer migration of oblique seismic reflection profiles: *Geophysics*, vol. 40, pp. 961-980.

Gabitzsch, J.H., 1978, Wave number migration: Application to variable velocity: presented at the 48th Annual International SEG Meeting, San Francisco.

Gal'perin, E.I., 1974, *Vertical Seismic Profiling*, translated by A.J. Hermont, Society of Exploration Geophysicists, Special Publication No. 12

Gardner, G.H.F., French, W.S., and Matzuk, T., 1974, Elements of migration and velocity analysis: *Geophysics*, vol. 39, pp. 811-825.

Garotta, R., and Baixas, F., 1979, Simulation directional acquisition as an aid to migration: presented at the 49th Annual International SEG Meeting, New Orleans.

Gazdag, J., 1978, Wave equation migration with the phase shift method: *Geophysics*, vol. 43, pp. 1342-1351.

Gazdag, J., 1980, Wave equation migration with the accurate space derivative method: *Geophys. Prosp.*, vol. 28, pp. 60-70.

Gazdag, J., 1981, Modeling of the acoustic wave equation with transform methods: *Geophysics*, vol. 46, pp. 854-859.

Gazdag, J., and Sguazzero, P., 1984, Migration of seismic data by phase shift plus interpolation: *Geophysics*, vol. 49, pp. 124-131.

Gibson, B., Larner, K., and Levin, S., 1983, Efficient 3-D migration in two steps: *Geophys. Prosp.*, vol. 31, pp. 1-33.

Godfrey, R.J., 1979, *A stochastic model for seismogram analysis:* Ph.D. thesis, Stanford University, SEP-17.

Godfrey, R. and Rocca, F., 1981, Zero memory non-linear deconvolution: *Geophys. Prosp.*, vol. 29, pp. 189-228.

Gonzalez-Serrano, A., 1982, *Wave equation velocity analysis:* Ph.D. thesis, Stanford University, SEP-31.

Gonzalez-Serrano, A., and Claerbout, J.F., 1984, Wave equation velocity analysis: *Geophysics*, vol. 49, pp. 1431-1456

Gray, W.C., 1979, *Variable norm deconvolution:* Ph.D. thesis, Stanford University, SEP-19.

Hale, D. and Claerbout, J.F., 1983, Butterworth dip filters: *Geophysics*, vol 48, pp. 1033-1038.

Hale, I.D., 1983, *Dip-Moveout by Fourier Transform:* Ph.D. thesis, Stanford University, SEP-36.

Hale, I.D., 1984, Dip-moveout by Fourier transform: *Geophysics*, vol. 49, pp. 741-757.

Hatton, L., Larner, K., and Gibson, B., 1981, Migration of seismic data from inhomogeneous media: *Geophysics*, vol. 46, pp. 751-767.

Hemon, Ch., 1978, Equations d'onde et modeles, *Geophysical Prospecting*, vol. 26, pp. 790-821

Herman, A.J., Anania, R.M., Chun, J.H., Jacewitz, C.A., and Pepper, R.E.F., 1982, A fast three-dimensional modeling technique and fundamentals of three-dimensional frequency-domain migration: *Geophysics*, vol. 47, pp. 1627-1644.

Hilterman, F.J., 1970, Three-dimensional seismic modeling: *Geophysics*, vol. 35, p. 1020.

Hood, P., 1978, Finite difference and wave number migration: *Geophys. Prosp.*, vol. 26, pp. 773-789.

Hubral, P., 1977, Time migration - some ray theoretical aspects: *Geophys. Prosp.*, vol. 25, pp. 738-745.

Hubral, P., 1980, Wavefront curvatures in three-dimensional laterally inhomogeneous media with curved interfaces: *Geophysics*, vol. 45, pp. 905-913.

Jacobs, A., 1982, *The pre-stack migration of profiles:* Ph.D. thesis, Stanford University, SEP-34.

Jakubowicz, H., and Levin, S., 1983, A simple exact method of 3-D migration: *Geophys. Prosp.*, vol. 31, pp. 34-56.

Jain, S., and Wren, A.E., 1980, Migration before stack - Procedure and significance: *Geophysics*, vol. 45, pp. 204-212.

Jones, W.B., and Thron, W.J., 1980, *Continued fractions: encyclopedia of mathematics and its applications:* American Mathematical Society, vol 11.

Judson, D.R., Lin, J., Schultz, P.S., and Sherwood, J.W.C., 1980, Depth migration after stack: *Geophysics*, vol. 45, pp. 361-375.

Kennett, B.L.N., 1983, *Seismic wave propagation in stratified media:* Cambridge University Press.

Kjartansson, E., 1979a, *Attenuation of seismic waves in rocks and applications in energy exploration:* Ph.D. thesis, Stanford University, SEP-23.

Kjartansson, E., 1979b, Constant Q - wave propagation and attenuation: *J. Geophys. Res.*, vol. 84, pp. 4737-4748.

Kosloff, D.D., and Baysal, E., 1983, Migration with the full acoustic wave equation: *Geophysics*, vol. 48, pp. 677-687.

Larner, K., Chambers, R., Yang, M., Lynn, W., and Wai, W., 1983, Coherent noise in marine seismic data: *Geophysics*, vol. 48, pp. 854-886.

Larner, K.L., Hatton, L., Gibson, B.S., and Hsu, I.C., 1981, Depth migration of imaged time sections: *Geophysics*, vol. 46, pp. 734-750.

Levin, F.K., 1971, Apparent velocity from dipping interface reflections: *Geophysics*, vol. 36, pp. 510-516.

Levin, F.K., Bayhi, J.F., Dunkin, J.W., Lea, J.D., et al., 1976, Developments in exploration geophysics, 1969 -1974: *Geophysics*, vol. 43, pp. 23-48.

Levin, S.A., 1984, Discussion On: "Dip limitations on migrated sections as a function of line length and recording time" by H.B. Lynn and S. Deregowski *(Geophysics*, v. 46, p. 1362, October, 1981). *Geophysics*, vol. 49, pp. 1804-1805.

Loewenthal, D., Lu, L., Roberson, R., and Sherwood, J., 1976, The wave equation applied to migration: *Geophys. Prosp.*, vol. 24, pp. 380-399.

Lynn, W.S., 1979, *Velocity estimation in laterally varying media:* Ph.D. thesis, Stanford University, SEP-21.

Lynn, W.S., and Claerbout, J.F., 1982, Velocity estimation in laterally varying media: *Geophysics*, vol. 47, pp. 884-897.

Ma Zaitian, 1981, *Finite difference migration with higher order approximation:* technical report of the China National Oil and Gas Exploration and Development Co.

Ma Zaitian, 1982, Steep dip finite difference migration: *Oil Geophysical Prospecting*, no. 1, pp. 6-15 (in Chinese).

Madden, T.R., 1976, Random networks and mixing laws: *Geophysics*, vol. 41, pp. 1104-1125.

May, B.T., and Covey, J.D., 1981, An inverse method for computing geologic structures from seismic reflections - zero-offset case: *Geophysics*, vol. 46, pp. 268-287.

Morley, L., 1982, *Predictive techniques for marine multiple suppression:* Ph.D. thesis, Stanford University, SEP-29.

Morley, L., and Claerbout, J.F., 1983, Predictive deconvolution in shot-receiver space: *Geophysics*, vol. 48, pp. 515-531.

References

Muir F., and Claerbout, J.F., 1980, Impedance and wave extrapolation: presented at the 42nd Meeting of the European Association of Exploration Geophysicists, Istanbul.

Ostrander, W.J., 1984, Plane wave reflection coefficients for gas sands at non-normal angles of incidence: *Geophysics*, vol. 49, pp. 1637-1648

Ottolini, R., Sword, C., and Claerbout, J.F., 1984, On-line movies of reflection seismic data with description of a movie machine: *Geophysics*, vol. 49, pp. 195-200.

Ottolini, R., and Claerbout, J.F., 1984, The migration of common midpoint slant stacks: *Geophysics*, vol. 49, pp. 237-249.

Ottolini, R., 1982, *Migration of reflection seismic data in angle-midpoint coordinates:* Ph.D. thesis, Stanford University, SEP-33.

Pan, P.H., 1983, Case history of the exploration of the Grand Isle 95 Field in the Gulf of Mexico: *Geophysics*, vol. 48, pp. 900-909.

Pann, K., Eisner, E., and Shin, Y., 1979, A collocation formulation of wave equation migration: *Geophysics*, vol. 44, pp. 712-721.

Phinney, R.A., Chowdhury, K.R., and Frazer, L.N., 1981, Transformation and analysis of record sections: *J. Geophys. Res.*, vol. 86, pp. 359-377.

Radon, J., 1917, Über die Bestimmung von Funktionen durch ihre Integralwerte längs gewisser Mannigfaltigkeiten, *Berichte Sächische Akademie der Wissenschaften.* Leipzig, Math. - Phys. Kl. 69, pp. 262-267. for English translation see Deans [1983].

Rice, R.B., 1962, Inverse convolution filters: *Geophysics*, vol. 27, pp 4-18.

Riley, D.C., 1974, *Wave equation synthesis and inversion of diffracted multiple seismic reflections:* Ph.D. thesis, Stanford University, SEP-3.

Riley, D.C., and Claerbout, J.F., 1976, 2-D multiple reflections: *Geophysics*, vol. 41, pp. 592-620.

Ristow, D., 1980, *Three-dimensional finite-difference migration:* Ph.D. thesis, University of Utrecht, The Netherlands.

Robinson, E.A., 1983, *Migration of geophysical data:* International Human Resources Development Corporation, Boston.

Robinson, J.C., and Robbins, T.R., 1978, Dip-domain migration of two-dimensional seismic profiles: *Geophysics*, vol. 43, pp. 77-93.

Rosenbaum, J.H. and G.F. Boudreaux, 1981, Rapid convergence of some seismic processing algorithms: *Geophysics* vol. 46, pp. 1667-1672.

Sangree, J.B., and Widmier, J.M., 1979, Interpretation of depositional facies from seismic data: *Geophysics*, vol. 44, pp. 131-160.

Sattlegger, J.W., 1975, Migration velocity determination: Part I. Philosophy: *Geophysics*, vol. 40, pp. 1-5.

Sattlegger, J.W., Stiller, P.K., Echterhoff, J.A., and Hentschke, M.K., 1980, Common offset plane migration (COPMIG): *Geophys. Prosp.*, vol. 28, pp. 859-871.

Schneider, W.A., Larner, K.L., Burg, J.P., and Backus. M.M., 1964, A new data-processing technique for the elimination of ghost arrivals on reflection seismograms: *Geophysics*, vol. 39, pp. 783-805.

Schneider, W.A., 1971, Developments in seismic data processing and analysis (1968-1970): *Geophysics*, vol. 36, pp.1043-1073.

Schneider, W.A., 1978, Integral formulation in two and three dimensions: *Geophysics*, vol. 43, pp. 49-76.

Schultz, P. S., 1976, *Velocity estimation by wave front synthesis:* Ph.D. thesis, Stanford University, SEP-9.

Schultz, P.S., and Claerbout, J.F., 1978, Velocity estimation and downward continuation by wavefront synthesis: *Geophysics*, vol. 43, pp. 691-714.

Schultz, P.S., and Sherwood, J.W., 1980, Depth migration before stack: *Geophysics*, vol. 45, pp. 376-393.

Schultz, P.S., 1982, A method for direct estimation of interval velocities: *Geophysics*, vol. 47, pp. 1657-1671.

Sengbush, R.L., 1983, *Seismic exploration methods:* International Human Resources Development Corporation, Boston.

Sheriff, R.E., 1980, *Seismic stratigraphy:* International Human Resources Development Corporation, Boston.

Sherwood, J.W.C., Adams, H., Blum, C., Judson, D., Jin Lin, and Meadours B., 1976, Developments in filtering seismic data: presented at the 46th Annual International SEG Meeting, Houston.

Sherwood, J.W.C., Schultz, P.S., and Judson, J.R., 1976, *Some recent developments in migration before stack:* released by Digicon Inc.

Slotnick, M.M., 1959, *Lessons in seismic computing:* Tulsa, Society of Exploration Geophysicists.

Stolt, R., 1978, Migration by Fourier transform: *Geophysics,* vol. 43, pp. 23-48.

Taner, M.T., and Koehler, F., 1969, Velocity spectra — digital computer derivation and applications of velocity functions: *Geophysics,* vol. 34, pp. 859-881.

Taner, M.T., and Koehler, F. (undated), *Wave equation migration:* released by Seiscom Delta Inc., Houston.

Taner, M.T., 1980, Long-period sea-floor multiples and their suppression: *Geophys. Prosp.,* vol. 28, pp. 30-48.

Taner, M.T., and Koehler, F., 1981, Surface consistent corrections: *Geophysics,* vol. 46, pp. 17-22.

Taner, M.T., and Coburn, K.W., 1980, *Surface consistent estimation of source and receiver response functions:* presented at the 50th Annual International SEG Meeting, Houston.

Tatham, R.H., and Stoffa, P.L., 1976, V_p/V_s — A potential hydrocarbon indicator: *Geophysics,* vol. 41, pp. 837-849.

Tatham, R.H., 1982, V_p/V_s and lithology: *Geophysics,* vol. 47, pp. 336-344.

Temme, P., 1984, A comparison of common-midpoint, single-shot, and plane-wave depth migration: *Geophysics,* vol. 49, pp. 1896-1907.

Thorson, J., 1984, *Velocity stack and slant stack inversion methods:* Ph.D. thesis, Stanford University, SEP-39.

Treitel, S., Gutowski, P.R., and Wagner, D.E., 1982, Plane wave decompositions of seismograms: *Geophysics,* vol. 47, pp.1375-1401.

Trorey, A.W., 1970, A simple theory for seismic diffractions: *Geophysics,* vol. 35, pp. 762-784.

Trorey, A.W., 1977, Diffractions for arbitrary source-receiver locations: *Geophysics,* vol. 42, pp. 1177-1182.

Tufekcic, D., Claerbout, J.F. and Rasperic, Z., 1981, Spectral balancing in the time domain: *Geophysics,* vol. 46, pp. 1182-1188.

Waters, K.H., 1981, *Reflection seismology, a tool for energy resource exploration* (second edition): Wiley, New York.

Wiggins, R.A., Larner, K.L., and Wisecup, R.D. 1976, Residual statics analysis as a general linear inverse problem: *Geophysics,* vol 41, pp. 922-938.

Yilmaz, O., 1979, *Pre-stack partial migration:* Ph.D. thesis, Stanford University, SEP-18.

Yilmaz, O., and Claerbout, J.F., 1980, Pre-stack partial migration: *Geophysics,* vol. 45, pp. 1753-1779.

Yilmaz, O., and Cumro, D., 1983, Worldwide Assortment of Field Seismic Records, released by Western Geophysical Company of America, Houston

Ziolkowski, A., 1984, *Deconvolution* International Human Resources Development Corporation, Boston.

Zoeppritz, K., 1919, Erdbebenwellen VIIIB: Uber Reflexion and Durchgang seismicher Wellen durch Unstetigkeitsflachen: *Göttinger Nachr.,* vol. 1, pp. 66-84.

Index

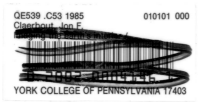

D